Building a Sustainable Transportation Infrastructure for Long-Term Economic Growth

Olga V. Smirnova
East Carolina University, USA

A volume in the Advances in Public Policy and Administration (APPA) Book Series

IGI Global
DISSEMINATOR OF KNOWLEDGE

Published in the United States of America by
IGI Global
Engineering Science Reference (an imprint of IGI Global)
701 E. Chocolate Avenue
Hershey PA, USA 17033
Tel: 717-533-8845
Fax: 717-533-8661
E-mail: cust@igi-global.com
Web site: http://www.igi-global.com

Library of Congress Cataloging-in-Publication Data

Names: Smirnova, Olga V., 1977- editor.
Title: Building a sustainable transportation infrastructure for long-term economic growth / Olga V. Smirnova, editor.
Description: Hershey, PA : Engineering Science Reference, [2019]
Identifiers: LCCN 2018026765| ISBN 9781522573968 (hardcover) | ISBN 9781522573975 (ebook)
Subjects: LCSH: Transportation--United States--Planning. | Transportation--United States--Finance. | City planning--United States. | Sustainable development--United States.
Classification: LCC HE203 .B85 2019 | DDC 388.0973--dc23 LC record available at https://lccn.loc.gov/2018026765

This book is published in the IGI Global book series Advances in Public Policy and Administration (APPA) (ISSN: 2475-6644; eISSN: 2475-6652)

British Cataloguing in Publication Data
A Cataloguing in Publication record for this book is available from the British Library.

For electronic access to this publication, please contact: eresources@igi-global.com.

Advances in Public Policy and Administration (APPA) Book Series

ISSN:2475-6644
EISSN:2475-6652

MISSION

Proper management of the public sphere is necessary in order to maintain order in modern society. Research developments in the field of public policy and administration can assist in uncovering the latest tools, practices, and methodologies for governing societies around the world.

The **Advances in Public Policy and Administration (APPA) Book Series** aims to publish scholarly publications focused on topics pertaining to the governance of the public domain. APPA's focus on timely topics relating to government, public funding, politics, public safety, policy, and law enforcement is particularly relevant to academicians, government officials, and upper-level students seeking the most up-to-date research in their field.

COVERAGE

- Government
- Law Enforcement
- Political Economy
- Politics
- Public Administration
- Public Funding
- Public Policy
- Resource Allocation
- Urban Planning

IGI Global is currently accepting manuscripts for publication within this series. To submit a proposal for a volume in this series, please contact our Acquisition Editors at Acquisitions@igi-global.com or visit: http://www.igi-global.com/publish/.

The Advances in Public Policy and Administration (APPA) Book Series (ISSN 2475-6644) is published by IGI Global, 701 E. Chocolate Avenue, Hershey, PA 17033-1240, USA, www.igi-global.com. This series is composed of titles available for purchase individually; each title is edited to be contextually exclusive from any other title within the series. For pricing and ordering information please visit http://www.igi-global.com/book-series/advances-public-policy-administration/97862. Postmaster: Send all address changes to above address.

Titles in this Series

For a list of additional titles in this series, please visit:
https://www.igi-global.com/book-series/advances-public-policy-administration/97862

Marketing Peace for Social Transformation and Global Prsperity
Alexandru-Mircea Nedelea (Stefan cel Mare University of Suceava, Romania) and Marilena-Oana Nedelea (Stefan cel Mare University of Suceava, Romania)
Information Science Reference • ©2019 • 304pp • H/C (ISBN: 9781522574644) • US $195.00

Mathematical Approaches to Understanding Democracy Emerging Research and Opportunities
Alberto Arteta (Troy University, USA)
Information Science Reference • ©2019 • 148pp • H/C (ISBN: 9781522575580) • US $165.00

Advanced Methodologies and Technologies in Government and Society
Mehdi Khosrow-Pour, D.B.A. (Information Resources Management Association, USA)
Information Science Reference • ©2019 • 726pp • H/C (ISBN: 9781522576617) • US $275.00

Handbook of Research on Promoting Peace Through Practice, Academia, and the Arts
Mohamed Walid Lutfy (London School of Economics, UK) and Cris Toffolo (Northeastern Illinois University, USA)
Information Science Reference • ©2019 • 553pp • H/C (ISBN: 9781522530015) • US $275.00

Handbook of Research on Military Expenditure on Economic and Political Resources
Ramesh Chandra Das (Katwa College, India)
Information Science Reference • ©2018 • 564pp • H/C (ISBN: 9781522547785) • US $325.00

Technology and the New Generation of Active Citizens Emerging Research and Opportunities
Paolo Beneventi (Independent Researcher, Italy)
Information Science Reference • ©2018 • 173pp • H/C (ISBN: 9781522537700) • US $135.00

Examining the Private Sector's Role in Wealth Creation and Poverty Reduction
Scott A. Hipsher (Webster University, Thailand)
Information Science Reference • ©2018 • 325pp • H/C (ISBN: 9781522531173) • US $185.00

Handbook of Research on Examining Global Peacemaking in the Digital Age
Bruce L. Cook (Chicago ORT Technical Institute, USA)
Information Science Reference • ©2018 • 435pp • H/C (ISBN: 9781522530329) • US $295.00

For an entire list of titles in this series, please visit:
https://www.igi-global.com/book-series/advances-public-policy-administration/97862

IGI Global
DISSEMINATOR OF KNOWLEDGE

701 East Chocolate Avenue, Hershey, PA 17033, USA
Tel: 717-533-8845 x100 • Fax: 717-533-8661
E-Mail: cust@igi-global.com • www.igi-global.com

Table of Contents

Detailed Table of Contents

Across the United States there is increased pressure for communities, especially in states like Nebraska, to engage in sustainable transportation infrastructure development. Through a case study of an ongoing statewide transportation initiative in nonmetropolitan Nebraska, this chapter examines transportation sustainability and planning from a regional and collaborative perspective. The Nebraska effort can be adapted to other states with significant rural and dispersed population centers. Funded by the state and the federal governments, Nebraska's transportation initiative, using an innovative public-private partnership, is creating and enhancing regional transit services in small urban and rural areas, using public transportation as a means to promote long-term economic growth and sustainability. Smaller urban and micropolitan communities, often serving as regional growth centers, frequently are overlooked when it comes to research on transportation planning and policy.

Transportation policymakers and planners have begun to realize the importance of sustainable transportation with regards to health, social implications, and the climate. Focusing on more active travel is one way that these officials are beginning to evolve cities in a way that supports these broader sustainability goals. In this light, this chapter focuses on how active transportation has evolved, and how policy and finance can support it. It also looks at emerging issues that may reshape transportation, such as connected and autonomous vehicles, and how we can maintain transportation sustainability in light of these innovations.

International trade requires efficient delivery between exporter and importer, usually by ocean vessels. Roughly 90% of cargo is transported in shipping containers, delivered for export by truck or rail, then received at the importer to be similarly discharged. This chapter discusses containerized trade, focusing on international chassis, the wheeled shipping container bed. The authors discuss the invention of containerization and give the historical context for US chassis provision. The chapter outlines the chassis logistical difficulties of US truckers. An overview of attempts to solve this by chassis pooling is provided. The chapter then addresses differences between chassis in the US and the rest of the world. Key chassis regulations are covered, followed by a discussion of the relation of chassis to sustainability. Antitrust issues concerning the main chassis providers and the three recently created ocean carrier alliances are covered. The authors conclude with three recommendations for improving US chassis access.

The lessons described in this chapter outline the mechanisms for cooperation through building new institutional designs for governance to build transportation construction projects. The scale of these projects included billions of federal, state, and local dollars invested in the 1970 through current day. Funding of transportation projects in Southern California during the period 1975 through 2010 addressed a range of challenges to economic growth. The chapter proceeds in four parts: one, a discussion of the environmental context; two, description of the institutional design for governance that developed; three, an overview of the projects developed and economic impact; four, applying the lessons learned to the emerging challenges of fiscal constraints, demographic change, and institutional re-design for transportation funding.

This chapter focuses on how investment in the American rail infrastructure has shaped changes in the population and residential patterns. Specifically, the chapter examines the association between commuter rail systems, urban rail transit systems, and the movement of the college-educated young into the inner city. Two hypotheses are proposed about the characteristics of rail systems and the relationship to the growth in the percentage of young college graduates residing in close-in neighborhoods. Using a sample of central cities within the 51 largest metropolitan areas in the U.S., the chapter compares the growth in young college graduates (ages 25 to 34 years) across cities with the different transit configurations. Using correlation analysis, the chapter explores the relationship between the presence of rail transit and the residential location choices of this population group. In the discussion and conclusion, the findings are summarized and implications for policy and sustainability are discussed.

This chapter analyzes the experiences with tolling in the Hampton Roads region of Southeastern Virginia to better understand residents' and drivers' support for tolls and behavioral responses to tolls. The Hampton Roads region, with its population of 1.7 million and extensive network of highways, roads, bridges, and tunnels, has a long history of toll facilities that date back to the 1920s. The most recent tunnel tolls, associated with the Elizabeth River Crossing Project and introduced in February 2014, are the focus of this chapter. This chapter analyzes two sets of survey data to provide insights that have implications for policies regarding tolling: (1) The Life in Hampton Roads Surveys which includes questions about support for tolls and toll avoidance behaviors; and (2) two surveys (pre- and post- toll implementation) commissioned by the regional transportation planning organization.

Fuel taxes have historically been the key revenue source for a great deal of transportation infrastructure, especially roads. For many reasons, such as reduced fuel tax receipts, governments at all levels have begun to explore additional financing options. This chapter explores an option that local governments have available to them in many states: local sales taxes earmarked for transportation projects. This chapter briefly discusses the literature on local sales taxes, the diversity of the laws regarding local sales taxes earmarked for transportation, potential consequences of increased reliance on local sales taxes earmarked for transportation, and briefly discusses a similar revenue source—the local fuel tax. This research is important to understanding the changing patterns of how public transportation is being financed in many states, and if the spread of non-earmarked local sales taxes are any indicator, how it is likely to be financed in many others moving forward.

In an effort to address financial constraints and environmental concerns states have increasingly turned to a combination of un-tolled (HOV) and tolled (HOT) lanes. Public-private partnerships (3Ps) are a popular mechanism for this more sustainable approach to highway infrastructure that couples environmental sustainability (efficient utilization of existing lanes, less congestion) with financial sustainability (private investment). This chapter offers an approach to 3P contract writing for HOV/HOT facilities that is structured by a stakeholder analysis of actors in the project accountability environment. By analyzing two Virginia 3P highway projects, the chapter shows it is possible to build into a contract a set of terms and conditions to enhance the likelihood of meeting the goals of multiple stakeholders. By necessity, such contracts cannot specify precise monetary returns and other stakeholder benefits, but they can be written to include trade-offs to minimize losses to one party at the expense of another.

While it is well known that cost is a politically salient issue, much less is known about the role of cost development and how costs get elevated into the political discussion. This chapter looks at the role of cost accounting and cost development in the recent takeover attempts of two airports by the state of North Carolina. The Charlotte-Douglas airport takeover was a failed attempt, and the Asheville Regional airport takeover succeeded. The chapter makes three important contributions to the study of sustainable transportation financial management. First, it shows that costs are important to the political discussion of the takeovers. Second, it provides a discussion of the relevant costs in each case. Third, it provides research on the political dimensions of cost in the United States in the management of transportation assets, which may limit sustainable transportation policy.

Since the Government Accounting Office report "Transit Agencies' Use of Contracting to Provide Service," there is a growing interest in contracting out and any measures increasing efficiency and cost-savings. This chapter looks at the results of a unique national survey of transit agency managers conducted in 2017 for a modern snapshot of the transit industry in the United States. While there are specific factors that make transit contracting easier (e.g., competition in the provision of services), there are also factors that require contracting out but make monitoring of contracts more difficult (e.g., no capacity to provide services and monitoring in-house). The authors discuss these factors and provide illustrative examples of factors that may enhance efficiency.

Preface

It is difficult to overstate the importance of the issue of transportation, since it touches everyday lives via commuting, shopping or delivery, and generally providing mobility to people and freight. The long-term economic growth will have even greater demand for transportation (BTS 2015). Vehicle congestion, according to some estimates, decreases GDP by 1 to 3 percent (Williams & Hammond, 2015). United States is spending about $245 billion a year to build and maintain infrastructure, yet estimates suggest that there is a gap of over $2 trillion that need to be closed (ASCE, 2017). At the same time, American Society of Civil Engineers (ASCE) gave the US infrastructure an overall grade of D+ in 2017 (ASCE, 2017), and large portion of that comes from transportation infrastructure. Fixing America's Surface Transportation (FAST) Act (Pub. L. No. 114-94) has provided $305 billion over fiscal years 2016-2020 for various surface transportation programs (FHWA, 2017). This book focuses on the sustainability and financing issues of long-term transportation planning for economic growth.

The main goal of this book is to fulfil the current gap in the literature that does not address transportation challenges facing American infrastructure in the 21st century in a comprehensive manner. The objectives include to provide a multi-modal approach by highlighting not only ground transportation and trucking, but also aviation (airports), marine ports and public transit. Besides transportation focus, the book will situate the challenges into a more global context of sustainability, especially financial aspects of it. The book will also bring in management and institutional aspects of contemporary transportation. The main value of the book is in its comprehensive nature, potentially serving as a reference handbook for the future generations.

The recent trends in US population also influence the demand for the transportation, such as lesser share of teenagers who obtain driver licenses across the country, the movement of young-adults into urban areas (where they can rely more on public transportation, for example), and growing senior population. Without mobility there can be no economic growth, and the long-term economic growth will demand more transportation options and better transportation infrastructure. The expanding transportation infrastructure ultimately influences the environment bringing another important dimension to contemporary transportation issues.

Transportation encompasses planning, public policy, engineering, environmental sciences, and features prominently in the discussions on sustainability, environmental pollution and climate change, urban planning, and economic development.

TARGET AUDIENCE

The book will be of interest to not only to those who are involved in transportation, but audiences interested in public infrastructure finance, overlapping jurisdictions management, and sustainable planning as well. Although the book is not intended to be a textbook, due to a variety of topics and transportation modes included, the book can be a supplemental reading material for any graduate level course/seminar focused on the contemporary transportation policy, public finance, and sustainability and planning.

Due to the nature of issues involved, and glacial pace of policy changes in transportation arena, we anticipate that most chapters of the book will remain relevant for an extended period of time.

BOOK OUTLINE

The book can be broadly divided into two major sections. The first section covers the broad issues of sustainability, transportation planning, public management, and planning issues in contemporary transportation. In essence, this part provides definitions of sustainability and active transportation policy as well as showing some of the pathways how transportation affects economic development. The second part of the book focuses on the transportation funding and cost efficiency challenges. This part underscores the politicization of costs in the discussion of transportation projects and the importance of accountability and authority in determining transportation funding. This section adds the challenges to how the transportation infrastructure can be financially supported for the long-term economic growth. While the book focuses primarily on the US context and examples, broader implications will be interested to a wider audience.

The first chapter in this book, "A Regional Approach to Mobility Management: Promoting Sustainability and Economic Growth Through Public Transit," by Morgan Vogel, Robert Blair, and Jerome Deichert, examines transportation sustainability from a regional and collaborative perspective on the example of non-metropolitan Nebraska initiative. The mobility management contributes to the long-term economic growth. In this respect, the authors identify the problems of lack of economic growth constraining transportation infrastructure, as well as rapid economic growth being constrained by inadequate transportation infrastructure. Ultimately, the regional approaches to economic mobility that connect rural areas to urban centers provide the largest promise in long-term economic development. The chapter provides economic definitions of sustainability, introduces mobility management concepts, and provide empirical case analysis not only of Nebraska initiatives but also comparing these initiatives to the neighboring Midwestern states.

The second chapter, "Evolving Transportation Sustainability: Climate Change, Transportation Planning, and Moves to Active Transportation Infrastructure," by William Griggs introduces the concept of active transportation planning policy or policy that focuses on biking and walking. This chapter provides an in-depth discussion of the sustainability issues of the built environment and how the infrastructure can frame the mobility options available for individuals. The chapter introduces the unique challenges posted by the climate change, and how local policies have been changing to answer this unique challenge. The chapter also expansively discusses emerging issues and technologies influencing sustainable transportation infrastructure. This chapter provides another look at the potential routes the transportation infrastructure can take to drive the positive changes in the future. Vulnerable populations are more affected by the lack of walking and biking as well as transit options than other population groups. The chapter showcases some innovative ways the cities can fund the active transportation policy in addition to the sales taxes and bond issuances. The chapter proposes three major managerial principles: embracing uncertainty and disruption, using data for decision-making, and weighing the full costs of transportation investment.

Chapter 3 by Bethany Stich, James Amdal, and Peter Webb, "Chassis: The Unsung Linchpin in the Global Supply Chain," provides details not only about trucking and chassis specifically, but also towards the ocean freight transportation. The chapter covers the historical development of chassis use in the USA. The chapter also introduces the important key players in the area as well as important policies that shape the use of chassis between ocean carriers, chassis lessors, and trucking companies. The chapter provides extensive details on the role chassis plays in the modern global supply chain with some technical details that will appeal to any reader interested in the freight transportation. The chapter also provides the details of the policy implications that are important for this area due to legacy contracts or other policy considerations in the US compared to the rest of the world. The chapter recommendations are focused on the "resilient dynamics of chassis" and improvements in transportation leading to further economic development.

Chapter 4 by Mark Pisano and Richard Callahan, "Regional Institutions for Transportation Sustainability and Economic Development: The Case of Southern California," provides extensive case study of the institutional involvement and development in Southern California (Los Angeles area). The authors specify details and names and other descriptions of how the transportation infrastructure development and financing of that development influenced the economic growth of the region. The chapter provides in-depth discussion of the connection of the transportation infrastructure development and economic development. The chapter discusses how the financing of the transportation projects can influence economic development along with the construction of the infrastructure itself. The chapter provides key figures and intimate details of the regional transportation development of institutions in the southern California. Besides such extensive analysis of one region, the chapter produces important definitions of sustainability as well as enumeration of modern challenges faced by the region. Similar challenges are faced by the country as a whole, and the chapter provides important pathways of how demographics or goods mobility can help or impede future economic growth given current development of the transportation infrastructure.

Chapter 5, "From College to the City: Implications of Urban Investments in Rail Transit Systems on the Movement of the Young, College Educated Into the Center of Large American Cities," by Lenahan O'Connell, Wie Yusuf, Timothy Brock, and Benjamin Blandford explains one of the links how transportation investment can generate central cities redevelopment. The authors focus their analysis on the largest 51 metropolitan areas and compare bus provisions with various rail projects such as light rail, heavy rail, and commuter rail. Besides details on various rail projects, the chapter provides background information on the central cities redevelopment and the potential of rail investments. This chapter concludes the first part of the book discussing major contemporary transportation issues in transportation and their link to the long-term economic growth.

The second part of the book focus on how such contribution of transportation infrastructure can be achieved financially. Given the sheer scope of the US infrastructure, even simple maintenance, the amount needed to maintain the infrastructure and not let it deteriorate, should be enormous. However, there are built-in deficiencies in the current funding mechanisms for transportation infrastructure investments as outlined in several chapters.

The sixth chapter of the book, "Support for and Behavioral Responses to Tolls," discusses not only the importance of alternative transportation options, but also highlights the importance of alternative transportation funding options. The chapter highlights the high costs associated with the failure to invest in the infrastructure. The chapter relies on multiple surveys, conducted before and after the implementation of toll bridges and roads in the Hampton Roads, VA region; this represents a quasi-experimental case for what people expectations are about their behavior changes combined with what they report they have actually done. The data is supplement by the previous reported numbers of traffic volumes and public transportation ridership to support chapter's main arguments. The chapter also provides a very good definition of financially sustainable transportation.

Chapter 7 by Whitney Afonso, "Financing Transportation Through Local Sales Taxes," provides extensive discussion of local sales taxes earmarked for transportation. The chapter provides a unique view of local governments involvement in infrastructure financing for economic development. The local sales taxes have a potential for a low rate and political feasiblity creating a more stable source than, for example, tolls, which do initially generate resistance from local residents. To provide a fuller discussion of local sales taxes, the chapter provides an in-depth discussion of state taxes, including list of states that permit local governments to have sales taxes earmarked for transportation and local fuel taxes. The chapter applies a unique framework which focuses on two key characteristics in describing variations in state legislation toward local sales taxes: jurisdictional eligibility and discretionary authority. The chapter outlines the benefits and drawbacks of local sales taxes, and provides a small section on the fuel taxes as a funding mechanism for the transportation projects.

Chapter 8 by Martin Mayer, Wie Yusuf, and Lenahan O'Connel on accountability in public-private partnerships (3P) ("Reducing Risk in Public-Private Partnership Contracts: Two Examples From Highway Tolling Projects Designed to Increase Sustainability") provides an in-depth analysis of the contract specifications that can ensure accountability of various stakeholders. The chapter provides unique challenges that are faced by 3P: uncertain outcomes over the long-term time frame. Such contracts cannot be specified as simple contracts, they need to provide accountability and flexibility at the same time. The chapter provides some illustrative examples as well as two specific case studies: the 3P contracts for I-495 HOV/HOT and I-95 HOV/HOT projects. The unique specifics of tolling projects are thoroughly discussed, but the authors also provide the common themes and recommendations that can be of use for anyone interested in the public private partnerships. The chapter also provides a discussion of congestion pricing.

Chapter 9, "Cost and Cost Politics in Airport Takeovers: Two Cases From North Carolina," by Zachary Mohr and Tonderai Mushipe showcases one successful and one unsuccessful examples of airport takeovers. The efficiency and lower cost may make airport infrastructure more sustainable, but the chapter shows how political factors play an important role in the discussion of the costs of the airports. The chapter discusses the fascinating topic of costs and their discussion in the public sphere, with illustrative details of the two airports takeovers in North Carolina providing details with dates, names, and numbers. But besides the details of these two airport cases, the chapter illustrate how indirect costs play a political role in the discussion on institutional authority over such key infrastructure pieces as airports. While this chapter will be of interest to those who are particularly interested in the details of aviation, those who pay attention the role costs play in the perceptions of efficiency and control is illustrated by the chapter.

As previous chapters have shown, the fiscal administration and funding mechanisms are important for the development of transportation infrastructure and for the economic development, and Chapter 10 by Leland and Smirnova, "Contracting Out in the Transit Industry: Recent Perspectives of Transit Agency Managers," represents the in-depth look at the perceptions of potential cost-savings in the transit industry. Chapters by Pisano and Callahan on Southern California has shown how transit development is important for the overall economic development and sustainability of the region. Previous studies find little to no cost-savings from contracting out, and the chapter indicates that their maybe certain path-dependency as those agencies that are able to capitalize on cost-savings and other benefits from contracting out tend to contract out, and those agencies that operate in the environments not conducive to contracting out (e.g. facing lack of eligible bidders and/or providers) do not contract out.

As can be seen from this preface, this book will appeal to the wide audiences of scholars and practitioners who are not only interested in the contemporary transportation issues, but also interest in innovative policy proposals, influences of modern technology on the future of the infrastructure, and unique contribution of the infrastructure to the long-term prosperity and economic growth. The scholars interested in the modern finance issues, especially those affecting transportation infrastructure, will also find useful material in this book. The book also shows the importance of institutional factors, including governance forms, for the sustainability and economic development. The authors provide either original analysis based on the area or industry wide surveys or in-depth case studies and contextual analyses to support their points.

Olga Smirnova
East Carolina University, USA

REFERENCES

American Association of Port Authorities (AAPT). (n.d.). *Exports, jobs, and economic growth*. Available online at http://www.aapa-ports.org/advocating/content.aspx?ItemNumber=21150

American Public Transit Association (APTA). (2017). *2016 Public Transportation Fact Book*. Available online at: http://www.apta.com/resources/statistics/Documents/FactBook/2016-APTA-Fact-Book.pdf

American Society of Civil Engineers (ASCE). (2017). *2017 Infrastructure Report Card*. Available online at https://www.infrastructurereportcard.org/americas-grades/

Bartle, J. (2006). The sustainable development of U.S. air transportation: The promise and challenge of institutional reform. *Public Works Management & Policy*, *10*(3), 214–224. doi:10.1177/1087724X06287494

Black, W. (2010). *Sustainable transportation: Problems and solutions*. New York, NY: Guilford Publications.

Bureau of Transportation Statistics (BTS). (2017). *US Carrier Traffic Statistics through May 2017*. Available online at https://www.transtats.bts.gov/TRAFFIC/

Bureau of Transportation Statistics (BTS). (2015). *Freight Facts and Figures 2015*. Available online at https://www.transtats.bts.gov/TRAFFIC/

Egilmez, G., & Tatari, O. (2012). A dynamic modeling approach to highway sustainability: Strategies to reduce overall impact. *Transportation Research Part A, Policy and Practice*, *46*(7), 1086–1096. doi:10.1016/j.tra.2012.04.011

Federal Highway Administration (FHWA). (2017). *Fixing America's Surface Transportation Act or "FAST Act"*. Available online at https://www.fhwa.dot.gov/fastact/

Jeon, C. M., Amekudzi, A. A., & Guensler, R. L. (2013). Sustainability assessment at the transportation planning level: Performance measures and indexes. *Transport Policy*, *25*, 10–21. doi:10.1016/j.tranpol.2012.10.004

Lindholm, M., & Behrends, S. (2012). Challenges in urban freight transport planning – a review in the Baltic Sea Region. *Journal of Transport Geography*, *22*, 129–136. doi:10.1016/j.jtrangeo.2012.01.001

Lutte, R., & Bartle, J. (2016). Examining the new role of public administrators in transportation sustainability: A case in air navigation modernization. *Public Works Management & Policy*. doi:10.1177/1087724X16679845

Stich, B., & Miller, C. (2012). A New Public Philosophy for a New Railroad Era: Applying the Blacksburg Manifesto to Contemporary Transportation Issues. *Administrative Theory & Praxis*, *34*(4), 602–621. doi:10.2753/ATP1084-1806340405

Swift, E. (2011). *The Big Roads: The Untold Story of the Engineers, Visionaries, and Trailblazers who Created the American Superhighways*. Boston: Houghton Mifflin Harcourt.

Szoltysek, J. (Ed.). (2012). *Developing of Transportation Flows in 21st Century Supply Chains*. Wydawnictwo Uniwersytetu Ekonomicznego, Katowicach. Available online at: http://www.sbc.org.pl/Content/74125/SE_121.pdf

Transportation Research Board (TRB). (2008). *Potential Impacts of Climate Change on U.S. Transportation*. Available online at http://onlinepubs.trb.org/onlinepubs/sr/sr290.pdf

Williams, R., & Hammond, P. (2015). *2015 Passenger Transportation Trends*. Retrieved from https://www.strategyand.pwc.com/trends/2015-transportation-trends

Chapter 1
A Regional Approach to Mobility Management:
Promoting Sustainability and Economic Growth Through Public Transit

Morgan D. Vogel
University of Nebraska at Omaha, USA

Robert Blair
University of Nebraska at Omaha, USA

Jerome Deichert
University of Nebraska at Omaha, USA

ABSTRACT

Across the United States there is increased pressure for communities, especially in states like Nebraska, to engage in sustainable transportation infrastructure development. Through a case study of an ongoing statewide transportation initiative in nonmetropolitan Nebraska, this chapter examines transportation sustainability and planning from a regional and collaborative perspective. The Nebraska effort can be adapted to other states with significant rural and dispersed population centers. Funded by the state and the federal governments, Nebraska's transportation initiative, using an innovative public-private partnership, is creating and enhancing regional transit services in small urban and rural areas, using public transportation as a means to promote long-term economic growth and sustainability. Smaller urban and micropolitan communities, often serving as regional growth centers, frequently are overlooked when it comes to research on transportation planning and policy.

DOI: 10.4018/978-1-5225-7396-8.ch001

INTRODUCTION

Across the United States, there is increasing pressure for communities, especially in states like Nebraska, to engage in sustainable transportation infrastructure initiatives in order to create, maintain, and secure economic and social prosperity for the future. In addition to physical infrastructure, many of these transportation initiatives include mobility management. In short, mobility management is a "strategic approach to transportation service coordination that improves efficiencies and increases transportation options to meet the needs of the public" (Mack & Ruse, n.d., p. 1). Along with improving service efficiency, mobility management also promotes sustainable and long-term economic growth by connecting small urban and rural areas with significantly dispersed populations to regional centers. This chapter demonstrates how mobility management can be used to promote sustainability and economic growth and, ultimately, facilitate enhanced rural transit opportunities among providers.

This chapter provides an overview of transportation literature and mobility management initiatives in several Midwestern states, followed by a case study of a comprehensive mobility management program in Nebraska. Special attention is given to Nebraska's regional coordination strategies used to implement a statewide mobility management initiative. A comparison is made between Nebraska's mobility management efforts and its peer-states in the Midwest with a discussion of the long- and short-term implications for nonmetropolitan communities. Opportunities for future research also are discussed.

BACKGROUND

As defined by the American Public Transportation Association, mobility management "is a strategic approach to service coordination and customer service that is becoming a worldwide trend in the public transportation sector" (2018, para. 2). When implementing mobility management strategies and networks, transit providers will transition from their established roles as fixed route or demand response providers in order to collaborate with other transit providers and partners. Thus, the concept of mobility management begins with the creation of partnerships among a variety of entities in order to increase transportation options for communities (American Public Transportation Association, 2018). Often, communities implementing mobility management strategies designate a mobility manager to assist with the coordination of transit services between providers, customers, and other state and local agencies (Nebraska DOT, 2017). In short, effective communication among mobility managers, providers, customers, and other parties involved in the transit process is essential. Communication between actors enables the success of mobility management programs.

Although mobility management can be accomplished in a variety of ways, implementation through a regional approach creates more manageable geographic areas while still connecting communities across political boundaries. Additionally, a regional approach to mobility management creates transportation networks that promote long-term economic growth and sustainability in small urban and rural areas that, without regional transportation services, might otherwise become economically disadvantaged and unsustainable communities. As we will see, several Midwestern states already have some form of a regional transportation network in place.

Promoting Long-Term Economic Growth for Micropolitan Communities

Getting people to and from these mid-sized urban areas is a critical challenge for rural transportation planning and policy. In many nonmetropolitan areas of the Midwest micropolitan areas constitute regional growth centers. Micropolitan communities range in size from 10,000 to 50,000 people. A micropolitan area includes the county in which the city is located and adjacent counties that are economically associated with it. They often serve as focal points for a variety of services (especially health and medical services) and retail options for area residents. As a result, the economic viability of micropolitan communities impacts a wide area.

Mobility management activities involve the identification of customer travel needs and the coordination of these needs with existing public transit providers. Additionally, inherent to the concept of mobility management is the implementation and coordination of strategies that are efficient for the taxpayer (Burkhardt & McLary, n.d.). Similar to the economic advantages of coordinating transportation service, the economic benefits of mobility management include the efficient use of transportation resources and lower costs in providing services by unit costs (Burkhardt & McLary, n.d.).

Another economic benefit of coordinated transportation and mobility management initiatives is the development of a larger pool of funding sources and overall increase in funding available to the entities involved (Burkhardt, Koffman, & Murray, 2003). Furthermore, increased transit service will benefit communities in a number of ways, including wages and benefits provided to transit employees and the purchases of goods and services that are made possible through regional transit systems (Burkhardt et al., 2003). Of course, these economic benefits are in addition to the primary motive of mobility management: increased and enhanced mobility for passengers.

In rural communities, transportation needs generally reflect one of two categories (a) slow (or negative) growth that compromises the growth-promotion aspect of transportation or (b) rapid economic and population growth that depends on transportation policy to effectively manage growth (Kidder, 2006). In the case of mobility management initiatives, many small urban and rural communities suffer from slow growth that limits the community's growth-promotion aspect of transportation. Furthermore, initiatives enhance mobility for marginalized or disadvantaged populations who physically or financially do not have access to vehicles. In Nebraska and other Midwestern states, these individuals often reside in small urban and rural areas. In this sense, transportation policy and poverty policy are intertwined (Kidder, 2006). While addressing concerns of social equity, mobility management initiatives can contribute to economic investment in these small urban and rural communities.

Infrastructure enhancements and improvements play key roles in community development and contribute to long-term economic stability. In the case of transportation infrastructure investment, for instance, policymakers favor high-productivity investments that increase connectivity, raise the standard of living and improve the overall quality of life, and ensure the best use of taxpayer dollars for society (Holtz-Eakin & Wachs, 2011). Given the trend in declining federal and state funding, practitioners need to find balanced investment opportunities that "can be achieved by identifying deficiencies and needs, developing long-range national and regional plans to address them, and harnessing an array of public and private resources to fund projects" (Majumdar, Sen, Highsmith, & Cherrington, 2013, p. 281). Regional mobility management strategies fit this description rather well. As Majumdar, Sen, Highsmith, and & Cherrington (2013) note, mobility management balances the "assessment of the transportation needs of the community, planning and coordinating with transportation infrastructure development, and evaluation of land-use policies to close the gaps in fixed-route [or demand response] transit services" (p. 281).

The social impacts of mobility management and coordinating transportation services are widespread and numerous, ultimately, promoting long-term economic growth for communities. First, improving service quality, increasing the accessibility of transit services to more people, and expanding transportation service areas have a significant and positive impact on the level of service provision, the number of passengers, and regional accessibility (Burkhardt et al., 2003). Secondly, since passengers now have access to more goods and services, individuals can obtain better jobs, better health care, and better shopping and retail services (Burkhardt et al., 2003). More importantly, individuals retain their independence as they gain access to these improved goods and services, which is a critical component of mobility management in terms of increasing social equity. Finally, local businesses will experience an increase in activity since more people can shop locally and visit local attractions (Burkhardt et al., 2003).

As the research shows, mobility management enables communities to remain economically viable and increase social capital within communities, which can lead to increased job opportunities and economic growth for communities. Given decreasing federal funding and resources, it is all the more important for small urban and rural communities to adopt a decentralized and regional approach to infrastructure investment in order to create sustainable communities. Regional mobility management initiatives are one way to do this.

Sustainability in Transportation

The viability of micropolitan and other nonmetropolitan communities depends on efficient, accessible, and affordable public transportation systems that can connect these communities to regional cores. In order for transportation systems to have an enduring effect on micropolitan communities, they must be developed with sustainability in mind. Regional mobility management adds this sustainable dimension.

While *sustainability* often is used in the environmental context, it also characterizes "long-term success or preservation of organizations, organizational plans, or economic prosperity that may not have an environmental or natural capital dimension" (Leuenberger & Bartle, 2009, p. 4). When conceptualizing sustainability from this perspective, one can see the centrality of economics to the concept of sustainability; sustainability is "the use of society's resources to provide the greatest satisfaction" (Leuenberger & Bartle, 2009, p. 27). From this perspective, we also can think of sustainability policy as providing the same level of services without decline in goods or values. In the fields of public administration and public policy, the ideas of sustainability and sustainable development are essential since they require public administrators and policymakers to consider both short-term and long-term implications of policy solutions. Sustainability in transportation policy, development, and infrastructure is no exception.

According to Leuenberger and Bartle (2009), the broad goal of sustainable transportation is balancing "the desire for mobility with the use of resources" (p. 76). Essentially, this description applies the same notion of sustainability from above but to the area of transportation policy and systems. Leuenberger and Bartle (2009) identify four dimensions of transportation sustainability: environmental, economic, financial, and social. For the purposes of regional mobility management structures, the economic, financial, and social dimensions of sustainable transportation are most important.

The economic dimension suggests that transportation is not the end in itself; rather, transportation is the means to an end with the goal being to achieve the highest social return on capital (Leuenberger & Bartle, 2009). Similar to the long-term economic growth investment in mobility management initiatives, economic sustainability is critical to investment in the project. Similarly, the financial dimension involves having sufficient funds to cover the costs of the project over the long term (Leuenberger & Bartle, 2009). This is also a component of investment in mobility management and, perhaps, a reason why additional funding from multiple sources is critical to the concept.

Finally, the social dimension of transportation sustainability encourages improvements in the standard of living and quality of life, in particular for marginalized populations who do not have access to existing service provisions (Leuenberger & Bartle, 2009).

In summary, the social dimension of creating sustainable transportation is integral to the concept of mobility management. As research shows, when transportation infrastructure and services promote long-term economic growth, investment in social capital, and long-term economic and social sustainability for communities, transportation development is a wise investment. Regional mobility management initiatives offer opportunities for communities to invest wisely in transportation infrastructure.

Regional Transportation Development in Midwestern States

In the Nebraska initiative, regional mobility management approaches appear to be critical to transportation development and one of the keys to establishing economically viable and sustainable micropolitan communities. When looking to Nebraska's peer-Midwestern states, it appears that regional collaborative structures and strategies related to transportation development are relatively common.

This chapter looked at other Midwestern states similar to Nebraska with significantly dispersed and rural population centers. Illinois, Indiana, Iowa, Kansas, Missouri, and Ohio were studied in order to get a sense of the scope and nature of regional collaborative structures related to transportation and, more specifically, the extent to which regional approaches are utilized for mobility management. Research on these states indicates that all of these states coordinate some degree of mobility management programs with regional planning organizations (RPOs) or regional transportation planning organizations (RTPOs). In short, RPOs and RTPOs "offer states a consistent statewide model for conducting planning that is continuing, comprehensive, and cooperative" (Kissel, Schwartz, James, & Dyer, 2016, p. 6). Many of the RPOs or RTPOs have policy boards and technical committees composed of local officials, planning staff from local governments, state DOT staff, representatives from different transportation modes, and other agency representatives or advocacy groups with an interest in transportation (Kissel et al., 2016). These regional planning organizations often act as an ongoing link between state DOTs, local officials, and other stakeholders. While these regional planning organizations may not be directly responsible for mobility management implementation, they are often involved in the planning process and long-term development.

Based on literature related to regional partnerships, the following proposition is offered.

Proposition One: Regional Approaches to Public Transit Promote Sustainability and Economic Development

Presently Iowa, Kansas, and Missouri appear to have the most developed mobility management networks in their states. Although stated as a priority by the state DOTs of Illinois, Indiana, and Ohio, there is no formal designation or creation of mobility management districts according to research into their state programs.

Iowa

As defined by the Iowa DOT, Iowa has 16 mobility management regions in the state with nine mobility managers (Iowa DOT, n.d.). Iowa also has 18 regional planning affiliations (RPAs) that work with Iowa DOT and metropolitan planning organizations (MPOs) to communicate policy updates, new tools and resources, and best practices (Kissel et al., 2016).

Kansas

Kansas is divided into 10 coordinated transit districts with only two mobility managers. They plan on increasing the number of mobility managers in the future (Spadafore, 2017). It is unknown if the Kansas mobility management program coordinates with any type of regional planning organization or entity.

Missouri

Missouri's mobility management initiative differs from Iowa and Kansas. Although Missouri's program is centralized under a program called *MO Rides*, the state's transit districts or regions follow the boundaries for the state's regional districts for Missouri's Regional Planning Commissions (RPCs) and Councils of Governments (COGs) (MO Rides, 2018). Many of the transit providers and services are coordinated by the RCPs and COGs.

Illinois

In Illinois, 96 out of 102 counties offer some form of transit service to communities (Illinois DOT, 2015). Illinois DOT has completed coordinated transportation plans for 11 regions in the state and has hired regional human service transportation plan coordinators who perform tasks that would be regularly associated with mobility managers (Kissel et al., 2016).

Indiana

Indiana does not explicitly mention statewide mobility management initiatives, though it does provide information and applications for 5310 and 5311 funding, in which applicants are encouraged to work with their regional planning organizations (Indiana DOT, 2018a & 2018b). In 2001, Indiana DOT launched the Small Urban and Rural Transportation Planning Program with five regional development organizations (RDOs) and four MPOs who assist the state in meeting federal transit requirements (Kissel et al., 2016).

Ohio

Ohio does not have statewide mobility management districts, but does have five regional transportation planning organizations (Kissel et al., 2016). Additionally, according to the Ohio DOT, 5310 funding is distributed directly to Ohio counties who are eligible and apply (Ohio DOT, 2018).

Though Illinois, Indiana, and Ohio do not have established mobility management districts like Iowa, Kansas, and Missouri, it appears that a significant amount of regional transportation planning and mobility management related tasks occur through regional planning organizations in all six Midwestern states.

Based on the descriptions of regional structures in different states, the following proposition related to mobility management initiatives is offered.

Proposition Two: Mobility Management Implementation Strategies That Reflect the Nature of Social Capital and Area Assets Specific to Each Region Promote Sustainability and Economic Development

An in-depth case study of rural transit planning and policy implementation in Nebraska follows. It illustrates the value of adopting regional strategies to implement statewide mobility management initiatives according to each region's social capital and other area and community strengths and assets.

CASE STUDY: MOBILITY MANAGEMENT IN NEBRASKA

Public Transit in Nonmetropolitan Nebraska

Nebraska contains less than two millions people, with the vast majority of the population concentrated in the major metropolitan areas in the eastern one third of the state. Population is widely dispersed among communities in the western two thirds of the state. Nebraska has 93 counties. Currently, there are nine counties in the state without any public transportation service and another five counties without rural area service, as illustrated below in Figure 1. In all but three cities, public transit is provided by demand response, where there are no fixed or regular routes. This includes eight of the nine micropolitan cities. In a demand response system, the customer calls in advance to schedule a trip. The transit provider then dispatches a vehicle to pick the passengers up and transport them to their destinations. Fees and schedules vary based upon location.

Much of the need for public transportation results from the lack of a vehicle in the household. According to the U.S. Census Bureau's American Community Survey, 5.7% of all households and 9.9% of households with a householder aged 65 or older lacked a vehicle in the 2012-2016 period (U.S. Census, 2017). Additionally, 8.8% of the state's unemployed did not have access to a vehicle (U.S. Census, 2018). By providing access to services and employment opportunities for these and other marginalized populations, public transportation strengthens the social equity and community capital of nonmetropolitan communities. This improves the likelihood of economic viability and long-term sustainability for these communities.

Figure 1. Locations in Nebraska with public transit service
Source: Compiled from information provided by Nebraska DOT

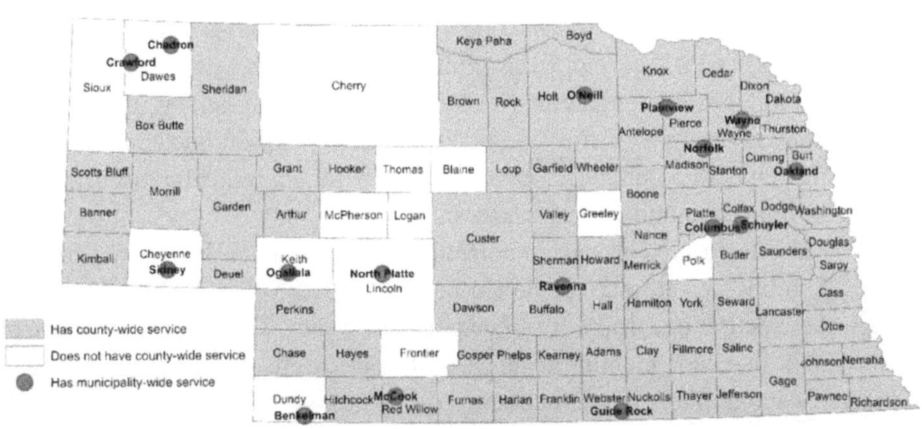

By statute, the Nebraska DOT (NDOT) is the principal state agency responsible for coordinating public transportation activities in the state and provides technical assistance to improve Nebraska's public transportation system when requested (Nebraska Public Transportation Act, 1993). The department also may contract to provide public transportation services.

NDOT provides funding to transportation providers in the state through three separate funding sources: Section 5311 (a federal program for rural public transportation); Section 5310 (a federal program for specialized transportation for the elderly and disabled in rural areas); and Program 305 (a state program for rural public transit).

The 5311 program provides capital, planning, and operating assistance to states to support public transportation in rural areas with populations of less than 50,000, where many residents often rely on public transit to reach their destinations. There are 58 rural public transportation providers delivering public transportation to 84 of Nebraska's 93 counties.

Advanced Partnership

To enhance its services and provide flexibility, NDOT formed a partnership called Advance with the Nebraska Safety Center (NSC) at the University of Nebraska at Kearney and the Center for Public Affairs Research (CPAR) at the University of Nebraska at Omaha. This partnership allows NDOT to extend its staff, bring in experts in content areas, add resources, learn about best practices, and benefit from the latest technology. Additionally, the partnership provides research and engagement opportunities for University faculty, staff, and students.

Through these activities, this partnership has provided the necessary flexibility to meet the varied needs of public transit providers. For example, the NSC was able to take driver training programs to locations throughout the state, making training more available. Using faculty specializing in e-governance and graduate students in computer science, CPAR developed an online portal that allows transit providers the ability to submit invoices, update service information, submit funding applications, and so forth. Other areas of collaboration among the partners include assistance with Federal Title VI compliance, safety management, analysis of transportation's community impact, and marketing and social media.

NDOT also integrated private organizations, along with its University partners, creating a public-private partnership to enhance travel options across the state through the creation of a Mobility Management Project. The goal of this project is to establish statewide mobility management strategies to address mobility needs and identify and implement alternatives to fill service gaps.

Figure 2. Mobility management regions in Nebraska
Source: Compiled from information provided by Nebraska DOT

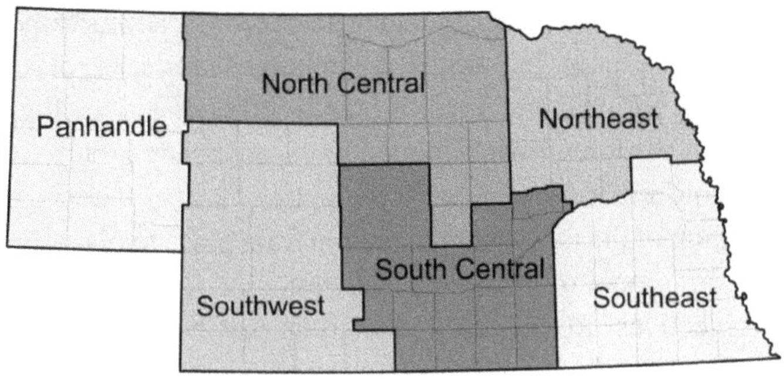

Regional Coordination

Nebraska is a large state geographically, and in order to create more manageable transit plans, and ensure local control, the Mobility Management Project identified six mobility management regions. The boundaries of the regions, as illustrated below in Figure 2, follow county boundaries, and the service areas of existing transportation agencies. The regions also generally conform to the borders of economic development districts, or regional planning organizations, whose staff participate in the project.

Each region formulates comprehensive transit plans to coordinate services and identify strategies to expand public transportation options for citizens in the area. The plans will vary from region to region, reflect on the geography of the area, and address mobility needs and gaps in public transit services identified in research conducted by the NDOT. The strategies will discover ways to leverage existing services, improve transit efficiency, identify service expansion opportunities, market to increase transit ridership, secure additional funding sources, and address other transit issues in the area.

Working with Advance staff, coordinating committees in each region will develop area-specific transit plans. Members of the committees include public, nonprofit and private transit service providers, social service agencies, local governments, and other public transportation stakeholders. Each region will have a mobility manager to coordinate transit services and implement strategies included in plans.

A coordinating committee was piloted in the southeast region. Based on information gathered in a series of meetings, service providers identified a number of collaborative transit strategies. Working with a private consultant and Advance, the committee also developed a mobility management handbook to assist other regional coordinating committees in the development of area transit plans and strategies.

Selected Regional Strategies

The Mobility Management Project included research on nonmetropolitan public transit needs and challenges in Nebraska. Researchers used a variety of data collection approaches: surveys, forums, and small group discussions at locations around the state. Social service agencies, governmental units, advocacy groups, and residents provided valuable information.

An understanding of the movement of people from place to place, or trips, forms the foundation for regional transit plans and strategies. Trips meet a multitude of transportation objectives. Research conducted as part of the Nebraska Mobility Management Project identified a range of nonmetropolitan trips, including: commuting workers to places of employment, transporting people to medical and health care facilities for appointments, getting shoppers to a variety of retail establishments, and connecting consumers to a range of personal services.

Research also identified trip destinations, or the places where people need to be transported. Destinations widely varied because they were generated by dispersed populations in nonmetropolitan locations in Nebraska. In other words, unlike most densely populated urban areas, the places that people need to get to in nonmetropolitan communities are often at distant locations. Trip destinations identified by research as part of the Mobility Management Project ranged widely. These destinations included moving people within nonmetropolitan locations; from rural to micropolitan areas; and to a lesser though important degree, from nonmetropolitan to metropolitan locations. In other words, there are inter-nonmetropolitan and intra-nonmetropolitan transit destinations.

The research found that many trip destinations often cross county and other governmental boundaries. That is significant since most transit agencies operate only within one county. In order to transport people across county lines, then, there must be some type of collaboration among the various governmental units. Governmental units use a range of collaborative instruments in partnering for the delivery of public services, including transit.

Researchers studying all types of intergovernmental initiatives have found that an "integral component of the interlocal cooperative process is the collaborative tool, or mechanism utilized to create the initial bond" (Blair & Janousek, 2013, p. 269). Blair and Janousek examined the use of collaborative mechanisms by local governments in Nebraska delivering a range of public services over a 15-year period. The researchers employed a classification system of eight types of mechanisms, placing them in two broad categories: informal administrative networks or formal organizational structures. They found that roughly two thirds of the collaborative mechanisms used by Nebraska local governments were informal in nature, and one third used formal collaborative structures, including tools such as agreements and federations (Blair & Janousek, 2013). The researchers in the Mobility Management Project also found a range of collaborative tools, though in the delivery of transit services the mechanisms of collaboration tended to be more formal in nature, likely due to the complexity of transit services and the involvement of nonprofit agencies in the transportation system.

While it is not possible to describe all of the regional transit strategies that have been developed or identified by the Mobility Management Project, it is informative to provide brief illustrations. These examples of regional transit strategies reflect diverse trip destination strategies and different collaborative mechanisms among the participating public, private and nonprofit entities.

Tri-City Road Runner Flexible Route Transit System

According to the Nebraska Public Transit website, Nebraska DOT provided the following description of the Tri-City Road Runner transit service:

One of the identified needs from the Mobility Management Project for the Panhandle Region was fixed route bus service in the Scottsbluff-Gering area. At the time, Scotts Bluff County Public Transportation provided demand response service in town and throughout Scotts Bluff County to all residents and visitors. On January 10th 2018, two flexible bus routes started operating along two established bus routes, which arrive and depart as determined by an established schedule, but deviate from the bus route for curb-to-curb pickups and drop-offs within ¾ mile of the bus route, and then return to the fixed route path. (Nebraska Public Transit, n.d., para. 6)

The two deviated fixed routes operate north/south within Scottsbluff, Gering, and Terrytown to provide connections throughout the community, with alignments primarily along 10th Street and Broadway. The routes serve grocery stores, higher density residential areas, major employment areas, medical facilities, and other activity centers. (Nebraska Public Transit, n.d., para. 7)

The collaborative structure employed in this transit strategy to move people around within a micropolitan area involved the creation of a multi-city and county-based transportation agency. The creation of this public transportation system required the participation of two moderately-sized neighboring municipalities and a county government. Because of the complexity of this regional transit strategy, the local governments employed a formal collaborative mechanism, using a series of formal agreements and structural arrangements by the governing bodies.

Tryon-North Platte Weekly Transit

This transit strategy was designed to provide transit service to and from North Platte (a micropolitan community) for residents of the neighboring McPherson County. Beginning in May 2017, "Hooker County Public Transportation made their demand response weekly transit service between Hooker County and North Platte available" to McPherson County residents, by adding a stop in Tryon, a small city located in McPherson County (Nebraska Public Transit, n.d., para. 8). This original service ran between Hooker County to North Platte, going through but not stopping in McPherson County. This coordination strategy improves the mobility of McPherson County residents who previously did not have public transit service available to them and leverages the existing services by adding a stop on a route that already existed.

While it would seem that this cross-county regional transit strategy would need to employ a more formal collaborative mechanism, in fact, this particular example used informal mechanisms. The Hooker County Public Transportation leveraged and expanded their services in an informal manner to transport McPherson County residents to another county to access needed services.

Tecumseh to Auburn Demand Response Weekly Transit

According to the Nebraska Transit website, Nebraska DOT provided the following description of the Tecumseh to Auburn transit service:

As part of the Mobility Management Project, several stakeholder group meetings were held to improve the coordination between transit providers to address the gaps and overlaps of service. In the southeast region, one such group meeting involved initiating communication between Johnson County Transit in Tecumseh and Blue Rivers Area Agency on Aging (BRAAA) for potential regional trip coordination. Johnson County Transit offers monthly regional trips to Nebraska City, Beatrice and Lincoln. In September 2017, Johnson County Transit identified a currently unmet transit need for regional service from Tecumseh to Auburn in Nemaha County. Since it was not feasible for Johnson County Transit to meet the demand at the time, coordination with BRAAA helped in addressing the need. BRAAA started a demand response Tecumseh-to-Auburn transit route for residents of Tecumseh. The service not only improves mobility for Tecumseh residents but also increases economic activity in Auburn. (Nebraska Public Transit, n.d., para. 13)

This regional transit strategy, which generally involved moving people among a number of nonmetropolitan and micropolitan communities, employed a range of informal collaborative mechanisms. In particular the transit strategy was based on improving the communication among the various providers to improve trip coordination and service efficiencies in an informal manner. The use of informal associations among providers to more efficiently deliver public services was found by Blair and Janousek (2013) to be a common collaborative mechanism in Nebraska local governments.

FINDINGS

Based on a review of Nebraska's mobility management initiatives compared to its peer-states in the Midwest, it appears that Nebraska has implemented mobility management initiatives based on somewhat different regional strategies and collaborative mechanisms. This inference is based on two observations. First, Nebraska's mobility management regions were specifically designated as transportation districts rather than adopting regional boundaries based on preexisting economic development districts or other issue-specific regional organizations. Second, mobility management strategies and partnerships in Nebraska's regions were created to reflect existing community relationships in each region.

Another inference pertains to the manner in which regional transit strategies are forged. In particular the collaborative mechanisms employed for these regional transit partnerships reflect the work by Blair and Janousek (2013). While formal organizations are often needed to implement area strategies, such as the case of the Tri-City Road Runner, informal mechanisms are also employed. Blair and Janousek (2013), for instance, learned that Nebraska local governments show "a tendency for the more indefinite, casual and need-based exchanges of resources in interlocal cooperation" (p. 280). This tendency for informal collaborative mechanism is consistent in two of the case studies discussed above and the research conducted for this Mobility Management Project. This suggests then that regional organizations such as area development districts which employ a range of collaborative tools in their interaction with local governments would be effective facilitators in the development and implementation of regional transit strategies.

Although Nebraska's peer-states exhibit some of these characteristics, Nebraska's case of mobility management is distinctive because it is developing a more comprehensive approach to mobility management through the creation of specific mobility management regions, regional coordinating committees, and regional partnerships as well as instituting regional mobility managers. The state's implementation of this statewide mobility management initiative through regional structures highlights the project's focus on creating sustainable transportation infrastructure and long-term economic growth that will enhance Nebraska's micropolitan communities by providing better integration with surrounding rural and small urban communities.

Even though Nebraska's mobility management initiative is still in the development stage, the state's commitment to mobility management initiatives is further in development than mobility management efforts in Illinois, Indiana, and Ohio in terms of regional strategies. While Illinois identifies the transit service offered in the state and coordinates transportation plans by region, Illinois has not assigned specific transportation districts. Indiana does not offer formally detailed mobility management initiatives, though it does have RDOs and MPOs that coordinate and assist with federal transit requirements. Ohio recently established regional transportation planning organizations, but does not address the specific coordination of mobility management. Like Nebraska, Indiana and Ohio DOTs assist with 5310 and/or 5311 funding. In general, while Illinois, Indiana, and Ohio offer some of the transportation infrastructure necessary for approaching mobility management from a regional perspective, their initiatives are not nearly as developed compared to other states in the Midwest.

Nebraska's model mirrors the mobility management initiatives in Iowa, Kansas, and Missouri since these states all approach mobility management from a regional perspective. Iowa, Kansas, and Nebraska's approaches to mobility management are similar, in that all three states are divided into transportation or mobility management districts with each region assigned a mobility manager or mobility coordinator. While Iowa's mobility management strategy has been implemented, Kansas and Nebraska's mobility management initiatives are still in the development and early implementation stages at the time of this writing.

Missouri's mobility management initiative, MO Rides, also has been implemented and appears to have increased service delivery and efficiency for passengers. Missouri's initiative has the added benefit of being organized according to the state's regional districts for COGs and RPOs; therefore, the mobility management implementation process for Missouri may have been easier since the Missouri DOT and/or mobility managers did not have to create transportation districts.

Iowa and Missouri's mobility management initiatives appear to be operationally successful, particularly from a regional perspective, providing increased mobility to marginalized populations. Nebraska and Kansas, albeit in the early stages, are laying the foundation for successful regional mobility management programs in their states similar to Iowa and Missouri. In the Midwest, Iowa, Missouri, Kansas, and Nebraska are among the leading examples of adopting regional approaches to mobility management.

DISCUSSION AND RECOMMENDATIONS

Based on research from peer-Midwestern states and the case study on Nebraska's approach to mobility management, there are several important points for discussion as well as recommendations.

The issue presented in this chapter is that, throughout the United States, there is increasing pressure for states with significantly dispersed and rural population centers to engage in sustainable transportation infrastructure development in order to promote long-term economic growth. In Midwestern states like Nebraska, many of these sustainable transportation initiatives include mobility management programs developed and implemented from a regional and collaborative perspective. Research on Midwestern states and the in-depth case study on mobility management in Nebraska presented in this chapter suggests that Midwestern states are developing mobility management programs from regional perspectives. This supports Proposition One as offered earlier in the chapter. These regional strategies have aspirations that may lead to transportation sustainability and economic development. However, additional research and data are needed to demonstrate the effectiveness of regional mobility management programs on these goals. Nonetheless, several of the Midwestern states, as discussed in this chapter show promise, particularly Missouri.

Since each state has developed a slightly different approach to regional coordination of mobility management, this suggests that guidelines on mobility management structures, goals, and indicators should be adaptable. For example, in Nebraska, the population density and composition differs dramatically from the western side of the state compared to the eastern side. Western Nebraska is composed of smaller urban and rural communities compared to the metropolitan communities of eastern Nebraska.

This demonstrates that there is no one size fits all approach to mobility management, at least in Nebraska. Rather, a regional approach to mobility management allows for mobility managers and stakeholders to draw on existing relationships and community assets in order to tailor mobility management programs and services to fit the needs of each region. This finding supports Proposition Two as offered earlier in the chapter. In the long term, regional mobility management initiatives are more sustainable and promote economic development and growth for micropolitan communities.

It should be noted that the propositions were not formally tested as a part of this chapter for two reasons. First, the purpose of this case study was not an evaluation of mobility management programs in Midwestern states. Rather, the purpose was to present a model for how to use regional approaches to develop and implement mobility management programs with the intent of promoting long-term economic growth and sustainability in micropolitan communities. Second, several of the mobility management initiatives studied in this chapter are still in the development and early implementation stages, therefore making it difficult to evaluate impacts or test the propositions based on ongoing and constantly changing implementation processes. Developing testable hypotheses is the next step in analyzing regional approaches to mobility management and is discussed later on in this chapter as an opportunity for future research.

Based on this research, there are two recommendations for developing mobility management programs.

1. It can be assumed regional planning and development structures exist in most states. These existing regional structures often reflect ongoing relationships among the range of potential stakeholders in the development of mobility management initiatives. It is recommended that, prior to developing mobility management programs or strategies, these existing regional structures and their relationships should be considered.
2. Closely related to the first recommendation is that these regional organizations can offer a great deal of knowledge and expertise in the development and implementation of regional mobility management programs. For this reason, it is recommended that members of these planning organizations should be included as stakeholders in regional mobility management coordinating committees.

FUTURE RESEARCH DIRECTIONS

A significant limitation of this research is that Nebraska's mobility management regions are still in the development stage; thus, the propositions set forth in this chapter cannot yet be tested. Although regional mobility management partnerships are occurring throughout the state of Nebraska, currently the regional mobility management structure is only operating in the state's Southeast region. This was designed as a pilot project in order to evaluate the regional coordination initiatives and progress prior to launching additional mobility management regions across the state. Since the pilot project remains under evaluation, there are not yet outcomes or results to present regarding the propositions offered in this chapter.

A major area for future research on this topic, especially in Nebraska, is evaluating the regional mobility management structures as they are developed, implemented, and at full operating capacity. As mentioned throughout the chapter, each region in Nebraska is unique in terms of its existing social capital and tendency toward collaboration. While Nebraska's Panhandle region may exhibit a history of collaboration, other regions may be less inclined to do so, at least without assistance from Nebraska DOT or other partners involved in the mobility management project. In this regard, gathering valid and reliable outcomes and results may take time as the mobility management regions are formed; nonetheless, the propositions offered in this chapter serve as guidance for transit practitioners who are implementing or considering undergoing regional mobility management initiatives.

Another limitation of this research is the inability to compare the effectiveness of mobility management across states due to the lack of performance measures (Majumdar et al., 2013). Without consistent indicators and measures of achievement for mobility management initiatives, it is difficult to compare accurately mobility management initiatives and the corresponding effectiveness across states. Future research on this topic should include the development of a set of performance indicators, subsets of which can be applied to any mobility management program. The purpose of these indicators would be to understand and explain the effectiveness of mobility management. They should be adaptable so that they can be made appropriate to a specific region and address its stated goals and objectives.

Five categories of performance measures to consider, as offered by the National Center for Mobility Management (n.d.), include:

1. **Inputs**: What was invested. Common measures could include items that currently are measured such as financial and capital resources.
2. **Outputs**: What was produced. These are quantifiable measures such as number of people served, number of trips, hours of service, and so forth.
3. **Outcomes**: What difference did the program make. These are short-term and long-term results that directly address the goals of the program.
4. **Stakeholder satisfaction**: How the program measured up to the satisfaction of the stakeholders. Often programs will measure rider satisfaction, but the satisfaction of other stakeholders, including service providers and funding agencies should be considered.
5. **Impacts**: What are the broad levels of change. This would look at how the community and individuals have changed as a result of the mobility management program. Has the program helped to secure economic and social prosperity of the community?

While these measures highlight the key components of constructing a mobility management initiative, they are measures to consider and, therefore, may not be applicable across states. A specific focus for future research is to test these five performance measures and encourage transit practitioners to adopt these measures nationwide as well as any additional performance indicators determined through future research. Until performance measures are nationally recognized, it is difficult to compare mobility management outcomes and results across states because there is not a valid and reliable method for evaluating overall impact. For this reason, an impact comparison of mobility management initiatives in Midwestern states to Nebraska's initiatives was not included in this chapter. Furthermore, given declining federal resources, establishing national performance measures for mobility management will be useful for transit providers and other transit practitioners as they compete for federal funding (Majumdar et al., 2013). Performance measures will aid providers in demonstrating the effectiveness of their mobility management programs and should be considered an important area for future research.

CONCLUSION

Increasingly, it seems small urban and rural areas are often forgotten when it comes to transportation planning and policy, despite the fact that significant portions of their residents are dependent on public transit services. The purpose of this research was to offer a regional model for how to use mobility management programs to expand and enhance transit services in these small urban and rural areas as well as promote long-term economic growth and sustainability for these communities.

There were two propositions presented in this chapter. First, in response to Proposition One, research from Nebraska and its peer Midwestern states suggests that utilizing regional transit strategies is an effective method for implementing mobility management programs. Drawing on regional strategies and approaches to public transit can tie a state's regions together by integrating the nonmetropolitan areas with regional micropolitan cores. Second, in response to Proposition Two, the case study of Nebraska suggests that regional mobility management structures which do not rely solely on a state's predetermined regional planning and development boundaries, such as economic development districts, may better reflect existing community relationships in each region. For instance, as Nebraska's mobility management regions develop, the regional mobility manager and regional coordinating committee stakeholders can develop specific mobility goals and indicators that reflect their region's strengths and assets.

Focusing on existing regional relationships and community assets will enable mobility management programs to contribute to a transportation infrastructure specific to each region. This research demonstrates that having a supportive and tailored regional transportation infrastructure promotes long-term economic growth and sustainable communities that will, in turn, create stronger relationships between micropolitan communities and regional cores.

ACKNOWLEDGMENT

Research reported in this publication is part of a Statewide Transit Initiative conducted by the University of Nebraska on behalf of the Nebraska Department of Transportation [Project RPT-C990(017)] supported in part by the Federal Transit Administration [PTE Federal Award PL1601 Subaward 17-002-01]. Any opinions, findings, and conclusions or recommendations expressed in this publication are those of the author(s) and do not necessarily represent the official view of the Federal Transit Administration.

REFERENCES

American Public Transportation Association. (2018). *Mobility management*. Retrieved from http://www.apta.com/resources/mobility/Pages/default.aspx

Blair, R., & Janousek, C. (2013). Collaborative mechanisms in interlocal cooperation: A longitudinal examination. *State & Local Government Review*, *45*(4), 268–282. doi:10.1177/0160323X13511647

Burkhardt, J., & McLary, J. (n.d.). *The business case for mobility management*. American Public Transportation Association. Retrieved from https://www.apta.com/resources/mobility/Documents/Business-Case-for-Mobility-Management.pdf

Burkhardt, J. E., Koffman, D., & Murray, G. (2003). *Economic benefits of coordinating human service transportation and transit services*. Transportation Research Board of the National Academies. Retrieved from http://onlinepubs.trb.org/onlinepubs/tcrp/tcrp_rpt_91.pdf

Holtz-Eakin, D., & Wachs, M. (2011). *Strengthening connections between transportation investments and economic growth*. Bipartisan Policy Center National Transportation Policy Project. Retrieved from http://bipartisanpolicy.org/wp-content/uploads/sites/default/files/NTPP%20Strengthening%20Connections%20Paper.pdf

IllinoisD. O. T. (2015). *Transit system*. Retrieved from http://www.idot.illinois.gov/transportation-system/Network-Overview/transit-system/index

Indiana D. O. T. (2018a). *Rural Transit Assistance Programs*. Retrieved from http://www.in.gov/indot/2816.htm

Indiana D. O. T. (2018b). *Enhanced mobility of seniors and individuals with disabilities*. Retrieved from http://www.in.gov/indot/2817.htm

Iowa D. O. T. (n.d.). *Iowa mobility management*. Retrieved from https://iowadot.gov/iowamobilitymanagement/home

Kidder, B. (2006). *The challenges of rural transportation*. Logan, UT: Western Rural Development Center.

Kissel, C., Schwartz, B., James, S., & Dyer, S. (2016). *Regional rural transportation planning: State models for local consultation, regional coordination, and regional transportation planning organizations*. NADO Research Foundation.

Leuenberger, D. Z., & Bartle, J. R. (2009). *Sustainable development for public administration*. Armonk, NY: M.E. Sharpe.

Mack, A., & Ruse, K. (n.d.). *Mobility management: Empirical evidence of fiscal benefits from multiple states*. Omaha: Nebraska Department of Roads and University of Nebraska at Omaha College of Public Affairs & Community Service. Retrieved from https://www.nebraskatransit.com/assets/pdf/Mobility%20Management%20White%20Paper.pdf

Majumdar, S. R., Sen, L., Highsmith, M. K., & Cherrington, L. (2013). The case of performance measurement in mobility management programs. *Public Performance & Management Review*, *37*(2), 280–301. doi:10.2753/PMR1530-9576370205

National Center for Mobility Management. (n.d.). *Performance measures for mobility management*. Retrieved from https://nationalcenterformobilitymanagement.org/wp-content/uploads/2014/09/Performance_Measures_Final.pdf

Nebraska D. O. T. (2017). *Nebraska mobility management: Mobility manager (Southeast Region)*. Retrieved from https://www.nebraskatransit.com/assets/pdf/Newsletter%201_MM_Oct2017_Final.pdf

Nebraska Public Transit. (n.d.). *Mobility management statewide project*. Retrieved from https://nebraskatransit.com/mobility-1.php

Nebraska Public Transportation Act, Nebraska Revised Statutes, Chapter 13 §1201-1214 (1993).

Ohio D. O. T. (2018). *Ohio mobility management program guide*. Retrieved from http://www.dot.state.oh.us/Divisions/Planning/Transit/Documents/Mobility%20Management%20Program%20Guide.pdf

Rides, M. O. (2018). *Mobility management*. Retrieved from http://morides.org/mobility-management/

Spadafore, M. (2017). *Mobility management: Successes and future plans*. Retrieved from https://s3.amazonaws.com/v3-app_crowdc/assets/a/ab/ab5808057ca15bba/Coordination_StatewideMobilityManagementInitiatives_MikeSpadafore.original.1510071701.pdf?1510071705

U.S. Census Bureau. (2017). *2012-2016 American Community Survey 5-Year Estimates Table B2045 Tenure by Vehicles available by Age of Householder*. Retrieved from https://factfinder.census.gov

U.S. Census Bureau. (2018). *2012-2016 American Community Survey 5-Year Estimates Public Use Microdata Sample for Nebraska*. Retrieved from https:factfinder.census.gov

ADDITIONAL READING

American Public Transportation Association. (2018). Retrieved from http://www.apta.com/Pages/default.aspx

Community Transportation Association of American. (2016). Retrieved from http://web1.ctaa.org/webmodules/webarticles/anmviewer.asp?a=23&z=2

Ellis, E. (2009). Mobility management. Retrieved from https://assets.aarp.org/rgcenter/ppi/liv-com/roundtable_091013_mobility.pdf

Federal Transit Administration. (n.d.). Retrieved from https://www.transit.dot.gov/

Litman, T., & Burwell, D. (2006). Issues in sustainable transportation. *International Journal of Global Environmental Issues*, *6*(4), 331–347. doi:10.1504/IJGENVI.2006.010889

National Rural Transit Assistance Program. (2018). Retrieved from http://www.nationalrtap.org/

Nebraska Transit. (n.d.). Retrieved from http://nebraskatransit.com/

Price, D., & Leather, A. (2011, November). Transport mobility management: Small changes – big impacts. *Journeys*, 20–30.

KEY TERMS AND DEFINITIONS

Collaborative Tools: A range of formal and informal mechanisms employed by local governments to coordinate the delivery of public services, including transit.

Demand Response: Type of public transit in which the customer calls in advance to schedule a trip.

Economic Development: A process that enhances the economic, social, and political well-being of individuals and communities through the adoption of new techniques and activities and overall improvement of living standards.

Fixed Route: Type of public transit in which the vehicle operates on a regular basis with a designated route and fixed schedule.

Micropolitan: An urban area, as defined by the U.S. Office of Management and Budget, containing a community with a population between 10,000 and 50,000 people.

Mobility Management: Strategic, efficient approach to transportation service coordination and transportation options in order to improve service delivery for members of the public.

Public Transit: A system of public transportation, using vehicles like buses, Vanpools, etc., operating by demand response or by fixed routes that are used by members of the public.

Regional Coordination: A development and implementation process across jurisdictions in order to address issues of regional significance and devise efficient and effective solutions.

Section 5310: A federal program for specialized transportation for the elderly and disabled in rural areas.

Section 5311: A federal program for capital, planning, and operating assistance to states in support of rural public transportation.

Chapter 2
Evolving Transportation Sustainability:
Climate Change, Transportation Planning, and Moves Toward Active Transportation Infrastructure

William Riggs
University of San Francisco, USA

ABSTRACT

Transportation policymakers and planners have begun to realize the importance of sustainable transportation with regards to health, social implications, and the climate. Focusing on more active travel is one way that these officials are beginning to evolve cities in a way that supports these broader sustainability goals. In this light, this chapter focuses on how active transportation has evolved, and how policy and finance can support it. It also looks at emerging issues that may reshape transportation, such as connected and autonomous vehicles, and how we can maintain transportation sustainability in light of these innovations.

INTRODUCTION

While transportation planning has been gradually moving toward prioritization of multimodal travel goals for quite some time (Frederick, Riggs, and Gilderbloom 2016), this change has been accelerated by issues related to climate change. Caused by a variety of human activities, scientists have observed a warming atmosphere, with the last three decades being successively warmer than any previous decade

DOI: 10.4018/978-1-5225-7396-8.ch002

since records began in the 1850's, with 2015 being the warmest on record (Brown et al. 2016). These unprecedented changes caused by human activity have spurred policy changes and has reframed the importance of active transportation and how it is financed and priorities for entities such as cities and counties.

This chapter traces the history and development of active transportation planning policy, e.g. policy that focuses on biking and walking (Sallis et al. 2004), and how it connects to health, (Corburn 2007) as well as the many other interrelationships between the built environment. These include land use, accessibility to parks and open spaces, public transportation, economic development and housing, along with health disparities that cross racial and ethnic lines (Krieger 2000; Maantay 2001). This ultimately concludes with recent transitions to climate action planning and efforts to establish not only healthy but "climate smart" transportation programs (Riggs 2014).

After an overview of transportation planning, the active transportation movement and climate action planning, the chapter dialogues ways that cities are using active transportation strategies and other urban policies to address climate-related transportation issues. This will include a review of policies and design strategies that are currently being used to improve the suitability of the environment for this kind of travel. Another review of financial best practice drawing on the work of Riggs and McDade (2016) will also be presented. Additionally, examples of how cities are pursuing programs that reframe sustainable transportation and travel will be shown. Each city has taken unique steps not only to plan climate-sensitive transportation solutions, but has also built financial vehicles to move from policy to action and implement those solutions.

Finally, the chapter discusses emerging issues related to sustainable transportation infrastructure and climate— autonomous vehicles in particular. While this emerging technology will reduce collisions and improve access to healthcare for those who need it most—particularly high-cost, high-need individuals at the younger and older ends of the age spectrum (Riggs and Boswell 2016), opportunities also exist to connect individuals to jobs and change the way cities organize space and optimize trips (Fagnant and Kockelman 2014; Guerra 2015). However, policy and critical dialogue has lagged in this area—particularly in the arena of both sustainability and dialogue about how innovation and technology will impact not only the way we commute but our efforts to address climate change.

BACKGROUND

The Climate Imperative

Over the last 100 years many parts of the US have become highly dependent on fossil fuels for economic and urban growth and transportation. Cities have experienced population and economic booms and have subsequently expanded -- all with the assumption that these resources were inexhaustible. This has brought to light apparent consequences for this consumption, based in part on the reliance on fossil fuels for transportation and the "compromised" livelihood this provides for future generations (Brundtland 1987). For example, massive species extinction and inundation of many low-lying countries around the world (Monbiot 2007).

What paradigm will cities follow in the future? What shape will they take as they transition to being more populous yet having a greater demand for economic prosperity bound by ecological sustenance? How might we rethink cities on the rise and those in decline? How do things like transportation, housing, and economics relate to ecological prosperity and livability for humans? These questions frame how many planners are thinking about the future of transportation around the world. Planners and policy makers are seeing unprecedented technological innovation at the same time as a high degree of uncertainty (Riggs and Boswell 2016a).

While these global implications and the issue of climate change are also real, many of the solutions, particularly those that deal with non-point sources related to transportation and the move to active travel initiatives.

Just what is "active travel"? It can mean walking, bike and even transit—anything that involves moving a body in a non-fossil-fuel, calorie-consuming manner. (While you may be quick to dismiss transit as "active" travel, work by Voss (2015) actually suggests that transit riders get a lot of exercise walking or biking to a stop, but they do not substitute that exercise for regular recreational activity—something that cyclists and walkers tend to do.) Research has shown correlations between places that support walking, cycling and transit and their environmental, social, and health benefits (Ewing 2005; Frank, Andresen, and Schmid 2004; Frank et al. 2005a; Saelens, Sallis, and Frank 2003; Environment 1993; Newman and Jennings 2008; Newman and Kenworthy 1999). Additional work has found denser urban areas to be correlated with improved environmental quality alongside increased population health—for example, having a lesser heat-island effect by providing more street trees on urban streets (Stone et al. 2014; Stone Jr 2005, 2014).

Given this background, policy moves should support sustainable systems, such as healthy buildings and transportation systems for bicycles and pedestrians. This can lead to improved economic and ecological resilience. For example, literature indicates that transportation investments increase local accessibility and improve economic productivity, thereby increasing local land and housing values (Landis et al. 1995; Dowall and Landis 1982; Cervero 2007; Nelson 1999; Diaz 1999). Research indicates that transit-connected areas weathered the 2008-2010 economic downturn better than more suburban areas and that, in city centers, housing prices either dipped less or went up (Cervero 2012; Glaeser, Gottlieb, and Tobio 2012). If features like walkability, bikability, or more energy-efficient buildings have environmental and financial impacts, as well as the potential to open up equitable places for people of all ages and abilities, whether on foot, on bikes, in wheelchairs, or in buses, cars, or streetcars (Dunham 2011; Dunham-Jones and Williamson 2011), then perhaps they are worth exploring with more aggressive policy and finance actions.

Moves Toward Active Travel Policy

In this light, shaping active transportation policy can help inform how we build more sustainable and healthy places. This is the approach guiding this chapter and links it to the idea of how we build a broad culture of health in the US. This is a topic of broad significance in the US given the increase in obesity and other chronic diseases in recent years. And, while many researchers in active transportation are eager to evaluate sustainable or healthy behaviors, this does not account for disparity in place / the built environment. In many locations, the built environment is not conducive for walking and biking. Many of the same locations there have cyclically had disproportionality high number of vulnerable individuals (Riggs 2011). These vulnerable individuals have historically dealt with disparity based on continued resource inequity, segregation and long-term exposure to adverse environmental conditions (Hu et al. 2003; Massey 2004; Thompson et al. 1999; Williams and Jackson 2005).

It is in these underprivileged locations that this chapter focuses on the concepts of 1) healthy places and practices, 2) vulnerable populations and 3) the idea of culture creation in the area of transportation behavior. This area of healthy transportation behavior is important but this chapter steps back and recognizes the disparity in transportation resources across geography place as a starting point. That being said, very little is known about how much or what kind of disparity there is with regard to active transportation infrastructure and resources in the US.

Yet active transportation research and policies are important because well-designed, complete streets are a seminal building block for influencing healthy behavior – especially in youth who have become less active and socially engaged in the recent past, based in part by the environment around them (Frank et al. 2007; Putnam 2001). They can help re-shape aspects of the built environment to contribute to walking and biking, and a 'culture of health' in transportation.

Studies have shown the health and active transportation benefits of living in a dense, urban, and connected environment (Cervero and Duncan 2003; Ewing 2005; Frank, Andresen, and Schmid 2004; Frank et al. 2005b; Glazier et al. 2014; S. Handy, Cao, and Mokhtarian 2005, 2006; Saelens, Sallis, and Frank 2003; Sallis and Glanz 2006; Sturm and Cohen 2004; Ewing et al. 2003; Cervero and Kockelman 1997) but the practical reality is that most communities are not designed for active travel (Sallis et al. 2004).

Given this, many have suggested using 'complete streets' policy to reshape urban streetscapes, or design the urban environment to accommodate and prioritize pedestrians and cyclists over cars (de Zeeuw and Flusche 2011). Yet this trend is not pervasive. Larger cities like San Francisco and New York have been implementing policies that promote active transportation policies but far fewer smaller, poorer communities are doing the same (Lillis 2013; Seskin and McCann 2012; Peters 2012). California, for example, has a *Complete Streets Act*, but its implementation is limited to new projects and plans, does not require a baseline Level of Service (LOS) for bicycles and pedestrians in plans and designs, and, most importantly, does not control land use decisions.

While integrating facilities for bicycling and walking may be easy in this context for larger and wealthier areas (who do many projects every year), it is not as easy or straightforward in smaller and poorer communities where there is less funding and fewer projects. In some of these locations, installing sidewalks near a local school would do little to promote walking to school since there is no housing within the proximity due to land use and zoning. Schools and places of community have been isolated from residential areas, presenting a structural constraint to environmental and behavioral health interventions. Studies about this have shown that environmental design alone is not the 'silver bullet' in influencing healthy behaviors (Forsyth et al. 2008).

Many communities, especially those that are rural and poor, do not have policies designed to improve the built environment for walking and biking, and facilitate a culture of health in active transportation—particularly in the infrastructural realm. The vast bulk of literature emphasizes this as a key impediment and obstacle to active travel. For many individuals it is not the behavioral aspects of active transportation that are a limiting factor but the fact that adequate places to walk or bike safely do not exist.

This is especially acute when considering youth trips to school and in many small towns in more suburban, rural and poor communities—which often have a higher concentration of minorities. These locations do not have the land use or built environment infrastructure (e.g. sidewalks and bike lanes) to facilitate active transportation nor do they have 'complete streets' policies that require bicycle and pedestrian facilities. Academics and practitioners know a lot about 'what' these locations look like and 'why' active transportation strategies are needed, but challenges remain on 'how' cities can achieve a more walkable and bikable infrastructure. This is a key policy, implementation and management challenge in cultivating healthier commutes to school and work.

Opportunities and Challenges

Given the background provided, there are both opportunities and constraints in many communities to become more 'transportation rich' with regard to active transportation. While the foundation of active transportation lies in infrastructure, there is a growing interest in how new and emerging mobility options can complement active transportation infrastructure. Literature shows that multimodality and transportation disruption hold promise not only in increasing total community health and active travel, but also in providing better and more egalitarian options for vulnerable populations in the future (Frederick, Riggs, and Gilderbloom 2016; Riggs 2018).

First, infrastructure and design matter. Since current roadway design is not planned for other modes of transportation such as bicycling and walking, context sensitive solutions can focus on physical infrastructure that supports active transportation and all users—irrespective of age or ability. Whether users are young or old, designing roadways that prioritize walking and biking and fit the context of the community can be a powerful opportunity. It can also be something that leads to increased walking and cycling behavior.

For example, Raleigh, North Carolina, was listed as having the state's most dangerous street (Hillsborough Street) for pedestrians and bicyclists. In 1999 a group of five hundred citizens and stakeholders joined for a charrette in making the Hillsborough Street safer for pedestrians and bicyclists through the implementation of a 'complete streets' policy. In October of 2010 the Hillsborough Street had been re-designed to include four round-a-bouts, a road diet, and streetscape improvements that had wider sidewalks and medians (Burden, 2011, p. 39). These changes are examples of context specific design improvements that make the pedestrian and bicyclist the focus of roadway construction.

One challenge to implementation was not just that existing current design of streets favored auto-mobility—it was funding and finance. While ample research now shows the benefits of walkability (Gilderbloom, Riggs, and Meares 2015) and even that things like converting streets from one-way to two-way or installing active transportation infrastructure (Riggs and Gilderbloom 2015, 2017) can yield greater economic sustainability, a key challenge is transportation funding.

Currently in the U.S., revenue streams for transportation projects are generated primarily through federal funding vehicle ownership, highway user fees, energy consumption, and beneficiary and local option fees (Vavrova, Chang, C., and Bina, L.

2017). While this federal funding to regions can boost the number of bicycle-related projects (Handy and McCann 2010; Cradock et al. 2009), it "provides insufficient funds to cover current transportation spending" (Laing 2013). With less gas tax money available to pay for projects, and with constitutional restrictions in some states on the allocation of that money, local communities face stiffer competition with one another for available state dollars to fund bicycle and streetscape infrastructure. Even those such as San Francisco, which has a sales tax measure, have reported gaps in funding (Coté 2013). With crumbling public infrastructure, communities look to alternate sources to find the money necessary to fund the implementation of planned alternative infrastructure.

This was summarized in some recent case examples from a study of bicycle and pedestrian program finance (Riggs and McDade 2016). The most widely used local funding mechanisms reveal that reliance on funding from a county sales tax measures was a highly popular mechanism. Bond issues were the second most widely used approach, primarily by larger cities but surprisingly also by the smallest. General fund allocations and transportation impact fees followed as the third most widely employed funding method. This case analysis suggested that even with a desire and plan for active transportation, many cities still struggle to find funding for their bicycle and pedestrian projects. Using general funds and even focusing on planning-related in-lieu fees was not common.

And this is where sales tax measures and bond issuances are becoming major ways of operationalizing active travel infrastructure come in to play. In a majority of cases these funding measures filled in the gap where the CIP did not deliver funding. Most plans indicated that they provided consistent revenue streams, if planned conservatively to account for market ups and downs. In most cases the cities indicated that the fees acted as a tool to capture the impacts of non-residents who also used streets, but they are were a necessity. They provided stable funding in a creative and unique way.

In this context, it is worth mentioning cases where cities were trying to be forward-thinking in order to reinvent transportation funding and thereby meet sustainability goals. Most considering private funding, such as the cities of Memphis, Denver and Christchurch (New Zealand) are all using private crowdsourcing to fund and implement certain components of their bicycle infrastructure (Andersen 2014; Anderson 2013). Others cited use of the parklet model, where certain portions of the street are in-effect "rented" to local businesses to create appealing infrastructure like parks, eating areas or other creative streetscape features (City of San Francisco 2013; Loukaitou-Sideris et al. 2012).

Likewise, many cities are using property-based measures, including London, Singapore, Stockholm, and Milan. Each has had success using cordon pricing systems to reduce vehicle congestion related delay in their financial and urban centers (May et al. 2002; Liu et al. 2014; Broaddus 2014; Broaddus, Browne, and Allen 2015). A cordon system is sometimes called a congestion zone and essentially creates a fee to access certain portions of an area based on time-of-day, day-of-week, etc. This invisible barrier can be used as tool to price trips in and out of an area, thereby impacting the number of trips and congestion.

The City of London began charging private vehicles this type of fee to enter Central London on weekdays as of 2003 (Litman 2006). In London, this pricing strategy has been combined with improvements to transit and improvements to safety and access for bicyclists and pedestrians. The result of this strategy is reduced congestion related delays in central city roadways during peak hours and a safer roadway environment for non-motorized travelers (Komanoff 2013). London has also had success using fees collected from their cordon pricing system to pay for enhanced bicycle and pedestrian infrastructure. As of 2006/2007, approximately 3% of net revenues ($4 million of $137 million in revenues) were spent on support for new pedestrian crossings and cycling initiatives (Transport for London 2008).

Into the Future

These opportunities provide a way that cities can attempt to continue to develop transportation sustainability and move to zero-emission, active transportation, yet they are not without challenges. Some of the fastest growing and most invested-in transportation industries in recent years have been networked and data driven personal travel. While it originated with peer-to-peer networks that allowed for car and ride-sharing, a sophisticated industry has arisen in the era of companies like Get-Around, Lyft and Uber. These companies offer variations of on demand services for travel via individual vehicles all from a networked, mobile platform. These companies connect riders with drivers through mobile smartphone apps, allowing point-to-point travel similar to a taxi service (Rayle et al. 2016).

Because such services allow for convenient point-to-point mobility, some say they have the potential to reduce automobile ownership in urban areas with the potential to active transportation (W. Riggs 2018). They may facilitate first and last mile connections via walking and biking, (Rayle et al. 2016; Shaheen and Chan 2016) and compliment transit (Rassman 2014). Other work has suggested that this smart and connected form of mobility actually reduces public transit ridership and increase vehicle miles traveled or VMT (Clewlow and Mishra 2017), and created complicated pick up and drop off issues (Riggs, Boswell, and Zoepf 2017).

The Uber and Lyft revolution (what transportation professionals classify as transportation network companies or TNCs) is compounded and underscored by moves toward autonomous vehicles (Larco 2017). As self-driving cars are getting smarter and merging with our devices, there are clear opportunities to shape advances in transportation, and to harness them to reshape cities and improve the socio-economic health of cities (Lipson and Kurman 2016). There are opportunities to reduce collisions and improve access to healthcare and to connect individuals to jobs and change the way cities organize space and optimize trips (Fagnant and Kockelman 2014; Guerra 2015).

That said, there are also challenges and little understanding of how travel behavior and streets will change in this brave new world. Some work has used chauffer-based experiments to try to test behavior (Harb et al. 2018)--providing the best simulation of this technology thus far--and many other high-level policy documents have made suggestions and predictions that range in type and scale—both time and geographic (Anderson et al. 2014; Isaacs 2016; Litman 2014; Airbib and Seba 2017). This work has shown that technological change is outpacing urban planning and policy. The 2018 resistance to the influx of scooters in many global cities provides one example of this. Cities have been unprepared to deal with disruption of the status quo and have established rules limiting the growth of this active (and non-emitting) mode of travel.

While scooters provide a bit of a paradox, with regard to autonomous vehicles, most planners and engineers estimate that, absent policy, vehicle miles traveled will increase (Sperling 2018; W. Riggs and Boswell 2016b). The cost of the ride and time spent in vehicles will go down thus the demand will go up. While some suggest fuel taxes can provide a solution, there are likely to be costs. In the U.S., fuel taxes have served as the primary revenue source for transportation projects. However, increasing fuel efficiency and alternative-fuel vehicles entering the market has led to a decrease in this primary funding source – so much so that, since 2001, expenditures have exceeded revenues generated by fuel taxes. With dwindling funding available for transportation, there is a very real need for new methods of funding for transportation projects.

In this light, focus, use or impact fees likely provide the key tool to increase funding and implementation capacity for active transportation projects. This is similar to what is already being done in London and Singapore, and advances and proliferation of mobile technology change our ability to reinvent this as a tool. As Wachs states in a 2010 Rand report,

Several technologies are nearly ready that would meter travel and charge fees for road use on a national scale. But larger-scale testing is required in order to compare the technical merits of each technology and test acceptability. In principle, systems almost ready for deployment can charge per-mile road use rates for travel that differs by jurisdiction, vehicle type, road type, time of day, and even current level of service or congestion. A proposed system of user fees could also charge for automobile insurance based on miles and location of driving and provide a technological pathway for the introduction of "congestion pricing," which is advocated by many to control the growth of urban traffic.

Likewise, Portland, Oregon's transportation system development charge (TSDC) is levied on new developments and on property-use changes. The fee is used to provide active transportation infrastructure and mitigate transportation impacts of new development or property use. Fees must be used within 10 years, and revenues are used only for qualified projects that are in the city's Capital Improvement Plan (CIP). Ironically they have also established a quite progressive scooter policy (City of Portland 2018), and some of this funding could go to infrastructure supporting that as a mode.

At the same time as these location strategies are implemented, leadership is needed more than money to deal with future challenges, particularly since many transportation advances involve partnership between the government and private industry. In this light, regional leadership is needed. History suggests that MPOs can be a force for regional change (Sciara 2017). Planners and policymakers could anchor future focus on reforms to MPOs' existing legal and administrative frameworks. Planners should revisit the membership and voting structures of MPO boards to ensure better stakeholder representation and permit some MPOs to generate and direct transportation funds at the local level.

CONCLUSION

As discussed throughout this chapter, there is a strong emissions, climate and health imperative for policy makers, planners and citizens to consider active transportation infrastructure to facilitate more green travel behavior. Yet this is not happening in a vacuum. Trends toward disruptive transportation will continue to present opportunities and challenges to meeting environmental targets. In light of this, the following general management principles are work considering for citizens, policy makers and citizens.

First, managers must *embrace uncertainty and disruption*. The transportation industry is going through a revolution, and it will be important to adapt active transportation goals (and infrastructure) to innovations we have not yet conceived. The disruptive mobility of tomorrow may not be the mobility of tomorrow. Our vision will change and, consistent with Moores Law,[1] the current disruption will be disrupted even faster than we anticipate. Yet one of the consistent principles that cities value most is walking and biking. Ensuring the disruption of the future does not prohibit or eliminate the possibility for active travel is essential—be it a scooter, a car, a Hyperloop train or a flying taxi.

Second, in light of this disruption, data is ever-more important. Policy makers, planners, elected officials, advocates, and citizens must be able to *"say it with data."* Whether we are talking about collisions or accessibility to bike lanes and sidewalks, quantification is an imperative, yet this is not only from a standpoint of arguing for additional infrastructure provisions. We are entering a future of owned user data from transportation companies like Uber and Lyft, who complement active transportation trips, particularly for poor and minority users. Should companies have a right to deny cities data about the vehicles they are running on city roads? Should they have a right to deny service and leave a city (or a portion of a city) without notice? These are important considerations going forward.

Third and finally, all of us must *weigh the full costs* of transportation investment. Be they investing in bike lanes or the social cost of collisions that could be mitigated by a fully-autonomous fleet of cars, we must share the burden as well as the benefit of transportation investment in the future.

For the sake of the planet and our health, policy, programs and management strategies should lead us not into a future solely built on digital transport, but one built to foster livability, sustainability and humanity. In doing so we can create streets and infrastructure that are not only livable but that prioritize the carbonless travel of tomorrow.

REFERENCES

Andersen, M. (2014). How Denver Got an Oil Company to Help Crowdfund a Protected Bike Lane. *People for Bikes*. Retrieved from http://www.peopleforbikes.org/blog/entry/how-denver-got-an-oil-company-to-help-crowdfund-a-protected-bike-lane?utm_source=twitterfeed&utm_medium=twitter

Anderson, M. (2013). *What Caused Portland's Biking Boom?* Retrieved from http://bikeportland.org/2013/07/02/what-caused-portlands-biking-boom-89491

Broaddus, A. (2014). Sustainable Transportation: Lessons from London. *Focus: Journal of the City and Regional Planning Department*, *11*(1), 10.

Broaddus, A., Browne, M., & Allen, J. (2015). Sustainable Freight: Impacts of the London Congestion Charge and Low Emissions Zone. In *Transportation Research Board Annual Conference*. Retrieved from http://andreabroaddus.com/wp-content/uploads/2015/01/Impacts-of-Sust-Transport-Policies-on-Freight-TRB-2015.pdf

Brown, D., Cabbage, M., McCarthy, L., & Norton, K. (2016). *NASA, NOAA Analyses Reveal Record-Shattering Global Warm Temperatures in 2015.* Press Release. Available at http://www. nasa. gov/press-release/nasanoaa-analyses-reveal-record-shattering-global-warm-temperatures-in-2015

Brundtland, G. H. (1987). World Commission on Environment and Development. *Our Common Future*, 8–9.

Cervero, R. (2007). Models for Change: Lessons for Creating Active Living Communities. *Planning Magazine,* 1.

Cervero, R. (2012). *Integrating Transit and Urban Development: Lessons and Challenges for Developing Countries.* Presented at the Urban Transport Pathways: Do the Experiences of the US and the EU Offer Lessons for the Developing World? Berkeley, CA.

Cervero, R., & Duncan, M. (2003). Walking, Bicycling, and Urban Landscapes: Evidence from the San Francisco Bay Area. *American Journal of Public Health, 93*(9), 1478–1483. doi:10.2105/AJPH.93.9.1478 PMID:12948966

Cervero, R., & Kockelman, K. (1997). Travel Demand and the 3Ds: Density, Diversity, and Design. *Transportation Research Part D, Transport and Environment, 2*(3), 199–219. doi:10.1016/S1361-9209(97)00009-6

City of Portland. (2018). *Shared Electric Scooter Pilot.* Retrieved from https://www.portlandoregon.gov/transportation/77294

City of San Francisco. (2013). *San Francisco Parklet Manual.* Retrieved from http://pavementtoparks.sfplanning.org/docs/SF_P2P_Parklet_Manual_1.0_FULL.pdf

Clewlow, R. R., & Mishra, G. S. (2017). Disruptive Transportation: The Adoption, Utilization, and Impacts of Ride-Hailing in the United States. In *Research Report UCD-ITS-RR-17-07.* Davis, CA: University of California, Davis, Institute of Transportation Studies. Retrieved from https://itspubs.ucdavis.edu/wp-content/themes/ucdavis/pubs/download_pdf.php?id=2752

Corburn, J. (2007). Reconnecting with Our Roots: American Urban Planning and Public Health in the Twenty-First Century. *Urban Affairs Review, 42*(5), 688–713. doi:10.1177/1078087406296390

Coté, M., & Cabanatuan, J. (2013). Task Force: Billions Needed to Fix Transit in S.F. *SFGate.* Retrieved from http://www.sfgate.com/bayarea/article/Task-force-billions-needed-to-fix-transit-in-S-F-5008878.php#src=fb

Cradock, A. L., Troped, P. J., Fields, B., Melly, S. J., Simms, S. V., Gimmler, F., & Fowler, M. (2009). Factors Associated with Federal Transportation Funding for Local Pedestrian and Bicycle Programming and Facilities. *Journal of Public Health Policy*, *30*(S1), S38–S72. doi:10.1057/jphp.2008.60 PMID:19190583

de Zeeuw, D., & Flusche, D. (2011). How a Bill Becomes a Bike Lane: Federal Legislation, Programs, and Requirements of Bicycling and Walking Projects. *Planning & Environmental Law*, *63*(8), 8–11. doi:10.1080/15480755.2011.604227

Diaz, R. B. (1999). Impacts of Rail Transit on Property Values. In *American Public Transit Association Rapid Transit Conference Proceedings*. Retrieved from http://www.rtd-fastracks.com/media/uploads/nm/impacts_of_rail_transif_on_property_values.pdf

Dowall, D. E., & Landis, J. D. (1982). Land-Use Controls and Housing Costs: An Examination of San Francisco Bay Area Communities. *Real Estate Economics*, *10*(1), 67–93. doi:10.1111/1540-6229.00258

Dunham, M. (2011). Where the Shoe Leather Meets the Road: Learning From Experience in Crafting a Complete Streets Ordinance. *Planning & Environmental Law*, *63*(8), 3–8. doi:10.1080/15480755.2011.604226

Dunham-Jones, E., & Williamson, J. (2011). *Retrofitting Suburbia, Updated Edition: Urban Design Solutions for Redesigning Suburbs*. John Wiley & Sons. Retrieved from http://books.google.com/books?hl=en&lr=&id=1xH4b4pQzOkC&oi=fnd&pg=PR6&dq=Dunham+2011+complete+streets&ots=VQxXk-4eTP&sig=-j4ZvAz2R8Rk_Bsvugby_qqQmQg

Environment, Department of. (1993). *Reducing Transport Emissions Through Planning*. London: HMSO.

Ewing, R. (2005). Can the Physical Environment Determine Physical Activity Levels? *Exercise and Sport Sciences Reviews*, *33*(2), 69–75. doi:10.1097/00003677-200504000-00003 PMID:15821427

Ewing, R., Schmid, T., Killingsworth, R., Zlot, A., & Raudenbush, S. (2003). Relationship between Urban Sprawl and Physical Activity, Obesity, and Morbidity. *Urban Ecology*, 567–582. PMID:13677962

Fagnant, D. J., & Kockelman, K. M. (2014). The Travel and Environmental Implications of Shared Autonomous Vehicles, Using Agent-Based Model Scenarios. *Transportation Research Part C, Emerging Technologies*, *40*, 1–13. doi:10.1016/j.trc.2013.12.001

Forsyth, A., Hearst, M., Oakes, J. M., & Schmitz, K. H. (2008). Design and Destinations: Factors Influencing Walking and Total Physical Activity. *Urban Studies (Edinburgh, Scotland)*, *45*(9), 1973–1996. doi:10.1177/0042098008093386

Frank, L., Kerr, J., Chapman, J., & Sallis, J. (2007). Urban Form Relationships with Walk Trip Frequency and Distance among Youth. *American Journal of Health Promotion*, *21*(4Suppl), 1–8. PMID:17465175

Frank, L. D., Andresen, M. A., & Schmid, T. L. (2004). Obesity Relationships with Community Design, Physical Activity, and Time Spent in Cars. *American Journal of Preventive Medicine*, *27*(2), 87–96. doi:10.1016/j.amepre.2004.04.011 PMID:15261894

Frank, L. D., Schmid, T. L., Sallis, J. F., Chapman, J., & Saelens, B. E. (2005). Linking Objectively Measured Physical Activity with Objectively Measured Urban Form: Findings from SMARTRAQ. *American Journal of Preventive Medicine*, *28*(2), 117–125. doi:10.1016/j.amepre.2004.11.001 PMID:15694519

Frederick, C., Riggs, W., & Gilderbloom, J. (2016). Multi-Modality and Public Health: A Multivariate Analysis of 148 U.S. Cities. Academic Press.

Gilderbloom, Riggs, & Meares. (2015). Does Walkability Matter? An Examination of Walkability's Impact on Housing Values, Foreclosures and Crime. *Cities, 42*(Part A), 13–24. doi:10.1016/j.cities.2014.08.001

Glaeser, E. L., Gottlieb, J. D., & Tobio, K. (2012). *Housing Booms and City Centers*. Retrieved from http://works.bepress.com/cgi/viewcontent.cgi?article=1007&context=joshua_gottlieb

Glazier, R. H., Creatore, M. I., Weyman, J. T., Fazli, G., Matheson, F. I., Gozdyra, P., ... Booth, G. L. (2014). Density, Destinations or Both? A Comparison of Measures of Walkability in Relation to Transportation Behaviors, Obesity and Diabetes in Toronto, Canada. *PLoS One*, *9*(1), e85295. doi:10.1371/journal.pone.0085295 PMID:24454837

Guerra, E. (2015). Planning for Cars That Drive Themselves Metropolitan Planning Organizations, Regional Transportation Plans, and Autonomous Vehicles. *Journal of Planning Education and Research*. doi:10.1177/0739456X15613591

Handy, S., Cao, X., & Mokhtarian, P. (2005). Correlation or Causality between the Built Environment and Travel Behavior? Evidence from Northern California. *Transportation Research Part D, Transport and Environment*, *10*(6), 427–444. doi:10.1016/j.trd.2005.05.002

Handy, S., Cao, X., & Mokhtarian, P. L. (2006). Self-Selection in the Relationship between the Built Environment and Walking: Empirical Evidence from Northern California. *Journal of the American Planning Association*, *72*(1), 55–74. doi:10.1080/01944360608976724

Handy, S., & McCann, B. (2010). The Regional Response to Federal Funding for Bicycle and Pedestrian Projects: An Exploratory Study. *Journal of the American Planning Association*, *77*(1), 23–38. doi:10.1080/01944363.2011.526537

Harb, M., Xiao, Y., Circella, G., Mokhtarian, P., & Walker, J. (2018). Projecting Travelers into a World of Self-Driving Cars: Naturalistic Experiment for Travel Behavior Implications. *Proceedings of the 97th Transportation Research Board.*

Hu, F. B., Li, T. Y., Colditz, G. A., Willett, W. C., & Manson, J. A. E. (2003). Television Watching and Other Sedentary Behaviors in Relation to Risk of Obesity and Type 2 Diabetes Mellitus in Women. *Journal of the American Medical Association*, *289*(14), 1785. doi:10.1001/jama.289.14.1785 PMID:12684356

Jr, S. (2005). Urban Heat and Air Pollution: An Emerging Role for Planners in the Climate Change Debate. *Journal of the American Planning Association*, *71*(1), 13–25. doi:10.1080/01944360508976402

Jr, S. (2014). *The City and the Coming Climate: Climate Change in the Places We Live*. Retrieved from http://www.jurareview.ro/2014_6_1/p_103_108.pdf

Komanoff, C. (2013). Lessons from London after 10 Years of the Congestion Charge. *Streetsblog*. Retrieved from http://www.streetsblog.org/2013/02/15/lessons-from-london-after-10-years-of-the-congestion-charge/

Krieger, N. (2000). Epidemiology and Social Sciences: Towards a Critical Reengagement in the 21st Century. *Epidemiologic Reviews*, *22*(1), 155–163. doi:10.1093/oxfordjournals.epirev.a018014 PMID:10939022

Laing, K. (2013). Bill would eliminate federal transportation funding. *TheHill*. Retrieved from http://thehill.com/blogs/transportation-report/infrastructure/190402-bill-would-eliminate-federal-transportation

Landis, J., Guhathakurta, S., Huang, W., Zhang, M., & Fukuji, B. (1995). Rail Transit Investments, Real Estate Values, and Land Use Change: A Comparative Analysis of Five California Rail Transit Systems. *EScholarship*. Retrieved from http://escholarship.org/uc/item/4hh7f652

Larco, N. (2017). *When Are AVs Coming? (10 Car Companies Say Within the Next 5 Years...) | Urbanism Next.* Retrieved from https://urbanismnext.uoregon.edu/2017/08/28/when-are-avs-coming-10-car-companies-say-within-the-next-5-years/

Lillis, R. (2013). Council Votes to Launch Bike Corrals, Plan Parklets in Sacramento. *The Sacramento Bee*. Retrieved from http://www.sacbee.com/2013/10/22/5844018/council-votes-launch-bike-corrals.html

Lipson, H., & Kurman, M. (2016). *Driverless: Intelligent Cars and the Road Ahead*. MIT Press.

Litman, T. (2006). London Congestion Pricing. *Implications for Other Cities*. Retrieved from http://www.mumbaidp24seven.in/reference/london_congestion_pricing.pdf

Liu, Z., Wang, S., Qu, X., & Shiwakoti, N. (2014). Congestion Pricing with Distance Tolls: A Review and New Developments. In *CICTP 2014: Safe, Smart and Sustainable Multimodal Transportation Systems*. American Society of Civil Engineers. doi:10.1061/9780784413623.331

Loukaitou-Sideris, Brozen, Callahan, Brookover, LaMontagne, & Snehansh. (2012). *Reclaiming the Right-of-Way: A Toolkit for Creating and Implementing Parklets*. Academic Press.

Maantay, J. (2001). Zoning, Equity, and Public Health. *American Journal of Public Health*, *91*(7), 1033–1041. doi:10.2105/AJPH.91.7.1033 PMID:11441726

Massey, D. S. (2004). Segregation and Stratification: A Biosocial Perspective. *Du Bois Review*, *1*(01), 7–25. doi:10.1017/S1742058X04040032

May, A. D., Liu, R., Shepherd, S. P., & Sumalee, A. (2002). The Impact of Cordon Design on the Performance of Road Pricing Schemes. *Transport Policy*, *9*(3), 209–220. doi:10.1016/S0967-070X(02)00031-8

Nelson, A. (1999). Transit Stations and Commercial Property Values: A Case Study with Policy and Land-Use Implications. *Journal of Public Transportation*, *2*(3), 77–93. doi:10.5038/2375-0901.2.3.4

Newman, P., & Jennings, I. (2008). *Cities as Sustainable Ecosystems: Principles and Practices*. Island Press.

Newman, P., & Kenworthy, J. (1999). *Sustainability and Cities: Overcoming Automobile Dependence*. Island Press.

Peters, S. (2012). *Impact Fees for Complete Streets*. Los Angeles, CA: University of California. Retrieved from http://164.67.121.27/files/Lewis_Center/CompleteStreetsInitiative/Peters_report.pdf

Putnam, R. D. (2001). *Bowling Alone: The Collapse and Revival of American Community*. Simon and Schuster.

Rassman, C. L. (2014). Regulating Rideshare without Stifling Innovation: Examining the Drivers, the Insurance Gap, and Why Pennsylvania Should Get on Board. *Pitt. J. Tech. L. & Pol'y*, *15*, 81.

Rayle, L., Dai, D., Chan, N., Cervero, R., & Shaheen, S. (2016). Just a Better Taxi? A Survey-Based Comparison of Taxis, Transit, and Ridesourcing Services in San Francisco. *Transport Policy*, *45*(January), 168–178. doi:10.1016/j.tranpol.2015.10.004

Riggs & Gilderbloom. (2015). Two-Way Street Conversion Evidence of Increased Livability in Louisville. *Journal of Planning Education and Research*. . doi:10.1177/0739456X15593147

Riggs, W. (2011). *Walkability and Housing: A Comparative Study of Income, Neighborhood Change and Socio-Cultural Dynamics in the San Francisco Bay Area* (Doctoral Dissertation). University of California, Berkeley, CA.

Riggs, W. (2014). Dealing with Parking Issues on an Urban Campus: The Case of UC Berkeley. *Case Studies on Transport Policy*. Retrieved from http://www.sciencedirect.com/science/article/pii/S2213624X14000431

Riggs, W. (2018). *Disruptive Transport: Driverless Cars, Transport Innovation and the Sustainable City of Tomorrow. Routledge Equity, Justice and the Sustainable City Series*. London: Routledge.

Riggs, W., & Boswell, M. R. (2016). *Thinking Beyond the (Autonomous) Vehicle: The Promise of Saved Lives*. Retrieved from https://works.bepress.com/williamriggs/71/

Riggs, W., & McDade, E. (2016). Moving from Planning to Action: Exploring Best Practice Policy in the Finance of Local Bicycling and Pedestrian Improvements. *Case Studies on Transport Policy*, *4*(3), 248–257. doi:10.1016/j.cstp.2016.06.004

Riggs, W. W., & Boswell, M. R. (2016a). *No Business as Usual in an Autonomous Vehicle Future*. Retrieved from https://works.bepress.com/williamriggs/53/

Riggs, W. W., & Boswell, M. R. (2016b). *Why Autonomous Vehicles Probably Won't Induce Sprawl*. Retrieved from https://works.bepress.com/williamriggs/60/

Riggs, W. W., Boswell, M. R., & Zoepf, S. (2017). A New Policy Agenda for Autonomous Vehicles: It's Time to Lead Innovation. *Planetizen*. Retrieved from https://works.bepress.com/williamriggs/75/

Riggs & Gilderbloom. (2017). How Multi-Lane, One-Way Street Design Shapes Neighbourhood Life: Collisions, Crime and Community. *Local Environment*. Retrieved from http://www.tandfonline.com/eprint/qZNFa3D9zyn9RD5chhvE/full

Saelens, B., Sallis, J., & Frank, L. (2003). Environmental Correlates of Walking and Cycling: Findings from the Transportation, Urban Design, and Planning Literatures. *Annals of Behavioral Medicine*, *25*(2), 80–91. doi:10.1207/S15324796ABM2502_03 PMID:12704009

Sallis, J. F., Frank, L. D., Saelens, B. E., & Kraft, M. K. (2004). Active Transportation and Physical Activity: Opportunities for Collaboration on Transportation and Public Health Research. *Transportation Research Part A, Policy and Practice*, *38*(4), 249–268. doi:10.1016/j.tra.2003.11.003

Sallis, J. F., & Glanz, K. (2006). The Role of Built Environments in Physical Activity, Eating, and Obesity in Childhood. *The Future of Children*, *16*(1), 89–108. doi:10.1353/foc.2006.0009 PMID:16532660

Sciara, G.-C. (2017). Metropolitan Transportation Planning: Lessons From the Past, Institutions for the Future. *Journal of the American Planning Association*, *83*(3), 262–276. doi:10.1080/01944363.2017.1322526

Seskin, S., & McCann, B. (2012). Complete Streets. *Policy Analysis*, *2011*. Retrieved from http://trid.trb.org/view.aspx?id=1212459

Shaheen, S., & Chan, N. (2016). Mobility and the Sharing Economy: Potential to Facilitate the First-and Last-Mile Public Transit Connections. *Built Environment*, *42*(4), 573–588. doi:10.2148/benv.42.4.573

Sperling, D. (2018). *Three Revolutions: Steering Automated, Shared, and Electric Vehicles to a Better Future*. Island Press. doi:10.5822/978-1-61091-906-7

Stone, B., Vargo, J., Liu, P., Habeeb, D., DeLucia, A., Trail, M., ... Russell, A. (2014). Avoided Heat-Related Mortality through Climate Adaptation Strategies in Three US Cities. *PLoS One*, *9*(6), e100852. doi:10.1371/journal.pone.0100852 PMID:24964213

Sturm, R., & Cohen, D. A. (2004). Suburban Sprawl and Physical and Mental Health. *Public Health*, *118*(7), 488–496. doi:10.1016/j.puhe.2004.02.007 PMID:15351221

Thompson, D., Edelsberg, J., Colditz, G. A., Bird, A. P., & Oster, G. (1999). Lifetime Health and Economic Consequences of Obesity. *Archives of Internal Medicine, 159*(18), 2177. doi:10.1001/archinte.159.18.2177 PMID:10527295

Transport for London. (2008). *Impacts Monitoring*. Retrieved from http://www.tfl.gov.uk/cdn/static/cms/documents/central-london-congestion-charging-impacts-monitoring-sixth-annual-report.pdf

Vavrova, M., Chang, C., & Bina, L. (2017). *A Framework to Analyze the Feasibility of Vehicle Miles Traveled Fees to Finance A Sustainable Transportation System*. Academic Press.

Voss, C., Winters, M., Frazer, A., & McKay, H. (2015). School-Travel by Public Transit: Rethinking Active Transportation. *Preventive Medicine Reports, 2*, 65–70. doi:10.1016/j.pmedr.2015.01.004 PMID:26793430

Williams, D. R., & Jackson, P. B. (2005). Social Sources of Racial Disparities in Health. *Health Affairs, 24*(2), 325–334. doi:10.1377/hlthaff.24.2.325 PMID:15757915

KEY TERMS AND DEFINITIONS

Active Transportation: Transportation that focuses on travel via modes that involve physical activity. Most often associated with walking and cycling, however sometimes includes travel via transit given that a transit trip is usually linked a bike or pedestrian trip at the origin or destination.

Complete Streets' Policy: Policy to reshape urban streetscapes or design the urban environment to accommodate and prioritize pedestrians and cyclists over cars.

Cordon System: Also sometimes called a congestion zone, a cordon is a spatially designated location of a city that carries a few to access it. It can be levied based on time-of-day or day-of-the-week, and is generally used to reduce the number of trips and congestion.

General Fund Allocations: Allocations from the primary operating fund that most governments. Many times, these funds are generated through fixed taxes such as property tax.

Impact Fees: Funding or fees tied to a direct impact from a project or use. For example, if new housing is expected to generate 100 new trips daily, a certain portion of funding to contribute to mitigating the negative impacts of those trips could be levied.

Sales Tax Measures: Special taxes raised by a local government based on local sales. Revenues are usually directed to a specific use (services, infrastructure, etc.) and such taxes are seen as a way to generate revenue for the impacts of non-residents of a city who use its' amenities but do not pay other local taxes like property tax. Can be used to borrow money from future revenue streams in the form of a municipal bond.

Transportation Network Companies: Also called TNCs these are peer-to-peer car networks and ride hailing services like Lyft and Uber that allow for on-demand travel from point to point across space.

ENDNOTE

1 Moore's Law is a principle popularized by George Moore one of the original founders of Intel. It hypothesized that chip speed would continue to double with every generation of technological advancement. This leads to exponential growth in speed and computing power. The same principle has been applied to most technological and computational devices that have emerged over the last 30 years, and the trend has generally held.

Chapter 3
Chassis:
The Unsung Linchpin in the Global Supply Chain

Bethany Stich
University of New Orleans, USA

James Amdal
University of New Orleans, USA

Peter Webb
University of New Orleans, USA

ABSTRACT

International trade requires efficient delivery between exporter and importer, usually by ocean vessels. Roughly 90% of cargo is transported in shipping containers, delivered for export by truck or rail, then received at the importer to be similarly discharged. This chapter discusses containerized trade, focusing on international chassis, the wheeled shipping container bed. The authors discuss the invention of containerization and give the historical context for US chassis provision. The chapter outlines the chassis logistical difficulties of US truckers. An overview of attempts to solve this by chassis pooling is provided. The chapter then addresses differences between chassis in the US and the rest of the world. Key chassis regulations are covered, followed by a discussion of the relation of chassis to sustainability. Antitrust issues concerning the main chassis providers and the three recently created ocean carrier alliances are covered. The authors conclude with three recommendations for improving US chassis access.

DOI: 10.4018/978-1-5225-7396-8.ch003

INTRODUCTION

Shipping containers are large, rectangular metal boxes in which cargo is transported by sea, rail, truck, and, rarely, air. When used for international shipping, they come in standard lengths of 20, 40, and 45 feet. This is known as containerized shipping. Prior to the advent of containerized shipping, cargo was shipped by a method known as breakbulk. In this method, large gangs of dockworkers known as stevedores and longshoremen loaded and unloaded cargo from ships. Stevedores usually owned the equipment for this unloading, and it was they who directly employed the longshoremen. This was the process whether the cargo was shipped in smaller boxes, crates, sacks, bales, etc.

Prior to the advent of containerization, break bulk cargo was loaded, lashed, unlashed and unloaded from a ship one piece at a time. This was a time consuming process at best. The cargos were also subject to theft and/or vandalism. Malcolm McLean revolutionized international shipping in 1956 by inventing a standardized method to transport cargo. Today roughly 90% of the world's trade in non-bulk goods utilizes containers for transport. Since its nascent beginning in 1956, over the decades containerization has grown to become the industry norm in international trade and transport. It offered a degree of standardization that allowed for intermodalism: the use of a standard storage unit (the container) to be used on a train, a truck (with chassis), or on various sized vessels: from small coastal ships to the mammoth container vessels of today. It also slashed the costs associated with international transport to a fraction of its former (pre-containerization) era.

It took some time for containerized shipping to really take hold of the shipping industry, but by the 1970s this method was rapidly revolutionizing it. Ships loaded high with containers (container ships) sailed up to the docks of ports, where the importing containers were unloaded by cranes, and the exporting containers waiting at the docks were simultaneously loaded onto the same ships. This new technology significantly impacted the labor forces of the stevedores and longshoremen. For example, according to Ron Brinson, director of the Port of New Orleans from 1986 to 2002, over the roughly twenty-year period from 1967 to 1986, the tonnage at the Port tripled, while the longshore labor force declined by a factor of 60 percent (Brinson, 2017).

The crucial transportation modes involved in getting shipping containers to or from an inland distribution hub or a port is by rail or by truck. Rail is important particularly for inland distribution hubs that are located remotely from ports or inland waterways. But for both distribution hubs and ports, trucks carry 64 percent of the tons and 69 percent of the value of freight moved in the United States (US).

By 2045, it is projected that trucks will move essentially the same percentage of the expected 25 trillion total tons of freight. Nowhere is the importance of trucking more obvious than in relation to North American Free Trade Agreement (NAFTA) freight flows. From August 2015 to August 2017, trucks carried 64.7 percent of US NAFTA freight. That amounts to $31.9 billion of the $52.0 billion of total imports (61.4%) and $31.1 billion of the $45.4 billion of total exports (68.5%) (Bureau of Transportation Statistics, 2017).

For these containers to be transported on trucks requires a chassis, on which the container is attached, and the combination of container and chassis hitched to the tractor truck. Trucking is key to shipping, and chassis are key to trucking. Therefore, the international chassis is, in turn, "…the linchpin of today's international commerce" (Lane, 2015).

BACKGROUND

Traditionally, ocean carriers (the companies which own and operate the container ships) provided international chassis in the US as part of their pricing package (Tioga Group, Incorporated, 2011). That is, the cost of the chassis was included in the price the ocean carrier charged the domestic trucking companies for bringing the containers to, or taking them from, the ocean vessels. However, recent recession-related shifts in international chassis management are creating gridlock at key transportation hubs, financially rewarding international ocean carriers while overburdening domestic motor carriers (trucking companies), and artificially inflating the cost of transport.

Due to the 2008 recession, international ocean carriers decided to divest themselves of their chassis to save money. This process started in 2009, and in 2010 Maersk was the first ocean carrier which stopped offering chassis (Connors, 2017). Most other ocean carriers did the same thing over the next three years (Connors, 2017). The chassis were sold to chassis leasing companies, also known as Intermodal Equipment Providers (IEPs). A chassis leasing company owns a fleet of chassis, and leases them to shipping companies, which then rent them to their various motor carriers. Collectively, the chassis leasing companies are known as chassis lessors. These lessors wanted guarantees that all the chassis they had purchased from the ocean carriers would be used. Exclusive contracts were put in place between the chassis lessors and the ocean carriers to ensure this (Connors, 2017). Renewal of these contracts has led to a legacy of exclusivity, whereby, in many cases, the ocean carriers are mandating the use of which chassis lessors can be used by the truckers, even though the ocean carriers themselves no longer own the chassis. That is why these contracts are called legacy contracts. The three main chassis lessors are TRAC Intermodal, Flexi Van Leasing, and Direct Chassis Link (DCLI). The following are a few examples of chassis leasing companies and supply types. Figure 1 illustrates the distribution of the different types throughout the country:

Chassis Pooling

"In co-op pools, members provide chassis that are centrally managed and shared by members. Other pools include neutral pools in which a leasing company or other third-party rents chassis to users by the day. There are also terminal-operated pools, which can be co-op or neutral pools or pools operated by truckers or logistics companies…Co-op pools are the primary chassis suppliers in all U.S. regions except the Northeast, where neutral pools dominate. However, all kinds of pools are present in each region." (Bonney, 2012b)

North American Chassis Pool Cooperative (NACPC)

NACPC was organized by motor carriers and provides premium chassis pools consisting of high end chassis with radial tires and LED lights. A chassis pool is a collection of different chassis fleets. Three NACPC locations in the Ohio Valley

Figure 1. Distribution of different of chassis supply types
Source: (Bonney, 2012b)

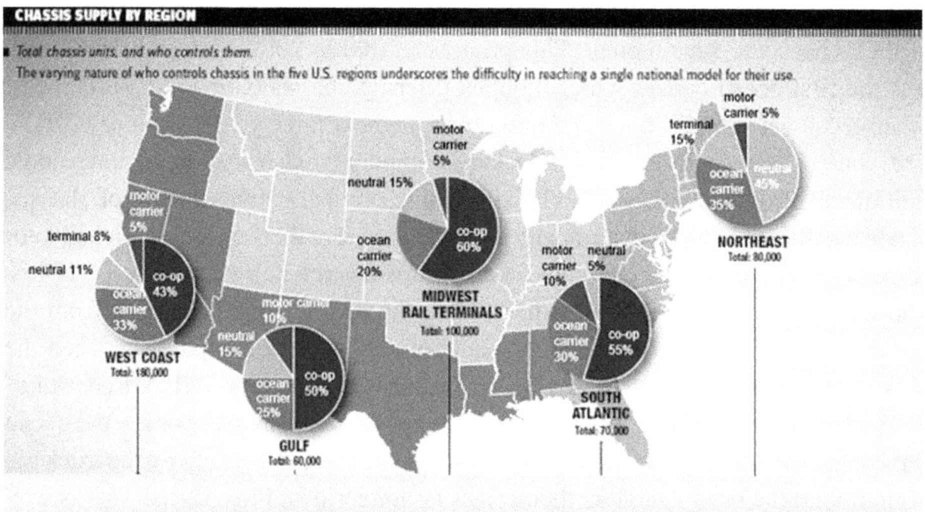

have a total of 300 chassis; Savannah, GA includes 150 chassis; and Houston 250 chassis. NACPC also provides 18,000+ chassis to five regional pools operated by Consolidated Chassis Management (CCM). CCM maintains six regional pools with a total of approximately 130,000 chassis (Morley, 2017a). In November 2017, NACPC announced they were adding another 1,200 new chassis and were planning future expansions with the goal of improving chassis quality and controlling costs.

Direct Chassis Link (DCLI): Charlotte, NC

DCLI's fleet consists of 136,000 chassis. Now that it has acquired TRAC intermodal's domestic chassis fleet as of January 2018, DCLI will own, lease, or manage approximately 136,000 international chassis, as well as approximately 80,000 other chassis, for a total chassis fleet of over 216,000. Additionally, the company manages over 86,000 domestic intermodal containers for third parties, via its REZ-1 asset management platform (DCLI, 2017; Morley, 2017b).

TRAC Intermodal: Princeton, NJ

Acquisition by DCLI of TRAC Intermodal's domestic chassis fleet was accomplished in January of 2018. TRAC Intermodal's international chassis fleet consists of 180,000 chassis. (Morley, 2017b).

Flexi Van Leasing: Kenilworth, NJ

Flexi Van's Fleet consists of 177,000 Chassis at 600 marine locations comprising nine (9) pools and sixty (60) depots (Morley, 2017a).

The shift in ownership of international chassis from ocean carriers to leasing companies occurred quite rapidly, and even though the ocean carriers no longer own the chassis, this can limit leasing options for international chassis users. Many stakeholders want the ocean carriers to be fully separate from the international chassis link of the supply chain (Federal Maritime Commission Bureau of Trade Analysis, 2015). The control of the leasing options by the ocean carriers means that many truckers are locked into using the international chassis leased by whichever IEP is legacy-linked to the ocean carrier for whom they are making any given leg of their journey; there is no choice for these truckers (Mongelluzzo, 2017). Generally, a motor carrier with permission to pick up a container from one ocean carrier would not have permission to use another ocean carrier's chassis to do so, despite complete physical interchangeability.

CHASSIS LOGISTICS

A chassis is defined as a special type of truck trailer/undercarriage developed specifically to facilitate roadway-based transportation of domestic and marine shipping containers. Chassis have a skeleton structure consisting of a frame, multiple axles (2 to 3) and several locking mechanisms (known as twist locks) to facilitate locking of the container on the container chassis. Chassis are designed specifically to transport shipping containers by truck between various shipping facilities, and are the primary means of transporting containers by roadway to and from ports and other shipping facilities.

For the purposes of this chapter, we are only discussing international chassis (unless otherwise specified), the standardized chassis designed specifically to be readily interchangeable for use with the internationally standardized 20', 40' and 45' shipping containers. Figure 2 illustrates the standard international chassis:

These international chassis are known variously as marine chassis, ocean carrier chassis, ocean liner chassis, and ocean container chassis. Other kinds of chassis include: domestic chassis, which are designed specifically for use with 48' and 53' length domestic shipping containers; and triaxle chassis, a chassis used in hauling 20' or 40' containers with three axles and a center that slides out, allowing for either size of container. This allows for better weight distribution and allows the hauling of heavier containers. International chassis are both the linchpin and chokepoint of global seaborne trade.

Global seaborne container trade is believed to account for approximately 60 percent of all world seaborne trade, which was valued at around 12 trillion US dollars in 2017. Containerized shipping links trading partners between the water, rail, and air modes as well as on-time distribution points and retail outlets. An on-time distribution point is a destination for freight which is then sent off to be further processed or

Figure 2. Standard international chassis
Source: (Cheetah Chassis, n.d.)

assembled with other components in such a way that all the necessary parts, the freight, arrive at the same needed time; thus on-time distribution. For example, the various parts of a car will be shipped to different on-time distribution points in such a way that they will arrive at the final assembly destination simultaneously. This entire process, from loading the container with a given freight cargo to final delivery is known as freight logistics.

Containers are how freight moves seamlessly from one mode to another and this system depends upon intermodal drayage. Drayage is the movement of freight via truck. Intermodal drayage refers to the movement of containers between ports and rail terminals, and an inland destination. Drayage typically includes either delivering an export container to the port or picking up an import container. It "is a hub-and-spoke system with the ports and rail terminals as the hubs and drayage providing the spokes" (NCFRP, 2011). According to the US Customs and Border Patrol, "…more than 26.3 million imported cargo containers [passed] through the nation's ports of entry [in 2015]" (U.S. Customs and Border Protection, 2016). If it takes about "…2.5 drayage trip legs for each container moved… due to the need for tractor-only moves and empty container repositioning" (Tioga Group, Incorporated, 2011, p. 1), then American truckers made almost 66 million truck trip legs in FY 2015." A trip leg is the movement of a truck tractor in intermodal drayage. It can be dropping off an empty container, loading or unloading a full container, or hooking/ unhooking a chassis. Trip legs take time for truckers to complete, and as the saying goes, "Time is money."

"The overall cost of driver and tractor time spent in marine container terminals is estimated at over $1 billion annually. The cost of obtaining international chassis at stacked terminals, as opposed to arriving with an international chassis, is estimated at upwards of $2-4 million... Congestion in the container yard is estimated to cost motor carriers about $37-47 million annually… Extra drayage associated with empty containers, bare chassis, or bobtail tractors (truck tractors with no chassis attached) costs about $11.5 million annually" (Tioga Group, Incorporated, 2011, p. 37 adjusted for inflation to 2017). Other costs include gate queuing (trucks waiting in line to get into the marine terminal), gate processing delays, exceptions and troubled tickets (instances where a truck is flagged because of repair issues with the chassis), and congestion on highways and streets. In situations where there are few main international chassis leasing companies, the cost of leasing from them could be as high as $20 per day, as opposed to $12-$15.50 per day when using a leasing company controlled by domestic motor carriers (Mongelluzzo, 2017). Ultimately, this results in higher prices for American consumers. If motor carriers had more freedom to choose their chassis lessors, or to own and use their own chassis, the cost of imported goods would be reduced, benefitting the US, not foreign ocean carrier lines. It is the legacy contracts which are the sticking point in the process.

For example, if a trucker picks up a Maersk ocean liner container, the trucker must use an international chassis leased by a Maersk legacy-linked IEP for that portion of their trip. Then, if they drop off the Maersk container to pick up a container from American President Lines (APL), they must first go and collect an international chassis from a different IEP legacy-linked to APL. The truckers who must do this are not compensated for these extra moves, or turns. Most truckers are paid by the turn, not the hour. Figure 3 shows the typical process a trucker goes through when obtaining a chassis:

This profusion of extra truck turns is obviously untenable. Therefore, to ameliorate this situation, as well as to coordinate equipment (storage and movement), regulatory compliance, and improve supply chain efficiencies, ports began "pooling" chassis. In pooling, chassis lessors negotiate an agreement whereby all lessor-owned international chassis are made available at terminals that are doing business with those lessors. At the port of Los Angeles-Long Beach (LA-LB), this took the form of a port-wide "pool of pools," in which the three largest international chassis lessors (TRAC Intermodal, Flexi Van Leasing, and DCLI) have arranged for truckers to now "...pick up and drop off chassis at any of the terminals served by the three chassis pools," instead of having to split the operation into two separate trips (Mongelluzzo, 2015a). This allowed truckers using TRAC, DCLI and Flexi Van chassis the full interchange of international chassis between the pools formerly managed separately within the port's jurisdiction. This was the first attempt to mitigate a major choke-point for truckers (Bonney, 2014).

Current Chassis Pools at Major US Container Ports

LA-LB (pool of pools). Participants include TRAC Intermodal, Direct Chassis Link Inc. (DCLI), and Flexi-Van Leasing. The total chassis pool is 100,000 and has been

Figure 3. International chassis sub-processes
Source: Tioga Group, 2011, p.14

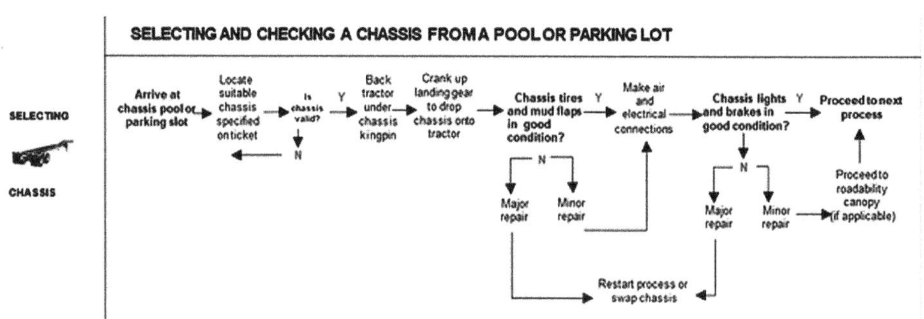

operational since May 2015. The pool incorporates fourteen (14) major marine terminals and four (4) major rail facilities with seventeen (17) start/stop locations.

The Port of New York-New Jersey (NY-NJ) tried to form a port-wide gray pool but the arrangement fell apart when Direct Chassis Link could not find a place to locate its part of the chassis pool (Morley, 2017a), as well as the labor issues associated with maintenance and repair (M&R) (Mongelluzzo, 2017).

The Port of Houston Authority (POHA) decided not to pursue chassis pooling given the complicated issues between the ocean carriers and unionized domestic labor. Private sector interests (motor carriers primarily) declined to pursue this option as well (Amdal, 2017).

Ports in the southeast belong to regional pools in which a truck driver can pick up and drop off the international chassis at the same terminal in any port within the regional pool (Hutchins, 2015). For example, in Memphis a cooperative pool (CO-OP pool) that incorporates 15,000 chassis from 11 major motor carriers has been put in place. Some, such as James Newsome of South Carolina Ports, have proposed a national chassis pool (Tirschwell, 2017a). In New Orleans, Gulf Consolidated Chassis Pool (GCCP) provides a list of common chassis facilities which include the Port of New Orleans, Ports America, and Burlington Northern Santa Fe, Norfolk Southern, and Union Pacific.

Presently, from a national perspective, there is no overarching management solution to chassis pooling or chassis ownership. It is this legacy of contract control by the ocean carriers which has led to ever-more vigorous calls for open choice of chassis on the part of truckers. Open choice refers to motor carriers and truckers being free to choose their own chassis based on market rates and convenience rather than following legacy contracts (Connors, 2017). Recently, the NACPC teamed up with Flexi-Van to offer an open choice pool at NY-NJ (Journal of Commerce, 2018; North American Chassis Pool Cooperative (NACPC), 2018).

Global Chassis vs. US Chassis

"Everywhere else in the world container chassis are supplied by customers, truckers, or off-terminal pools, and are brought to the marine terminal by the drayage driver. Drivers in other countries do not interchange chassis with the ocean carriers or terminal operators. Costs or delays in obtaining a chassis are therefore an internal drayage company issue in those countries, and of no concern to the marine terminals…" (Tioga Group, Incorporated, 2011, p. 64). In the US, however, chassis leasing companies are the predominant model.

When discussing the differences between chassis access in the US and the rest of the world, it is important to understand the distinction between "carrier haulage" and "merchant haulage". Carrier haulage is when the charge made for truck transportation (drayage) including the chassis, is handled by the ocean carrier (J.-P. Rodrigue & Booth, 2013, p. 2). Merchant haulage refers to instances where drayage is arranged by the cargo owner, known as the beneficial cargo owner (BCO) using their standard trucking company, in which case the trucking company is responsible for arranging the chassis (J.-P. Rodrigue & Booth, 2013, p. 2).

In the US, larger shipping companies use carrier haulage, and do not pay separate chassis fees in most cases (Tirschwell, 2017b). Small-to-medium sized shipping companies, however, rely on merchant haulage (Mediterranean Shipping Company, 2018a). These smaller companies are in essence footing the chassis bill for the larger ones, and the smaller companies' rates for ocean-carrier-mandated leasing companies are rising (Tirschwell, 2017b).

"Motor carriers, logistics companies…and to a lesser extent…long term [chassis] leasing companies" are the main providers of chassis in European and Asian shipping (J.-P. Rodrigue & Booth, 2013, p. 2). The chassis and truck are managed as a single resource separate from the leasing companies or ocean carriers (J.-P. Rodrigue & Booth, 2013, p. 3). A similar model is being tried on the shipping market in the US, and has been responsible for the surge in demand for new, state-of-the-art chassis with LED lights, radial tires, and anti-lock brakes (Bonney, 2017).

Apart from the US, shipping containers are utilized at terminals via the "*grounded*" method, in which they are stacked without chassis attached. This means more space at the terminal yard. In the US, "wheeled" containers are those which are stored on chassis, and although this can make for easier movement of containers onto ships and rail cars, this method leads to a much higher usage of chassis, as well as requiring more chassis storage space. "Usually there is also more yard tractor time and mileage driving to and from the storage area" (J.-P. Rodrigue & Booth, 2013, p. 1). This in turn generates more emissions from the trucks, negatively impacting air quality.

Merchant haulage is predominant in Europe (J.-P. Rodrigue & Booth, 2013, p. 2). "China, Hong Kong, and Japan [practice carrier haulage at a rate of 70 to 75 per cent]," but in other parts of Asia merchant haulage is also more common (J.-P. Rodrigue & Booth, 2013, p. 2). Regarding the specifics of chassis economics in Asia, "A possible explanatory factor behind the lower supply of chassis in Asia concerns drayage distances and the nature of economic activities. The export-oriented economic development model has favored the location of factories close to marine terminal facilities. Drayage distances are relatively short and containers are loaded/unloaded immediately, with the tractor remaining hooked to the chassis and the driver waiting until the next move. In this context, the utilization level of chassis assets is therefore higher" (J.-P. Rodrigue & Booth, 2013, p. 2).

Key Chassis Regulations

Containers and chassis are governed by a set of laws, regulations, conventions and standards both internationally and nationally. The first instance of international chassis regulation was the Customs Convention of 1972, which was entered into force on December 6, 1975. Oversight is provided by the World Customs Organization (WCO), An intergovernmental body which guards and enforces the legitimacy and legality of international trade. Shortly thereafter the International Maritime Organization (IMO), An agency of the United Nations which oversees the safety of global shipping, put the International Convention for Safe Containers in force on September 6, 1977 with a two-fold purpose:

- To maintain a high standard of safety for human life in the transport and handling of containers;
- To facilitate the international transport of containers by providing uniform international safety regulations applicable to all forms of surface transport.

The International Organization for Standardization's International Standards for freight containers and chassis has adopted 30 international standards for all manner of containers: air, surface, intermodal; containers on-board vessels, tank containers, platform and platform-based containers (The Institute of International Container Lessors, 2017).

The Uniform Intermodal Interchange & Facilities Access Agreement (UIIA initially) is the standard contract governing the interchange of intermodal equipment between ocean carriers, railroads, equipment leasing companies and intermodal trucking companies. It covers facility access, equipment interchange procedures, equipment usage rules, liability and insurance requirements, administrative processes, and dispute resolution (Tioga Group, Incorporated, 2011). At the terminal the driver will go through terminal sub processes that include:

- Verifying the identity of the driver and motor carrier
- Verifying the transaction is legitimate (inbound and outbound)
- Checking the condition of equipment (inbound and outbound) and issuing an Equipment Interchange Report (EIR)
- Performing the exchange of container and chassis with the container yard

Regarding roadability, that is, safe usage of chassis on the roads, legislation was passed in 2009 with oversight provided by the US Federal Motor Carrier

Safety Administration (FMCSA). This established a regulatory requirement for safe operation, inspection, repair and maintenance of intermodal chassis in the US. Requirements include:

- Single IEP for each chassis (December 2009)
- IEP establishment of inspection, maintenance, repair, and recordkeeping program (December 2009)
- Standardized audit trail of driver Roadability Component Defect (RCD) reports
- Standardized audit trail of Driver Vehicle Inspection Reports (DVIR)
- US Department of Transportation (USDOT) number applied to all chassis (December 2010)

"The key effect is to hold IEPs responsible for maintaining chassis to FMCSA standard and establish a corresponding audit trail… The burden has thus been placed disproportionately on the drivers and motor carriers, who must either find a good chassis or wait to have one fixed" (Tioga Group, Incorporated, 2011, p. 68). These regulations and the resulting standardized audit trail were created to ensure that IEPs actually maintain international chassis on schedule and repair defects noted by drivers, thereby establishing a shared safety responsibility among intermodal equipment providers, motor carriers and drivers (J. P. Rodrigue, 2012). Given the average age of an international chassis in the US is 19 years, this legislation was notable for maintaining safety and roadability across chassis operations.

The terms under which chassis are provided according to the current system is governed by the Ocean Carriers Equipment Management Association (OCEMA), a US association of 15 main ocean carriers who oversee the operational safety of US intermodal ocean freight transportation. It is licensed by the Federal Maritime Commission (FMC), the federal body charged with oversight of international ocean freight in the US. The terms of chassis provision outlined by OCEMA are paraphrased as follows:

- Individual ocean carriers can coordinate the formation of chassis pools and the upkeep of the chassis, although this is now the purview of separate chassis leasing companies;
- Chassis can be exchanged among individual ocean carriers (although technically they chose to divest themselves of them) and among marine terminal operators and all other parties involved in the shipping process, i.e., "…rail terminal operators, container yard operators, rental companies, shippers, inland carriers, and logistics providers…"

- Individual ocean carriers can form, own, and operate chassis pools or pool-owning companies, although no ocean carriers technically own any of the pool companies anymore. The pools can be managed by carrier-formed chassis leasing companies or neutral parties (gray pools);
- The ocean carriers collectively delegate inspection and maintenance and repair (M&R) of chassis and compliance with state and federal chassis-related law

(Ocean Carrier Management Equipment Association, 2017, pp. 9–10).

US Federal chassis regulations are related to roadability requirements. The original legislation for establishing this safety responsibility came about after much acrimony between the Teamsters (truckers) Union, the American Trucking Associations (ATA), the International Longshoremen's Association (ILA) and the International Longshore and Warehouse Union (ILWU). These parties all favored the legislation. Terminal operators, ship lines and railroads opposed it (Beadle, 2004). It was the Safe, Accountable, Flexible, Efficient Transportation Equity Act: A Legacy for Users (SAFETEA-LU) of 2005.

Specifically, the legislation specifies in "Section 4118. ROADABILITY…a requirement to identify intermodal equipment providers (chassis lessors) responsible for the inspection and maintenance of intermodal equipment (chassis) that is interchanged or intended for interchange to motor carriers in intermodal transportation" and "a requirement that an intermodal equipment provider… systematically inspect, repair, and maintain, or cause to be systematically inspected, repaired, and maintained…" the intermodal equipment (Rep Don Young (R) Alaska, 2005, p. 1730). The ILA and the ILWU have frequently sparred with the chassis lessors at union terminals to make sure that the unions are the ones "caused" by the lessors to perform the maintenance and repair (M&R) of chassis (Bonney, 2015; Mongelluzzo, 2015b).

Chassis and Sustainability

The present chassis supply regime hampers sustainable freight transportation, by contributing to avoidable congestion and the resulting emissions. To better understand this, a working definition of sustainability is in order. "The term 'sustainability' represents a pattern of human activity that aims to use the planet's resources in a manner that meets the needs of the world's population now and in the future, while achieving a balance between environmental conservation, economic development, and livability (including consideration of social equity and justice)" (Federal Highway Administration, 2017). When related to transportation, sustainability entails a discussion of the harm caused by vehicle emissions.

Emissions

Freight diesel is a main generator of particulate matter (PM), oxides of nitrogen (NOx), which generate ozone, and greenhouse gas (GHG) emissions, particularly carbon dioxide (CO2). PM and NOx can cause respiratory failure, cancer, and early death in children and older adults. Freight accounts for 29% of all transportation-related GHG emissions, and 68% of that number is due to freight trucking (Federal Highway Administration, 2017). Aside from PM, NOx, and CO2, diesel freight trucking also releases Hydroflourocarbons (HFC), Nitrous oxide (N2O) and Methane (CH4) (Federal Highway Administration, 2017). All these materials are hazardous to human health, and CH4 is another major GHG. Diesel freight trucking, then, is a contributor to climate change. It makes sense to seek ways both to reduce freight trucking emissions and to limit the amount of freight trucking activity to the greatest possible extent.

One way to achieve this is through technology. Diesel trucks can be retrofitted with filters which reduce emissions, and Port LA-LB has made significant investment in liquefied natural gas (LNG) fueled freight trucks as part of its Clean Truck Program (John A. Volpe National Transportation Systems Center, 2018, p. 44). Legislation is another means. The Federal Congestion Mitigation and Air Quality Improvement (CMAQ) program, funded by the USDOT, has been utilized by the state of Ohio to fund the replacement of diesel trucks with LNG vehicles (John A. Volpe National Transportation Systems Center, 2018, p. 41). European experiments with subsidized electric varieties of the largest over the road freight trucking vehicles, however, have not proved to be cost effective thus far (John A. Volpe National Transportation Systems Center, 2018, p. 46). In addition to CMAQ, other federal funding sources for emission reduction technologies, focused on diesel retrofitting and the use of alternative fuels for large freight trucks, include the National Highway Freight Program (NHFP), the EPA's National Clean Diesel Campaign, and the US Department of Energy's Clean Cities Program (John A. Volpe National Transportation Systems Center, 2018, pp. 55–56).

Chassis and Inefficiency

Chassis, however, remains a sticking point in optimizing sustainability at American ports and intermodal terminals. Inefficiencies within the current chassis provision regime lead to excess truck moves, thereby producing more emissions, as well as increasing those from congestion-related idling. The differentiation in chassis provision models means that there is "...no consistent prevailing strategy" (Kellaway,

2014, p. 11). Co-op pools, neutral pools, third-party chassis pools (Flexi Van, TRAC, DCLI), legacy contracts, and trucker-owned chassis are all mixed together in an uncoordinated way that results in increased turn times for truckers (Kellaway, 2014, p. 11). When chassis were uniformly provided by the ocean carriers as part of the business cost, turnaround time at ports and intermodal terminals was about one hour. Under the modern chassis regime, turnaround time can be from two to six hours, and off-terminal pools can exacerbate this problem (Kellaway, 2014, pp. 11, 13–14). The aging chassis fleet is increasing maintenance times for truckers, and "...chassis flips [the mounting of grounded containers on chassis] and pool transfers [are] adding additional delays" (Kellaway, 2014, p. 12). All of this means more time for trucks to drive within, and to and from, terminals, increasing emissions and last mile congestion (the buildup of freight traffic at the approach to ports and intermodal terminals). The overarching solution to the chassis problem is to make the entire chassis process uniform, with the ocean carriers, truckers, shipping companies, rail companies, ports, and terminals working together towards a common paradigm (Kellaway, 2014, p. 23). The solution should focus on neutral chassis pools, as these have been shown to generate less truck movement and emissions at the American-style stacked-chassis terminals (Tioga Group, Incorporated, 2012, p. 14).

Antitrust Issues

Chassis Lessors

The deregulation of the 1980s, specifically with an eye towards loosening antitrust legislation, ultimately led to the evolution of the three dominant chassis leasing companies, Flexi Van, Direct Chassis Link Intermodal (DCLI), and TRAC Intermodal (Magnier, 1988). Antitrust refers to "legislation against or opposition to trusts or combinations; specifically: consisting of laws to protect trade and commerce from unlawful restraints and monopolies or unfair business practices" (Merriam-Webster, 2018). Part of the leasing companies' evolution was the Federal Maritime Commission (FMC)-scrutinized formation of Consolidated Chassis Management (CCM) in 2012. This chassis consortium consisted of 20 ocean carriers who "managed" but no longer technically owned their chassis, facilitating the use of these chassis in "pools" so that truckers could use them interchangeably at marine terminals (Bonney, 2012a). In 2014, the Department of Justice's antitrust division ruled that the Ports of LA-LB international chassis providers could share their international chassis, roughly 100,000, in a gray pool. This ruling set the precedent for the creation of additional chassis pools.

Ocean Carriers

In April of 2017, the major ocean carriers consolidated into three ocean carrier shipping alliances with Vessel Sharing Agreements (VSAs). These three alliances are:

The Ocean Alliance

CMA CGM (Compagnie Maritime d Affretement Compagnie Generale Maritime), COSCO (China Ocean Shipping Company), OOCL (Orient Overseas Container Line Ltd.), APL (American President Lines) and Evergreen (APL is now owned by CMA CGM);

The Transport High Efficiency Alliance (THE) THE Alliance

NYK Group (Nippon Yusen Kabushiki Kaisha Line and NYK Ro Ro [Roll on/Roll off, referring to vessels designed for wheeled cargo, such as cars, to roll directly aboard]), MOL (Mitsui O.S.K. Lines), O.S.K. (Osaka Shosen Kaisha), "K" Line (Kawasaki Kisen Kaisha, Ltd.), Hapag Lloyd, UASC (United Arab Shipping Company Co. and Yang Ming (UASC has merged with Hapag Lloyd); and

The 2M Alliance

Maersk Line and MSC (Mediterranean Shipping Company), with HMM (Hyundai Merchant Marine Co., Ltd.) and Hamburg Sud (Hamburg Sud is now owned by Maersk Line). HMM is not officially in the alliance, but they have slot purchases and exchanges with MSC as well as Maersk (American Export Lines, 2017).

These three alliances now account for "...nearly 80% of global container trade and roughly 90% of container capacity on major trade routes. The main trade lane that is highly affected by this change and the main reason for the new alliances is the North America-Asia a.k.a. "East-West" trade lane between the Far East and North America which will represent 96% of East-West trade" (American Export Lines, 2017). As may be expected, this level of consolidation has brought about concerns expressed by the FMC of antitrust violations by the alliances, or, more specifically, abuses by them of antitrust immunity provisions (Bonney, 2016b).

Under 46 U.S.C. §§ 40307(a)(1), (2), "Exemption from antitrust laws":

- The ocean carriers may act in concert if doing so doesn't negatively impact US commerce)
- The ocean carriers may act in concert when moving in foreign nations during trade with the US;

- The ocean carriers may act in concert during marine terminal operations which take place outside the US;
- The ocean carriers may act in concert according to any agreements made before the 1984 amendments to the Shipping Act of 1916 (United States Congress, 2014).

This is related to chassis, for example, in the case of THE Alliance wanting to purchase chassis and containers as an alliance per se, rather than these purchases being made by individual ocean carriers. The chassis lessors interpret federal regulations as prohibiting this, saying that "...section 10(c)(4) of the Shipping Act of 1984 makes it illegal for ocean carriers to collectively discuss the cost of services with truck, rail, or air carriers that are not regulated by the FMC" (Bonney, 2016a).

This means that a group of ocean carriers cannot negotiate chassis rates or services with non-ocean shipping companies, unless by doing so they do not violate existing anti-trust laws. This prohibition holds as long it does not affect the total charge of "moving freight from beginning to end" by a group of ocean carriers (Mediterranean Shipping Company, 2018b; United States Congress, 1999).

This section of the Act can clearly be interpreted as meaning that chassis ownership by the new ocean carrier alliances is off limits, unless the ocean carriers' antitrust exemptions are significantly extended. Such an extension is strongly opposed by the FMC as well as the US Justice Department, to the extent of subpoenaing some of the heads of the alliances' members, even though the ocean carrier alliances have made attempts to remove the appearance of collusion from their VSAs (Bonney, 2016b; Hutchins, 2017a; Journal of Commerce, 2016). American shipping companies and chassis lessors are in fear of the ocean carrier alliances' collective buying power, and this concern has been taken seriously by Congress, which is moving to give the FMC more enforcement power for the protection of US shipping interests (Bonney, 2016a; Hutchins, 2017b, 2017c).

Solutions and Recommendations

Per Lane 2015, the international chassis is indeed "...the linchpin of today's international commerce". Wherever the practice of legacy contracts is still in place, the international chassis is also a chokepoint for international commerce. In this chapter, we have seen part of the array of chassis provision arrangements in various forms of pools: pool of pools, gray pools, regional pools, co-op pools, and common chassis facilities. These arrangements are designed to mitigate the effects of legacy contracts on the number of turns a truck must make when utilizing a chassis. To further this end, we offer the following recommendations:

1. Adopt a national chassis pool model: Open Choice

Throughout the nation, the availability of suitable chassis at ports is a vexing problem for truckers, terminal operators and even shipping lines. At several large ports, on both the East Coast and the West Coast, chassis leasing companies have formed “pools” where chassis are offered to truckers regardless of their affiliation to a shipping line. This ameliorates the problem of double truck turns because truckers can use any available chassis but truckers now view “open choice” as the next logical improvement. All these iterations increase the speed of truck turns and improve terminal efficiency.

However, in order to address this issue throughout the US port system, a national program of “open choice” needs to be advocated by the Federal Maritime Commission.

2. Mandate a modern chassis fleet nationwide

Another issue critical to chassis operation is their “roadability”. The average age for chassis in the U.S. is 19 years. Efforts should be advanced to require all chassis to be state-of-the-art: LED lights, radial tires, and anti-lock brakes.

Provision should also be made that the inspection, maintenance and repair of chassis be mandated with the respective unions as part of their contract negotiations.

3. Investigate chassis utilization analytics and chassis fleet rightsizing for typical US port sizes

Recently, investigators have attempted to predict the availability of chassis based upon the size of ship discharging cargo. To date, their efforts have been challenged by the extreme fluctuation of ship sizes per port. Nonetheless, these investigations should be supported and expanded. They should also be paired with studies on the rightsizing of chassis fleets at pilot ports (small, medium and large ports on all US coasts).

REFERENCES

American Export Lines. (2017, May 8). *Understanding the 3 New Ocean Carrier Shipping Alliances | Learn How the New Carrier Alliances Affect Your Ocean Freight Contracts and/or Shipments*. Retrieved March 12, 2018, from https://www.shipit.com/archives/2017/05/08/understanding-the-3-new-ocean-carrier-shipping-alliances/

Chassis

Beadle, A. D. (2004, June 27). *Is there life to chassis legislation?* Retrieved March 3, 2018, from https://www.joc.com/trucking-logistics/there-life-chassis-legislation_20040627.html

Bonney, J. (2012a, January 16). *"Open Pools" for Chassis*. Retrieved March 14, 2018, from https://www.joc.com/maritime-news/open-pools-chassis_20120116.html

Bonney, J. (2012b, April 23). *Fixing Broken Chassis System Proves Elusive*. Retrieved May 22, 2018, from https://www.joc.com/maritime-news/international-freight-shipping/fixing-broken-chassis-system-proves-elusive_20120423.html

Bonney, J. (2014, September 23). *DOJ won't challenge LA-LB 'gray chassis' plan*. Retrieved November 21, 2017, from https://www.joc.com/port-news/us-ports/port-los-angeles/doj-won%E2%80%99t-challenge-la-lb-%E2%80%98gray-chassis%E2%80%99-plan_20140923.html

Bonney, J. (2015, December 7). *NY-NJ 'gray' chassis pool awaits deal with ILA*. Retrieved October 5, 2017, from https://www.joc.com/port-news/us-ports/port-new-york-and-new-jersey/ny-nj-%E2%80%98gray%E2%80%99-chassis-pool-awaits-deal-ila_20151207.html

Bonney, J. (2016a, November 23). *Lessors: THE Alliance members shouldn't jointly buy equipment*. Retrieved March 12, 2018, from https://www.joc.com/maritime-news/container-lines/alliance/lessors-alliance-members-shouldnt-jointly-buy-equipment_20161123.html

Bonney, J. (2016b, November 29). *FMC's Doyle warns of alliance's contracting power*. Retrieved March 12, 2018, from https://www.joc.com/regulation-policy/transportation-regulations/us-transportation-regulations/fmc-doyle-warns-alliance-contracting-power_20161129.html

Bonney, J. (2017, December 4). *Trucker-led pool adds chassis, plans more expansion*. Retrieved March 7, 2018, from https://www.joc.com/trucking-logistics/trucking-equipment/trucker-led-pool-adds-chassis-plans-more-expansion_20171204.html

Bureau of Transportation Statistics. (2017, October). *August 2017 North American Freight Numbers | Bureau of Transportation Statistics*. Retrieved October 30, 2017, from https://www.bts.gov/newsroom/august-2017-north-american-freight-numbers

Cheetah Chassis. (n.d.). *Standard international chassis image*. Retrieved from http://www.cheetahchassis.com/media/1189/blue-20-40-45-tandem-ps.jpg

Connors, P. (2017, November 13). *Shippers need "open choice" for chassis*. Retrieved March 1, 2018, from https://www.joc.com/port-news/port-equipment/evolution-chassis-business-model-must-continue_20171113.html

DCLI. (2017). *DCLI to Purchase TRAC Intermodal's Domestic Chassis Fleet | Direct ChassisLink Inc*. Retrieved April 19, 2018, from https://dcli.com/about/news/dcli-to-purchase-trac-intermodals-domestic-chassis-fleet/

Federal Highway Administration. (2017, February 1). *FHWA Freight and Land Use Handbook: Section 3.0 - FHWA Freight Management and Operations*. Retrieved September 19, 2018, from https://ops.fhwa.dot.gov/publications/fhwahop12006/sec_3.htm

Federal Maritime Commission Bureau of Trade Analysis. (2015). *US; Container Port Congestion & Related International Supply Chain Issues: Causes, Consequences & Challenges*. Retrieved from http://www.fmc.gov/assets/1/Page/PortForumReport_FINALwebAll.pdf

Hutchins, R. (2015, October 31). *Can Miami be a major gateway for Asian imports heading out of Florida?* Retrieved October 13, 2017, from https://www.joc.com/port-news/us-ports/port-miami/can-miami-be-major-gateway-asian-imports-heading-out-florida_20151031.html

Hutchins, R. (2017a, March 23). *Container lines defend against collusion suspicion*. Retrieved March 12, 2018, from https://www.joc.com/maritime-news/container-lines/container-lines-defend-against-collusion-suspicion_20170323.html

Hutchins, R. (2017b, May 3). *Fearing alliance collusion, Congress eyes US supplier protection*. Retrieved March 12, 2018, from https://www.joc.com/regulation-policy/transportation-regulations/us-transportation-regulations/fearing-alliance-collusion-us-congress-considers-oversight-expansion_20170503.html

Hutchins, R. (2017c, June 9). *Legislation would give FMC more alliance muscle and money*. Retrieved March 12, 2018, from https://www.joc.com/regulation-policy/transportation-regulations/us-transportation-regulations/legislation-would-give-fmc-more-alliance-muscle-and-money_20170609.html

John A. Volpe National Transportation Systems Center. (2018). *Primer for Improved Urban Freight Mobility and Delivery - Operations, Logistics, and Technology Strategies* (Performimg Organization Report No. FHWA-HOP-18-020). Cambridge, MA: Federal Highway Administration.

Journal of Commerce. (2016, December 6). *THE Alliance strips joint-contracting language from VSA*. Retrieved March 12, 2018, from https://www.joc.com/maritime-news/container-lines/alliance/alliance-strips-joint-contracting-language-vsa_20161206.html

Journal of Commerce. (2018, March 7). *US and Global Trade: TPM 2018 Tracker - The latest news, insights, and analysis from the JOC's TPM 2018 conference in Long Beach*. Retrieved March 16, 2018, from https://www.joc.com/international-logistics/tpm-2018-tracker-news-insights-analysis_20180307.html

Kellaway, K. (2014, April). *State of the Union in the Drayage Industry and the impact on Green Fleet Initiatives*. PowerPoint presented at the Port Stakeholders Summit, Baltimore, MD. Retrieved from https://www.epa.gov/sites/production/files/2014-07/documents/summit-kellaway.pdf

Lane, J. C. (2015, October). *NY-Appellate-Division-rules-Graves-Amendment-extends-to-intermodal-chassis*. Retrieved October 12, 2017, from http://www.thelanelawfirm.com/ny-appellate-division-rules-graves-amendment-extends-to-intermodal-chassis.html?no_redirect=true

Magnier, M. (1988, June 16). *Container leasing enters new era*. Retrieved March 14, 2018, from https://www.joc.com/container-leasing-enters-new-era_19880616.html

Mediterranean Shipping Company. (2018a). *Intermodalism | MSC*. Retrieved March 5, 2018, from https://www.msc.com/gin/help-centre/guide-to-international-shipping/new-to-international-trade-intermodalism?lang=ru-ru

Mediterranean Shipping Company. (2018b). *Shipping Glossary of Terms | MSC*. Retrieved May 7, 2018, from https://www.msc.com/usa/help-centre/interactive-shipping-glossary

Merriam-Webster. (2018). *Definition of antitrust*. Retrieved April 19, 2018, from https://www.merriam-webster.com/dictionary/antitrust

Mongelluzzo, B. (2015a, May 5). *Drayage community praises new LA-LB gray chassis pool*. Retrieved October 5, 2017, from https://www.joc.com/port-news/us-ports/port-los-angeles/drayage-community-praises-new-la-lb-gray-chassis-pool_20150305.html

Mongelluzzo, B. (2015b, October 21). *ILWU chassis trouble likely to be resolved in court*. Retrieved October 5, 2017, from https://www.joc.com/port-news/longshoreman-labor/international-longshore-and-warehouse-union/ilwu-chassis-trouble-likely-be-resolved-court_20151021.html

Mongelluzzo, B. (2017, June 9). *New chassis business model would cut costs, trucker says*. Retrieved October 5, 2017, from https://www.joc.com/port-news/us-ports/new-chassis-business-model-would-cut-costs-trucker-says_20170609.html

Morley, H. (2017a, September 14). *NY-NJ joint chassis pool collapses*. Retrieved October 5, 2017, from https://www.joc.com/trucking-logistics/drayage/ny-nj-joint-chassis-pool-collapses_20170914.html

Morley, H. (2017b, October 25). *DCLI snags TRAC's domestic chassis fleet*. Retrieved November 7, 2017, from https://www.joc.com/trucking-logistics/trucking-equipment/dcli-snags-trac%E2%80%99s-domestic-chassis-fleet_20171025.html

North American Chassis Pool Cooperative (NACPC). (2018). *About*. Retrieved March 16, 2018, from http://www.nacpc.org/about/

Ocean Carrier Management Equipment Association. (2017). *Ocean carrier equipment management association* (4th ed.). Federal Maritime Commission Agreement.

Rep Don Young (R) Alaska. Safe, Accountable, Flexible, Efficient Transportation Equity Act: A Legacy For Users, Pub. L. No. 109–59, § 4118. ROADABILITY, 1729. (2005). Retrieved from https://www.gpo.gov/fdsys/pkg/PLAW-109publ59/pdf/PLAW-109publ59.pdf

Rodrigue, J. P. (2012). *Guidebook for Assessing Evolving International Container Chassis Supply Models*. Transportation Research Board. Retrieved from http://www.trb.org/Publications/Blurbs/168158.aspx

Rodrigue, J.-P., & Booth, M. (2013, February 26). *Grounded and chassis container terminal operations*. Retrieved March 7, 2018, from https://people.hofstra.edu/jean-paul_rodrigue/downloads/Booth%20&%20Rodrigue_PT57_V5.pdf

The Institute of International Container Lessors. (2017). *The Institute of International Container Lessors (IICL): About the Industry - Laws, Regulations, Conventions and Standards*. Retrieved November 21, 2017, from https://www.iicl.org/aboutIndustry/laws.cfm

Tioga Group, Incorporated. (2011). *Truck Drayage Productivity Guide* (Vol. 11). Transportation Research Board. Retrieved from http://www.trb.org/Publications/Blurbs/165528.aspx

Tioga Group, Incorporated. (2012). SmartWay DrayFLEET, Truck Drayage Environment and Energy Model Version 2.0 User's Guide (Final Report No. EP 11H000338). Transportation and Climate Division Office of Transportation and Air Quality U.S. Environmental Protection Agency.

Tirschwell, P. (2017a, October 9). *Demise of chassis pool an ominous sign for NY-NJ*. Retrieved October 9, 2017, from https://www.joc.com/port-news/us-ports/demise-chassis-pool-ny-nj-has-many-shaking-heads_20171009.html

Tirschwell, P. (2017b, November 17). *It is time to give shippers a choice on chassis*. Retrieved March 1, 2018, from https://www.joc.com/maritime-news/container-lines/it%E2%80%99s-time-give-shippers-choice-chassis_20171117.html

United States Congress. The Shipping Act of 1984 as Modified by The Ocean Shipping Reform Act of 1998, Pub. L. No. S. 414, § 10. (1999). Retrieved from http://www.shippers.com/Shipping_Act.asp

United States Congress. (2014). *U.S.C. Title 46 - SHIPPING*. Retrieved March 14, 2018, from https://www.gpo.gov/fdsys/pkg/USCODE-2014-title46/html/USCODE-2014-title46-subtitleIV-partA-chap403-sec40307.htm

U.S. Customs and Border Protection. (2016, March 4). *CBP Releases Fiscal Year 2015 Trade and Travel Numbers- Trade Facilitation and Enforcement Supports Economic Prosperity*. Retrieved December 1, 2017, from https://www.cbp.gov/newsroom/national-media-release/cbp-releases-fiscal-year-2015-trade-and-travel-numbers

KEY TERMS AND DEFINITIONS

Anti-Trust Legislation: Laws to protect trade and commerce from unlawful restraints and monopolies or unfair business practices.

Beneficial Cargo Owner (BCO): The owner of a cargo which is shipped.

Bobtail Tractor: A tractor truck with no chassis attached.

Breakbulk: Cargo which is not moved in shipping containers, but rather transferred to and from ships piece by piece by gangs of dockworkers.

Carrier Haulage: When the charge made for drayage, including the chassis, is handled by the ocean carrier. In the US, larger shipping companies use carrier haulage, and often do not pay separate chassis fees. The smaller ones do.

Chassis: A special type of truck trailer/undercarriage developed specifically to facilitate roadway-based transportation of domestic and marine shipping containers.

Chassis Leasing Companies: Owners of fleets of chassis, which lease them to shipping companies, which then rent them to their various trucking companies.

Chassis Lessors: The collective term for chassis leasing companies. The three main chassis lessors are TRAC Intermodal, Flexi Van Leasing, and Direct Chassis Link Incorporated (DCLI).

Container Ships: Ocean and river-going vessels specifically designed to carry shipping containers. The ocean-going vessels are getting larger every year.

Containerized Shipping: The practice of transporting cargo internationally in shipping containers of standard lengths of 20, 40, and 45 feet.

Domestic Chassis: Chassis designed specifically for use with 48' and 53' length domestic shipping containers.

Drayage: Moving goods by truck.

Equipment Interchange Report (EIR): The report a driver issues at a terminal when the driver checks the condition of inbound and outbound equipment, especially chassis.

Federal Maritime Commission (FMC): The federal body charged with oversight of international ocean freight in the US.

Freight Logistics: The process from loading the container with a given freight cargo to final delivery.

Gray Pool: A neutral chassis pool, in which several chassis lessors all provide the chassis together under a neutral management agency. Truckers doing business with a gray pool are free to lease any chassis belonging to any of the lessors in the pool. This gives truckers some degree of choice of which chassis to use.

Grounded Containers: Shipping containers utilized at terminals in which they are stacked without chassis attached. This is the global practice, and differs from the US practice of Wheeled Containers.

Intermodal Drayage: The truck portion of movement of containers between the port and an inland destination

Intermodal Equipment Providers (IEPs): Lessors of chassis and other shipping equipment.

Intermodalism: Use of a standard storage unit (the container) to be used on a train, a truck (with chassis), or on various sized vessels.

International Chassis: Standardized international chassis designed to be readily interchangeable for use with the internationally standardized 20', 40', and 45' shipping containers.

International Maritime Organization (IMO): An agency of the United Nations which oversees the safety of global shipping.

International Ocean Carriers: The companies which own and operate container ships.

Legacy Contracts: Exclusive contracts put in place between chassis lessors and international ocean carriers to ensure that all the carriers' chassis would be used. Renewal of these contracts has led to a legacy of exclusivity, and in many cases, the international ocean carriers are mandating which chassis lessors can be used by the truckers.

Longshoremen: A type of dockworker. Longshoremen work for Stevedores, loading and unloading ships and maintaining/repairing chassis

Marine Chassis: Another name for international chassis.

Marine Terminal Operators (MTOs): Those who provide the necessary docking, wharf operations, warehousing, etc. to ocean carriers.

Merchant Haulage: Drayage arranged by the cargo owner, using their standard trucking company, which is responsible for arranging the chassis.

Motor Carriers: Trucking companies which are hired by shipping companies.

North American Chassis Pool Cooperative (NACPC): A chassis leasing company organized by motor carriers and providing premium chassis pools consisting of high-end chassis with radial tires and LED lights.

Ocean Carrier Chassis: Another name for international chassis.

Ocean Carriers: The oceangoing vessel lines which carry shipping containers. They have recently merged into three alliances.

Ocean Carriers Equipment Management Association (OCEMA): A US association of 15 main ocean carriers who oversee the operational safety of US intermodal ocean freight transportation. It is licensed by the FMC.

Ocean Container Chassis: Another name for international chassis.

Ocean Liner Chassis: Another name for international chassis.

On-Time Distribution Point: A destination for freight which is then sent off to be further processed or assembled with other components so that all the freight arrives simultaneously.

Open Choice: Motor carriers and truckers are free to choose their own chassis based on market rates and convenience rather legacy contracts.

Pooling: Chassis lessors negotiate an agreement where all lessor-owned international chassis in the pool are available at terminals that are doing business with it.

Roadability: The condition of a chassis being safe for road use.

Shipping Companies: Legal entities in the business of moving freight.

Shipping Containers: Rectangular metal boxes in which cargo is transported by sea, rail, truck, and rarely, air.

Stevedores: A type of dockworker who usually owns the equipment for breakbulk vessel loading/unloading, and maintaining/repairing chassis, and who employs longshoremen for these purposes.

Triaxle Chassis: Chassis for 20’ or 40’ containers with three axles and a center that slides out, allowing for either size of container. This allows for better weight distribution and hauling of heavier containers.

Trip Leg: A movement of a truck tractor in intermodal drayage. It can be dropping off an empty container, loading or unloading a full container, or hooking/unhooking a chassis.

Turn: A complete move of a truck to drop off and pick up a container and/or a chassis. Truckers are paid for turns picking up or dropping off containers, not switching out chassis.

Twist Locks: Locking which fastens a container on the chassis.

Uniform Intermodal Interchange and Facilities Agreement (UIIA): Standard contract governing the interchange of intermodal equipment between ocean carriers, railroads, equipment leasing companies, and intermodal trucking companies.

Vessel Sharing Agreements (VSAs): Arrangements whereby the major ocean carriers consolidated into three alliances. Each carrier in an alliance shares its vessels with the other members of that alliance.

Wheeled Containers: Containers stored on chassis. This facilitates movement of containers onto ships and rail cars but leads to much higher use of terminal space and more emissions.

World Customs Organization (WCO): An intergovernmental body which guards and enforces the legitimacy and legality of international trade.

APPENDIX

Figure 4. Key chassis developments timeline
Source: Authors

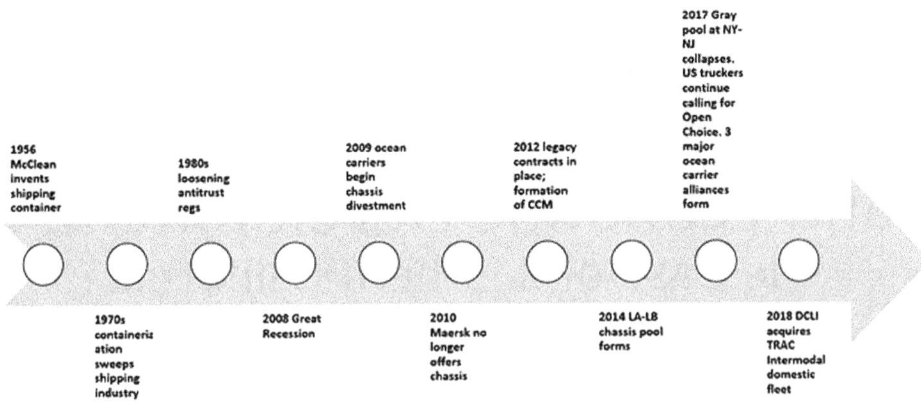

Chapter 4
Regional Institutions for Transportation Sustainability and Economic Development:
The Case of Southern California

Mark Pisano
University of Southern California, USA

Richard F. Callahan
University of San Francisco, USA

ABSTRACT

The lessons described in this chapter outline the mechanisms for cooperation through building new institutional designs for governance to build transportation construction projects. The scale of these projects included billions of federal, state, and local dollars invested in the 1970 through current day. Funding of transportation projects in Southern California during the period 1975 through 2010 addressed a range of challenges to economic growth. The chapter proceeds in four parts: one, a discussion of the environmental context; two, description of the institutional design for governance that developed; three, an overview of the projects developed and economic impact; four, applying the lessons learned to the emerging challenges of fiscal constraints, demographic change, and institutional re-design for transportation funding.

DOI: 10.4018/978-1-5225-7396-8.ch004

Figure 1. California population, travel, and highway expenditure trends
Source: Southern California Association of Governments. 2008. Comprehensive Regional Plan. Los Angeles, CA: SCAG., Page 69.

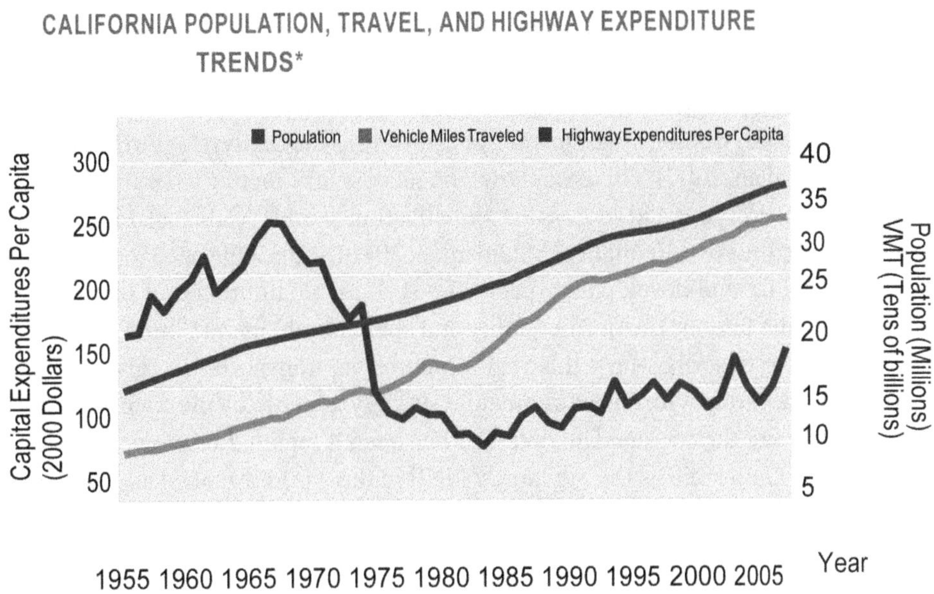

INTRODUCTION

Lessons learned from past successes in funding transportation construction are increasingly valuable given the current state of crisis in infrastructure funding. For example, the issuance of tax-exempt municipal bonds fell by 10 percent in 2017 (London Financial Times, 2018). The state of the current crisis is sharply illustrated in the State of California where *per capita* expenditures for highway in 2010 are nearly 40 percent less than in 1970, while the vehicle miles traveled in the same time frame increased by nearly 90 percent (SCAG, 2008; 69).

The range of contemporary infrastructure and transportation challenges can be addressed by drawing on the strategic lessons learned to advance economic development from the region served by the Southern California Association of Governments (SCAG). Six million people called this region home in 1950, dispersed throughout the seven-county, multiple-centered region. The population grew to nearly 12 million residents by 1975, almost doubling into an urban form many called "sprawl." For those who lived in the region, it was region of multiple small town centers. Even in Los Angeles, the largest city, people identified with the section of the city where they lived. The population growth occurred after the dismantling of a regional “Red-Car” rail system that was removed to make way for freeway construction as part of the relentless dynamics of growth at any cost.

Rather than frame any specific infrastructure project solely as a transportation project, SCAG, as the formal metropolitan planning organization (MPO), linked transportation infrastructure to growing the regional economy. The story that unfolds is one of moving people and moving goods to move the economy. The Los Angeles regional case study demonstrates institutional design as the catalyst for infrastructure building (Callahan, 2007). This case shows the success in attracting federal, state, and local funding to improve and address the globalization of trade (Callahan, Pisano, and Lindor, 2010), improve air quality (Mazmanian, 2010), reduce the need for freeways (Edelman, 2000), and develop all aspects of a sustainable future for a Los Angeles region with “the famous traffic, air quality and sprawl problems” (Elkind, 2014, 5).

This chapter describes how this region altered its transportation, growth, and economic base, along with an environmental strategy, to mitigate the environmental challenges facing the region. The outcome has been a regional economy essential to the state of California as the 5th largest GNP in the world if ranked as a separate nation. One measure of the institutional success has an over 60 percent increase in the growth of personal income since 1972, despite the constraint of no increase in the amount of gas tax available for infrastructure building and the pressure of steady population growth coupled with doubling of the vehicle miles traveled (SCAG, 2008;78)

This chapter focuses on the years 1975 through 2010. In that timeframe, funding of transportation projects in Southern California addressed a range of challenges to economic growth. Researching the SCAG region through case study methodology provides an opportunity for in-depth explanation the “how” of an issue chosen because of national significance (Yin, 2009; 9). For governance research, case studies provide needed texture that derives from field research (Heinrich, Hill and Lynn, 2004; 13), with the opportunity to identify the significant intervening variables (George and Bennett, 2005; 20-21). The case study approach adds to a wide range of research “conducted following a neoclassical approach using various forms of aggregated production function” with inconsistent findings across varied time frames and geography (Chen and Haynes, 2015).

This chapter explores the infrastructure response to those challenges in the following four sections: one, a discussion of the historical context; two, a description of the institutional design for governance; three, an overview of the projects that addressed problems of economic development in the region; and four, a discussion of the future challenges.

THE HISTORICAL CONTEXT

The region served by the Southern California Association of Governments (SCAG) was one of the fastest growing regions in the nation, almost doubling in size from 10.5 million to 18.8 million people. Immigration growth was also the fastest in the country since the turn of the last century. Air quality was the poorest, not only in the US, but globally. Global trade was on the upswing: The region served as the primary port of entry from the Pacific Rim. National military policy drastically reduced defense spending leading to a decline in the economic base of the region. These challenges occurred in the most congested and transportation challenged region in the country (SCAG, 2004).

This rapid growth, coupled with sudden defense spending deceleration, raised fundamental questions on environmental, economic, and infrastructure sustainability. The concept of sustainability offers analytical leverage that applied to environmental issues (Fiorino, 2010). The concept has been applied to development (Rodriquez, et al., 2016) and extended to demonstrate the connection in public institutions between environmental and fiscal sustainability (Callahan and Pisano, 2014). Sustainability has been applied in research on fiscal and administrative policy (Chapman, 2010; Dadayan and Ward, 2009), as well as globally (Rodríguez et al., 2014). Sustainability in funding the construction of infrastructure to promote environmental protection and advance economic development has been the enduring challenge in Southern California.

The wide range of environmental and demographic pressures were created by the process of the second half of the 20th century. The national economic growth strategy sought to make up for the lost ground of the depression and two world wars. Los Angeles and the southern California region was the poster child for this determination: It was home to a substantial portion of the dynamic military industrial complex that propelled victory over both events. During the first half of the century the region become a manufacturing center, rivalling and exceeding the Great Lakes states and the eastern seaboard. Many of the soldiers stationed or trained at the numerous military facilities and ports returned with wives and friends due to the availability of vast areas of land and the desirable climate. These favorable conditions started both a migration of people from within and outside the United States. These trends continued throughout the rest of the century by attracting populations throughout the world, including Asia and South America.

It was affordable and alluring. New freeways coupled with the automobile set the region off to the races. Nowhere else in the United States did the construction of freeways match the population and economic growth of the region. The development of the Federal Interstate System in the 1950s through the 1970s not only made coast to coast a reality, but it propelled the development of regional spaces in America.

Freeway construction spurred the ubiquitous production of cookie cutter subdivision homes on the expansive alluvial plains from the San Gabriel and Santa Monica mountains.

Nationally, a cultural, political, and economic consensus overcame the impediments to growth experienced in the first half of the century, driving the growth of the next 25 years. This consensus formed the political will in Southern California and other regions in the country creating the growth impacts of the post war ebullience.

The federal response to these dynamics and complexities was to form new institutional designs that created instruments for cities and counties and states to work together differently in regional structures and new intergovernmental systems. Federal legislation enacted regional entities called Metropolitan Planning Organizations (MPOs) in 1962, followed by Air Quality Districts in the Clean Air Act of 1970 and Area-wide Waste Management Districts in the Clean Water Act of 1972. At a minimum, legislative authority to implement federal provisions were enacted by all states. Not to be outdone, states sometimes acted before the nation: Many of the first air quality districts and water basin plans were created in California. Today there are over 400 MPOs. Along with these regional efforts, the Advisory Commission on Intergovernmental Relations was established to manage this new overlay of governance and financial assistance to the country.

What emerged in the last part of the 20th century was collaborative problem solving. Various federal agencies stimulated investment but involved states and local governments aimed at dealing with the problems that growth had created. The Environmental Protection Agency was created to lead efforts to control and manage the problems emerging from the development of the most powerful economy in the world. The Department of Housing and Urban Development was established to set forth provisions for area-wide planning provisions (APO) and a clearing house for coordinating federal programs and regulatory actions that impacted housing and urban development. The Department of Transportation invested in various rail transit projects.

INSTITUTIONAL DESIGN FOR GOVERNANCE

In the Southern California region, federal efforts were operationalized through varied regional agencies. At the core of the transportation solutions to these problems was a transformation of how the region organized itself to address these changes. Understanding infrastructure projects and economic development starts with the study of the regional organizations created: The Southern California Association of Governments (SCAG, 1994).

California enacted its own statute to enable MPOs to be formed by local elected officials within the state. The legislation outlined the criteria for their formation and started a highly-charged discussion around the state, particularly among the elected officials in southern California. This discussion, which erupted at times into a conflictual debate, focused on local control. At the core of the issue was the capacity for jurisdictions to maintain the old order of being able to act independently to control their own future. The debate stirred the local populations' longstanding belief in the value of independence, which aligned with de Tocqueville's belief—described a century earlier—in the proven success of local problem-solving.

The leadership of Los Angeles City Councilman Tom Bradley was instrumental in the establishment of the Southern California Association of Governments (SCAG). Bradley became the second president of SCAG, continuing his early support of SACG throughout his four terms as mayor of the largest city in the region and a pivotal city for global trade. In 1967, six counties, Los Angeles, Ventura, Orange, Riverside, San Bernardino, and Imperial, (San Diego remained separate) along with 123 cities, formed SCAG as a regional joint power agency, not as an agency established by state government. San Diego, which is part of the economic mega-region, elected to stay separate from the SCAG region. Both agencies were designated as the Metropolitan Planning Agency (MPO) by the state and the U.S. Secretary of Transportation. They were also designated as the Area-wide Planning Agency by the federal Housing and Urban Development Department (HUD) to carry out the intergovernmental review functions of the Advisory Council of Intergovernmental Relations (ACIR). SCAG's and SanDAG's formation addressed the intense political struggle between independent action of individual jurisdictions versus coordinated action needed to advance regional economic development and environmental protection.

In its first decade, SCAG focused on two policy arenas: growth and transportation. First, it set out to reconcile the growth targets or assumptions contained in the general plans of each jurisdiction. In 1970, the first regional area-wide growth plan estimated over 50 million people would live in region by 2000. The jurisdictions in the SCAG region ultimately agreed on a 50 percent reduction in the 1976 Growth Plan. This started a growth reconciliation process that continues today. The tension and debate over the size of individual growth plans was not only contentious but also important. Over the next several decades, sensitivity analysis on growth projections and distribution were found to be the most significant of all the policy assumptions in terms of cost implications and mitigation impacts for air quality, congestion, housing, and other issues. Land use and growth distributions impacts were quantified and found critical to resolving the policy issues of the region.

SCAG modified its planning processes in the reconciliation process for its members' plans and regional forecasts. Successive plans led to refinements. For example, the 1982 plan reduced the regional growth targets for the region to 12.8

million by the year 2000. However, the consistency between the regional growth target adopted by the regional board and each jurisdiction's actual plan remained a problem. SCAG's response in 1987 was to modify its institutional structure to include more jurisdictions on its board. This institutional redesign increased the size of the existing board of 27 elected officials comprised of six county supervisors and 21 mayors and council people to 76 elected officials comprised of seven county supervisors and 69 city officials. In effect, the body became a legislative body for the region with increased voices and participation in the process.

Additionally, SCAG developed a system of sub-regions comprised of 11 historical growth areas throughout the region, and all cities and counties had seats on the board of these sub-regions. The planning processes evolved to formally incorporate the input from the sub-regions into the regional plans for growth, transportation, and air quality. The primary reasons for this change in the institutional design of SCAG was to achieve buy-in from the multiple separate and independent acting jurisdictions.

Over the next two decades the region accomplished a reconciliation between local and regional targets, with regional forecasts closely tracking actual growth as it occurred. As a result, the growth vision and the realities of the dynamics driving growth evolved into a political consensus. In 2001, the *Compass Program* in SCAG capped a decade long regional dialog with significant data findings. SCAG showed that changes in the land use policies of two percent of the land area of the vast six county region would dramatically improve the region's performance. The most startling result was the success in both the highly-populated portions of the region and the less populated rural parts of the region. Also, the new institutional design of SCAG, which included sub-regions in the planning process, produced results in the region's performance. Transportation congestion was reduced. Air quality mitigations were visibly and quantitatively improved. While difficult to achieve, the results of this innovation in governance addressed economic development while changing the infrastructure landscape of the region.

FACILITATING ECONOMIC DEVELOPMENT

As a regional planning organization, SCAG worked with various other regional agencies designed for specific infrastructure investment and for creating public policy that impacted economic development. These regional organizations developed strategies that changed the way the region supported private sector development and undertook public infrastructure construction. The public agencies listed below are a partial list of the many regional partners that advanced both missions, but were newly designed and significant in the time period in discussion:

- South Coast Air Quality Management District (AQMD)
- Los Angeles County Transportation Commission (LACTC)
- Southern California Regional Rail Authority (Metrolink)
- Orange County Transportation Authority (OCTA)
- Alameda Corridor Transportation Authority (ACTA)

Transportation Funding

The national consensus to pay for needed resources to maintain infrastructure investment as well as to sustain continued economic growth started to fray in the period between 1950 and 1975. The needs outpaced the availability of the federal and state gasoline and diesel sales tax systems that provided significant funding to implement plans. The regional transportation plan in1978 began to note this decline, showing that only 82 percent of the plan was funded.

An institutional design provided a breakthrough in funding for highway and transit development in 1977 with the passage of state legislation that created County Transportation Commissions in all the urbanized counties in the state. The commissions filled a gap between regional planning and implementation of projects.

A significant institutional design feature of the commission in the legislature provided landmark taxing authority to the commissions. This authority enabled each to raise sales taxes if a majority voted to approve an increase. Subsequently, approval required a two-thirds vote due to the passage of Proposition 13 in 1980. This funding mechanism grew to be the major source of funding in most of the counties in the SCAG region, providing up to 60 percent of all the fiscal resources spent on transportation in the region to this day. Other sources included federal funding at 10 percent, state assistance at 16 percent, and local jurisdictions providing the remainder. Only two of the member counties of SCAG—Ventura County and Imperial County—have not enacted sales tax measures.

Transit Infrastructure

The center piece of SCAG's first transportation plan was a starter rail plan that would revitalize the central business center. This plan looked to directly advance economic development through constructing subway infrastructure. The vision of the newly elected Mayor of the City of Los Angeles, Tom Bradley, in 1973 proposed building rail to connect the downtown businesses with the region. Supported by the Los Angeles Chamber and the international headquarter firms in the region, the mayor was able to coalesce the energy and support of the region, leading the U.S. Department of Transportation to designate this proposal the New Start Program for southern California. The Los Angeles Central Business District

(CBD) was the symbolic center pole of the sixty-mile circle of this project, an area that encompassed the 16th largest economy in the world. The banks and financial community were posed to take their place in the national and international world. The rail construction advanced the region as the gateway to the newly emerging Pacific Rim global economy.

The 11-mile heavy rail, mostly underground, was the first step in rebuilding the rail system that had been dismantled in the first half of the century to make way for the automobile. Geographically, the proposed line would link the regional rail at Union Station with the Civic Center and business areas. The proposed line would traverse downtown as a regional core, extend through the Sepulveda Pass to end in Northern Hollywood. This proposal would allow future rail construction to then extend through the San Fernando Valley which comprised almost half of the population of the City of Los Angeles. The region started actual construction of the largest new rail program in the nation in the early 1980s.

The plan encountered tension between transit and freeway development immediately, a tension that continues and will likely continue for decades to come. Symbolically, automobility had become an icon for the American spirit given the freedom that it provides. However, from a political perspective, building transit provided an important alternative to building increasingly unpopular and expensive freeways (Edelman, 2000). Transit and multiple occupant vehicles provided an alternative to constructing freeways that would need to displace homes in the built urban environment.

The SCAG plan outlined how the transit system of buses from the Regional Transportation District (RTD), the second largest bus fleet and system in the nation would support this hub. The bus system would connect the rail lines to the vast and horizontal Los Angeles county. Initially buses were thought to be the answer in this horizontal region. They could roam on the vast freeway system, and the plan described additional lanes on freeways and major arterials for their exclusive use. But congestion on both systems not only slowed autos and trucks but buses as well. Relying on a freeway-based strategy with transit for the large centers would be necessary but not enough.

Two additional transit systems developed over the past 30 years to support the movement of people to the business core and throughout the region. These projects were seen as essential to avoiding freeway gridlock and to support growing the economy. The regional transportation agency, the Los Angeles County Transportation Commission (LACTC), along with Supervisor Kenneth Hahn of Los Angeles County, led the effort for passage of a half-cent sales tax increase as a new funding source for transit. The LACTC provided the regional strategy for a comprehensive rail transit system and oversaw Metro construction in Los Angeles County. The LACTC started building its first light rail line in the mid-1980s, concurrent with the construction of the downtown subway. The LACTC oversaw construction of the 22-mile blue line that linked the county's two largest cities—Los Angeles and Long Beach. Today the 111-mile system is the second largest light rail system in the nation with over 200,000 trips per day. The substantial investment was made possible by the passage of three separate sales tax increases, and the absence of a sunset provision on the initial measure. Combined the taxes yielded over $120 billion to build infrastructure and support transit operations over four decades.

In 1986, a regional commuter system of 534 miles was included in the SCAG Regional Transportation Plan. The plan proposed transit binding together all parts of this expansive region. The LACTC expanded their focus to acquire the regional right of ways from existing rail lines. These railroad lines were disposed of due to commercial railroad company mergers and the deregulation of the rail industry due to the Stagger Act. The planning of this system resulted from SCAG's analysis of the possible uses for these abandoned lines for both passengers and goods movement in the region. A new instructional design, the Southern California Regional Rail Authority (Metrolink), was a partnership arrangement among the County Transportation Commissions in the SCAG region. This strategy pilot developed a regional network for a commuter rail system in only five years from idea to operating system. The commissions purchased a large portion of the right away and entered operating agreements with the railroads for use of tracks still handling freight. Ridership on the system reached 42,265 passengers in 2012 but continues to be limited in size by the agreements with the freight rail systems.

Freeways

Regional freeway construction was altered with the development of the Century Freeway. Stalled by litigation for over 20 years, the settlement agreement transformed freeway construction into a housing, economic development, and freeway project to regenerate an economically depressed part of the region. The Century Freeway construction resulted from the settlement of 20-year lawsuit based on the process for the Environmental Impact Statement (EIS). The settlement by Judge Henry Pregerson involved not only building the 18 miles of freeway but adding a transit line that linked Norwalk on the east to El Segundo on the west, along with a carpool lane. Finally, to replace the housing removed by the right of way clearance of the freeway, 10,000 housing units were to be constructed. His settlement was a metaphor for the new era of infrastructure investment in the region. Highway capacity would be provided only if it was accompanied with mitigation: transit and carpool lanes to offset air quality and housing to replaced displaced individuals. Growth, capacity, and a single purpose investment, even if funded with federal and state monies, were no longer sufficient.

Concurrently, SCAG facilitated the annual regional transportation planning process to consider the impact on economic development, housing, and environment. New freeways construction shifted from the Los Angeles core region of SCAG to the southern and eastern parts of the region in Orange County and Riverside County to address population growth and the need for economic development in these areas. The links in Riverside and San Bernardino were among the last to accomplish the difficult task of receiving annual appropriations of federal and state funding.

Parts of these new freeways relied on tolls and a system of value capture from land development. A new system of HOV lanes that blanketed the region was created to alter the single occupant car mania. The Orange County freeway system was built using a new institutional design: collecting funding streams to pay for the system coupled with a new approach to building the projects. The north south projects, the San Joaquin Hill Corridor and the Foothill Corridor, were contained in the 1982 SCAG RTP as new corridors to be built by a new agency set up under state law, the Orange County Transportation Authority (OCTA). The authority had three funding streams to issue debt sold in the financial markets. For building and operations, OCTA contracted with the state of California's transportation department—Caltrans. The first major funding stream was the first toll-funded project in the state since the start of the interstate program. The second source was payment from large land development projects, including the Irvine Company, Mission Viejo. This funding source was developed and led by Bruce Nestande, an elected supervisor in the County of Orange and a SCAG board member. The third source of funding was federal financial assistance in the form of grants along with wetland and open space agreement land contributions.

The institutional design was combination of policy agreements that enabled the financing and approval process for these innovative projects. These infrastructure projects provided more than highway and carpool capacity. The projects supported housing, open space, and wetland protection coupled with funding provisions that retired debt. Not unlike other debt-funded projects of the past several decades, the project has been refinanced and has been restructured because of economic cycles and fluctuating usage patterns.

In the southern and eastern region of SCAG, the State Route 91 Corridor was built for additional freeway and carpool capacity connecting Riverside County and Orange County. Developed through a concession agreement between Caltrans and a private developer—a public-private partnership (P3) agreement—this project illustrates another institutional design to build infrastructure for freeway/carpool capacity. It did not utilize the typical approach of giving a public entity responsibility for the decision-making and financing for the project with contracts. Instead, legislation granted decision-making on tolls, financing, and implementation to a private firm. The decision-making on tolls led to differences between the counties, in part due to increases in fares by the concessionaire to cover costs that resulted in a lawsuit filed by Riverside County against Orange county. Poor financial performance and the differences among the political bodies led the private concession holder to sell the project to the Orange County Transportation Commission.

Carpools and Information Technology

If there were too many vehicles with only one person per vehicle carrying the "limousine cowboys" of the west, then why not encourage more passengers and create carpool lanes? Caltrans built the first lanes in the early 1970s to widen Interstate 10 in the San Gabriel Valley—east of Los Angeles. SCAG had included it in the Regional Transportation Plan (RTP) of 1975. The carpool lanes increased the number of required occupants in vehicles to an average of nearly three persons per car in return for travelling in a newly separated lane. Since more people would be riding in fewer vehicles it was also included as a transportation control measure in the Air Quality Management Plan.

The next freeway experiment was on Interstate 10 travelling west of Los Angeles. This time a lane was taken away from motorists and the result was a total failure. Motorists almost rioted on the freeway itself, and the newly appointed Executive Director of SCAG, Mark Pisano, had to field calls on a radio talk show from irate residents questioning the wisdom of reducing their freeway lanes with impolite language. The lesson learned was clear: Carpool lanes could be added to new freeways or via widening existing freeways, but the number of lanes could not be reduced. This limitation slowed the construction of building carpool lanes to only a few miles of freeway a year over the course of five decades.

To complement the investment program of building carpool lanes, government and business entered into a partnership to promote voluntary carpooling among workers. A public-private partnership known as Commuter Computer became an early version of data-based ridesharing matching. The board was comprised of public officials from state and local entities along with business leaders from throughout the region. The database matched employees of the region's employers, public and private, in geographical locations so that individuals could form carpools. The program provided information to individuals about location and convenience of matches, as well as information about the cost savings of this approach to transportation along with air quality improvements. The matching effort also become part of the Air Quality Plan. This approach to trip making was demonstrated to be more effective than facility construction.

Goods Movement

The economic reality of the SCAG region in the last half of the 20th century changed from manufacturing, defense, finance, and business headquarters to two new industrial bases. The region became the capital of the entertainment/content industry not only for the United States but the world. The second major economic change was the region's transformation to become the logistics and supply chain center for the United States. As a major port of entry for the Pacific Rim, the goods movement upsurge of the last quarter of the 20th century created a transportation and environmental challenge that has been overwhelming for the region. As evidence of the economic development of the region, over 60 square miles of warehouse space was constructed in this period. The region now handles more than 40 percent of the nation's goods. This new economic base constitutes 30 percent of the employment in the region.

As the governance of the LACTC focused on moving people, the economic development of the region became increasingly linked to the global economy. The growth of containerized shipping through the ports of Long Beach and Los Angeles literally created bottleneck on moving goods throughout the region and the nation. In response, the SCAG facilitated an inclusive planning process that led to the development of the Alameda Corridor Transportation Agency (ACTA). The innovative institutional design of ACTA become the use of a surcharge per container paid by the railroads. This surcharge funded $800 million of the debt for the $2.4 billion total project costs for constructing a 20-mile rail corridor to speed the shipment of containerized goods. The corridor linked the ports with the intersection of the east west rail. The issuance of the debt was made possible by a loan from the Federal Department of Transportation, and the loan structure provided a credit enhancement needed by creditors. In effect, private sector dollars funded economic development through global goods movement. At the time of construction completion and initial operations through 2007, this infrastructure project became a national model for public-private partnerships (Callahan, Pisano, and Lindor, 2010). It also became a model for the establishment of the State of California's Transportation Infrastructure Finance Investment Act (TIFIA).

The South Coast Air Quality Management District (SCAQMD) developed new institutional designs for improving air quality. The air quality improvement program in the region transformed the livability of the region (Mazmanian, 2010). The core element of the plan was an alteration of the controls on cars and industry. The AQMD designed a system of air quality offsets and cap and trade for stationary sources. This approach become an early model for the state and regional effort on climate change. Equally important, the AQMD partnered with SCAG on the development of the Air Quality Maintenance Plan (AQMP) that linked land use, transportation, and energy in the planning processes for the region. The linkage can be seen most vividly in the goods moved by truck and rail.

Truck traffic is now crushing the regional freeway system, while emitting small scale particulates and nanoparticles that are undoing air quality gains in the region. The connection between goods movement encouraging economic development and emissions from diesel combustion particularly impact lower income and disadvantaged communities. Opposition from these communities have stopped investment programs to expand goods movement facilities within the region, which remain a significant infrastructure challenge for expanding the economic base of the region.

Growth

The issue of growth limitation and growth management became not only a policy debate within the region but a state led initiative, complicating the politics of the issue. Then Governor Jerry Brown started the process with a 1976 Office of Planning Research Report titled "Growth Management." While multiple laws were introduced, the conflict between environmentalists and developers blocked significant legislation. This stalemate continued until 2008 with the enactment of legislation SB 375, authored by Daryl Steinberg. The SB3 75 Sustainable Growth stature spelled out steps to reconcile transportation, environmental mitigation, and growth demands. It included the polices of the air quality plan in state legislation but the funding for plan implementation was left unanswered. Under state law, the SCAG region is required to set growth targets, with local governments required to show consistency with the targets. In addition, both the region and local governments are required to align with air quality plan requirements, including climate change targets, transportation plans, and fair share housing requirements. The legislation required sustainability plans developed by local governments in their own plans for population, housing, and employment goals.

The requirement for sustainability plans connected to growth developed from 30 years of experiments within the regions of the state, particularly by SCAG and SACOG (Sacramento Region). The direct stimulus for these experiments was the "Blueprint Planning" initiative of the State of California, Business, Housing and Transportation Agency led by then Secretary Sunne McPeake. The initiative called for each region to develop vision statements that demonstrated how the state goals of economy, environment, and equity could be maximized through their planning processes. The vision needed to be achieved by integrating land use, transportation, air quality, and energy. The approval process of these strategies within the regions created a new political understanding of sustainability in each jurisdiction. As part of an expected dialog, the individual jurisdictions discussed acting in a mutually interdependent strategy rather than acting separately, creating a foundation for cooperation for the region's future.

EMERGING CHALLENGES

The core strategy lesson from Los Angeles was to frame transportation infrastructure projects as assets to generate economic development. The Los Angeles experience created assets as new public institutions, new sources of revenue, and new rail systems that moved people and goods. The success of the Southern California region in past four decades now faces a new set of issues that challenge sustainable economic development through transportation infrastructure. Despite these innovations and successes, the region is not able to keep up with the growth and the environmental needs of the region. Despite the region's success in self-help, financing and implementation innovations gaps remain. Funding of infrastructure continues to be the most significant issue facing the region. The funding analysis of the 2017 Regional Transportation Plan (RTP) reported that 50 percent of the maintenance and capital construction needs were unfunded. Another tension is that the sources of funding in the RTP are 75 percent locally derived revenues. The state provides 15 percent of the funds. The remaining 10 percent is federal funding.

Demographics

Demographics is more than a trajectory of the population growth over time, but reflect cultural, social, economic, and even political behavior (Pisano, 2017). The growth in working-age population drives two-thirds of the growth in the economy. The region's population substantially increased in the later decades of the century, enabling it, and the country, to have an added economic growth spurt. A small shift in age cohort sizes can have a dramatic impact on the economy when you calculate these changes for over 300 million people. The major finding of the research is that these changes had a more significant impact on the economy and taxes than national monetary and fiscal policies (Pisano, 2017). In the last decade of the 20th century, an analysis of the southern California region showed the impacts were positive. The year 2001, when the baby boomers began to turn fifty-five, marked a reduction in the growth rate of incomes and taxes paid by individuals. These changes accelerated in the first decade of the 21st century, contributing to the arrival of the Great Recession. Unfortunately, these negative effects will last for decades.

Another change in demographic make-up is the movement of people via immigration from other countries and migration from other parts of the country. Each movement contributed to the development of the region beginning from World War I to the turn of the century. Immigration from the South and Midwest, along with the Pacific Rim, turned southern California into the immigration capital of the country. Immigration population contributed 72 percent of the population growth. Not only were the numbers significant, but the fertility rate of the immigrant women enabled the region, and the nation, to have a fertility rate above 2.1 for the last two decades. But now the region's fertility rates are below replacement. Not only are growth in income and taxes paid declining, but the growth in the working age population is 30 percent lower than the growth in the last three decades of the 20th century. These demographic changes will impact the future financing of public goods, including the SCAG region infrastructure.

Adaptations for Sustainable Growth Policy

The sustainable growth strategy that is both legislatively and locally enacted will be tested. Recognizing fiscal deficiencies in transportation, economic, and environmental performance goals, the state has enacted legislation creating new institutions of Enhanced Infrastructure Financing Districts (EIFD). These newly enacted entities fund the sustainability city strategies of the individual jurisdictions throughout the state as well as the multiple jurisdictions that have multiple investment objectives.

Southern California's institutional designs suggest a pathway for the legislature of the State of California for the needed next sets of infrastructure investments. For example, the visioning and public policies established by the interactive process of the "Compass Vision" were written into the path breaking California State Legislation SB 375 Sustainable Cities statute. The next step was to create the mechanism for capturing the fiscal wealth these plans created, which was accomplished through two statutes: SB 628 and AB313 which established Public Financing Authorities (PFA). These governmental entities have revenue generating authorities when there is a link between revenue and expenditure. The PFAs also have the authority to enter contractual agreements with private sector parties and collect fees. The nexus of public sector strategy and public sector business plans drive the governance of this process.

The SCAG region is now continuing the evolution of new institutions arrangement and new revenue sources to address the infrastructure challenges in the region using the new legislative authority. A new goods movement system is being developed through an Infrastructure Fund Alliance involving SCAG, Edison, and the State of California Clean Air Board, the Ports and the shippers a regional district that would look at the fuel source issue of trucks and trains, with an option for incentives to effect use pricing to change the behavior of how the urban systems operate. Pricing would evolve from how the PFA collect revenues to capture a portion of the increased value of land use policies and economic development (value capture) to add funding streams to pay for debt issued by these new districts to fund the regions' needs. The lessons learned can be applied to addressing the emerging issues of fiscal constraints (US GAO, 2013), demographics (Pisano, 2017), and to a new inter-governmental system redesign.

Regional Economic Challenges

Along the way several storm clouds have gathered that the region needs to overcome to take advantage of these assets. The recession in the 1990s for the country was a depression for southern California and the impacts were severe. Today the regional economy is 26th out of 27th in terms of income per capita. For transit, the ride share economy (Technology of New Transportation, TNTs) via Uber and Lyft has led to a ridership decline of over 20 percent in the past decade despite the massive investments made in transit. For goods movement, despite truck lane proposals in the RTP, the high cost of the facilities and lack of revenues leave investments unfunded. Truck traffic now utilizes over 30 percent of the many corridors that travel west to east in the region. Trucks and trains are overburdening the geography east of the I-5 freeway in the region where most of the new warehouses and affordable housing stock for workers in all sectors of the economy. The high cost has made housing unfordable for the majority of those entering the housing market. Currently less than 30 percent of new entrants can afford housing in the region, including the eastern part of the region.

The commuter traffic west of the I-5, the location of the entertainment/content economy and the growing higher income service economy, is now so severe that people are not able to leave their neighborhood in rush hour—which has expanded to include every day of the week. A vivid example of the increased geographical separation in the region is the development of the I 405 HOV improvements separating the new economic base around Santa Monica from the urban core. Some have referred to this as the "Berlin Wall" that separates Los Angeles.

CONCLUSION

The Southern California infrastructure building institutions became the foundation for the economic might of the region. The core "sixty-mile" circle became one of the most dynamic regions in the country. The region has made the most of capturing the benefits of the national consensus that existed midway through the last century to grow into a leading economic performer. The region also innovated to complement the traditional paradigm with self-help funding of new infrastructure.

The current challenges illustrate the dynamic system of infrastructure building and economic development. Past success only positions a region to address future challenges. The 40 years of investment in construction of transportation systems in Los Angeles suggest the strengths of a diversified portfolio of projects advanced through varied institutional designs. The complexity of supporting economic development calls for addressing related but disparate goals. Confronting the complexity invited thoughtfully designed governance structures as catalysts for effective collective decision-making and collection action. The range of new institutions addressed the movement of people and goods, improving air quality, and reducing freeway reliance.

The process of governing these projects included a subsequent recognition of the need for adaptation and change, that governance structures designed in an earlier decade earlier needed to evolve to meet new challenges. The variety of institutions provides evidence that one size does not fit all transportation construction projects. The varied economies, demographics, and political jurisdictions in Los Angeles region called for a variety of institutions The SCAG region's governance successes demonstrates an iterative process, with structures evolving over time, and with new governance designs emerging to address new challenges. Progress on transportation funding and construction was incremental and innovative as infrastructure construction extended over a long-time period. The Southern California regional experience found that all the solutions need not be present at the moment of creation, and it is reasonable to expect adaptation and evolution over time.

REFERENCES

Callahan, R. (2007). Governance: The Collision of Politics and Cooperation. *Public Administration Review*, (67), 299–301.

Callahan, R., & Pisano, M. (2014). Aligning Fiscal and Environmental Sustainability. In D. Mazmanian & H. Blanco (Eds.), The Elgar Companion to Sustainable Cities: Strategies, Methods and Outlook (pp. 154–165). Academic Press.

Callahan, R., & Pisano, M. (n.d.). Fiscal Sustainability, Demographics, and the Social Determinants of Health Driving Intergenerational Equity. In D. L. S. María (Ed.), *Financial Sustainability and Intergenerational Equity in Local Governments. IGI Global*. doi:10.4018/978-1-5225-3713-7.ch006

Callahan, R., M. Pisano, & A. Linder. (2010). Leadership and Strategy: A Comparison of the Outcomes and Institutional Designs of the Alameda Corridor and Alameda Corridor East Projects. *Public Works Policy and Management*, (14), 263-287.

Chapman, J. I. (2008). State and local fiscal sustainability: The challenges. *Public Administration Review*, *68*, S115–S131. doi:10.1111/j.1540-6210.2008.00983.x

Chen, Z., & Haynes, K. (2015). Regional Impact of Public Transportation Infrastructure: A Spatial Panel Assessment of the U.S. Northeast Megaregion. *Economic Development Quarterly*, *29*(3), 275–291. doi:10.1177/0891242415584436

Dadayan, L., & Ward, R. B. (2009). State and local finance: Increasing focus on fiscal sustainability. *Publius*, *39*(3), 455–475. doi:10.1093/publius/pjp014

Elkind, E. N. (2014). *Railtown: The Fight for the Los Angeles Metro Rail and the Future of the City*. Los Angeles, CA: University of California Press.

Fiorino, D. (2010). Sustainability as a Conceptual Focus for Public Administration. *Public Administration Review*, *70*(68), S78–S88. doi:10.1111/j.1540-6210.2010.02249.x

George, A., & Bennett, A. (2005). *Case studies and theory development in the social sciences*. Cambridge, MA: MIT Press.

Heinrich, C. J., Hill, C., & Lynn, L. E., Jr. (2004). Governance as an organizing theme for empirical research. In *The Art of Governance: Analyzing Management and Administration* (pp. 3-19). London Financial Times.

Mazmanian, D. (2009). Los Angeles Clean Air Saga—Spanning the Three Epochs. In D. Mazmanian & M. Kraft (Eds.), *Toward Sustainable Communities: Transitions and Transformations in Environmental Policy* (pp. 88–113). MIT Press. doi:10.7551/mitpress/9780262134927.003.0004

National Academy of Public Administration. (2018). *Federal Systems Redesign*. NAPA.

Pisano, M. (2017). *The Puzzle of the American Economy*. Santa Barbara, CA: Praeger.

Pisano, M., & Callahan, R. (2013, October). Developing effective mechanisms that promote fiscal sustainability. *Government Finance Review*, 74.

Rodriguez, B., Pedro, M., Navarro Galera, A., Alcaide Munoz, L., & Deseada Lopez Subires, M. (2014). Factors influencing local government financial sustainability: An empirical study. *Lex Localis: Journal of Local Self-Government*, *12*(1), 31–54. doi:10.4335/12.1.31-54(2014)

Rodriguez, B., Pedro, M., Navarro Galera, A., Alcaide Munoz, L., & Deseada Lopez Subires, M. (2016). Analyzing forces to the financial contribution of local governments to sustainable development. *Sustainability*, *8*(9), 925–943. doi:10.3390u8090925

Southern California Association of Governments. (1994). *Regional Comprehensive Plan*. Los Angeles, CA: SCAG.

Southern California Association of Governments. (2004). *Compass: Charting the Course for a Sustainable Southland*. Los Angeles, CA: SCAG.

Southern California Association of Governments. (2008). *Comprehensive Regional Plan*. Los Angeles, CA: SCAG.

Tang, S. Y., Callahan, R. F., & Pisano, M. (2014). Using common-pool resource principles to design local government fiscal sustainability. *Public Administration Review*, *74*(6), 791–803. doi:10.1111/puar.12273

U.S. Government Accountability Office (GAO). (2013). *State and Local Governments' Fiscal Outlook: April 2013 Update*. Retrieved from http://www.gao.gov/assets/660/654255.pdf

Yin, R. K. (2009). *Case study research: Designs and methods* (4th ed.). Thousand Oaks, CA: Sage.

KEY TERMS AND DEFINITIONS

Cross-Sectoral: Involving two or more of the following sectors: the public, private, and nonprofit.

Demographics: The study of populations and subpopulations over time.

Economic Development: The financial performance in a specified geographic region, affecting wages, housing, businesses, goods movement, including a range of metrics on individual and aggregate changes over time.

Fiscal Sustainability: The availability of budget and other financial resources over time.

Governance: The exercise of public authority through public agencies and partners.

Institution Design: The rules in a society for governance of each of the public, private and nonprofit sectors, for the exercise of public authority, for the oversight of markets, for the organizational structures and partnerships initiating, overseeing, and implementing public policy.

Strategy: Aligning the internal operations of an organization or the design of institutions with the external demands and changes in the environment.

Chapter 5
From College to the City:
Implications of Rail Transit on the Movement of the Young, College Educated Into the City Center

Lenahan L. O'Connell
University of Kentucky, USA

Juita-Elena (Wie) Yusuf
Old Dominion University, USA

Timothy J. Brock
Independent Researcher, USA

Benjamin Blandford
University of Kentucky, USA

ABSTRACT

This chapter focuses on how investment in the American rail infrastructure has shaped changes in the population and residential patterns. Specifically, the chapter examines the association between commuter rail systems, urban rail transit systems, and the movement of the college-educated young into the inner city. Two hypotheses are proposed about the characteristics of rail systems and the relationship to the growth in the percentage of young college graduates residing in close-in neighborhoods. Using a sample of central cities within the 51 largest metropolitan areas in the U.S., the chapter compares the growth in young college graduates (ages 25 to 34 years) across cities with the different transit configurations. Using correlation analysis, the chapter explores the relationship between the presence of rail transit and the residential location choices of this population group. In the discussion and conclusion, the findings are summarized and implications for policy and sustainability are discussed.

DOI: 10.4018/978-1-5225-7396-8.ch005

INTRODUCTION

American cities experienced disinvestment and economic decline in the early decades of the post-World War II era when many in the middle class moved to the suburbs, leaving behind low-income residents and a deteriorating inner city. This produced an ongoing crisis in central cities as American cities, which tend to rely on property taxes to meet many of their responsibilities (Garvin, 1996), saw their housing stock and tax base deteriorate.

The mass migration of affluent urbanites to the suburbs had other adverse consequences. It fostered the widespread purchase of automobiles for commuting, which ultimately produced today's transportation bottlenecks, congestion, and infrastructure breakdowns. It consumed large amounts of farm and wild land, requiring the building of many miles of roadway with detrimental effects on the landscape and increased greatly the number of miles Americans drive (Gillham, 2002). Many of the affluent, college-educated, middle-class Americans now endure the wasted time and frustration of long commutes and time spent in traffic when driving into the city centers where many professional jobs are located.

In recent years, many cities have responded by adopting economic development policies to improve accessibility to the city center by fostering development closer to the city center through investments in light rail, transit-oriented development, and various downtown amenities. Presumably, these would attract talented, affluent residents and improve the tax base. Urban theorists assert that transportation accessibility to the urban core is one of the key generators of economic growth (Glaeser, 2011; Polèse, 2013). In their research on American cities, Florida and Adler (2018) found that the most advantaged Americans, primarily those with college degrees, tend to "locate in the most economically functional and aesthetically desirable places—in and around the urban core, along transit routes, close to universities and knowledge institutions, and along waterfronts and other natural amenities." (p. 610)

Rail transit, in particular, is often advanced as a means to attract high-income residents to downtown areas. Presumably, higher income residents will upgrade the housing stock and the overall inner-city tax base, rendering the inner city more prosperous and attractive to higher income residents. In addition to its contribution to sustainable city finances, rail transit can generate environmental sustainability, especially when it fosters high density development around transit stops. It can reduce the average urbanite's carbon footprint along with the number of miles he or she drives each year (Glaeser, 2011). In contrast to suburban sprawl, new lane miles may not be needed. All of this can enhance the quality of life and even contribute to a healthier populace because urban transit riders tend to get more physical activity as they walk to and from rail stops (Frank, Engelke, & Schmid, 2003).

This chapter begins with a background discussion of the American rail system to differentiate the commuter rail and urban rail transit system that are the chapter's focus. The chapter examines the association between commuter rail, urban rail transit, and the movement of the college-educated young into the cities. An overview of the literature on transportation and population movement provides the context for understanding how investments in rail systems fit within urban economic development and changes in population and residential patterns. The literature is used to set up two hypotheses about characteristics of rail systems and effects on the population of young college graduates residing in close-in neighborhoods. Using a sample of central cities within the 51 largest metropolitan areas in the U.S., the chapter compares the growth in young college graduates (ages 25 to 34 years) across cities with the different transit configurations. Using correlation analysis, the chapter explores the relationship between the presence of rail transit and the residential location choices of this population group. In the discussion and conclusion, the findings are summarized and implications for policy and sustainability are discussed.

BACKGROUND

Four Categories of Rail Transport

American rail systems can be organized into four broad categories: freight rail, passenger rail, commuter rail, and urban rail transit (Brock & Souleyrette, 2013), as shown in Table 1. These categories include freight rail, passenger rail, commuter

Table 1. Categories of U.S. rail systems

	Sub-categories	Geographic Scale
Freight Rail	• Class I • Regional Rail • Shortline	• National/Regional Network
Passenger Rail	• Amtrak • Alaska Railroad • High Speed Rail	• National Intercity Connectivity
Commuter Rail	• Legacy • New Start	• Greater Metropolitan Commuter Shed
Urban Rail Transit	• Light Rail (Street Cars) • Heavy Rail (Subway or Metro)	• Intra-Urban • Downtown

Note: Commuter Rail and Urban Rail Transit are the focus of this chapter.
Source: Adapted from Brock and Souleyrette (2013)

rail, and urban rail transit. The focus of this chapter is on commuter rail and urban rail transit, what will be referred to throughout the chapter as rail transit.

Commuter rail encompasses both regional and suburban rail systems. These systems connect the downtown of a major city to the surrounding suburban communities. Operating on a frequent and regular schedule, they tend to run most often at traditional peak commuting hours. Commuter rail systems operate on shared track corridors with freight rail carriers and Amtrak passenger rail. These shared commuter corridors usually range between 30 and 200 miles of track, although the very largest systems in the country have up to several hundred miles of track. Commuter rail systems serve the metropolitan areas of many of the nation's largest cities including New York City, Chicago, Philadelphia, and Washington, D.C. (Brock & Souleyrette, 2014). With their wide service areas, commuter rail systems tend to serve affluent professionals and managers who live in the low-density, outer suburbs and commute to work in the central city. Their service quality is high, and in the wake of the post-war suburban expansion, many of the old commuter rail lines have been upgraded and extended into new and up-scale communities.

Urban rail transit systems are defined as electric-powered, fixed guide way rail systems that transport passengers within the city center and inner suburbs. One type of urban rail transit (also sometimes referred to as rail rapid transit) is heavy rail (subway or metro), which operates on a separated right-of-way and moves large numbers of passengers at once. With more cars coupled together, these heavy rail systems can move large volumes of passengers. Another type is light rail transit, also called streetcars. These transit systems operate on separated right-of-way, reserved corridors along highway medians or at-grade with street traffic. Although both are on city streets, 'light rail' commonly refers to a separated right-of-way, while 'streetcar' usually refers to at-grade vehicles that mix with traffic. Many of the larger and older cities built their heavy transit systems in the early years of the 20th century, such as New York, Boston, and Chicago. Some new systems have come on-line more recently, for instance, in the Washington, D.C. and San Francisco Bay areas.

Transportation and Population Movement: An Overview

Heavy and light rail systems were built in the early decades of the 20th century with rail ridership exceeding bus ridership through 1945. Altshuler and Luberoff (2004) reported that in the post-war era rail ridership, especially for light rail or streetcars, took a nosedive, falling from 9,246 million passengers in 1945 to 3,904 million in 1950. Consequently, many urban rail transits systems were discontinued, especially streetcars in midsize cities. Ridership on such systems reached its nadir in 1980 when ridership was only 122 million. Between 1945 and 1975, total ridership on bus and rail dropped from 23,254 million passengers to 7,219 million passengers (Altshuler & Luberoff, 2004).

Americans had turned to the automobile for transportation. The number of automobiles on the roads rose sharply and cities experienced increases in congestion along with a shortage of available and low-cost parking. During the 1950s and continuing into later decades government at all levels invested heavily in new roadways, but the limits of building highways were soon recognized. Most large urban areas, especially those on the coasts, had a limited amount of land suitable for the construction of highways and parking facilities. The rapid rise in the population in urbanized areas gave rise to a search for another approach to congestion.

By the 1970s, many community leaders, politicians, and downtown business owners were calling for investment in rail infrastructure. They were joined by environmentalists, rail manufacturers, and rail operators (Altshuler & Luberoff, 2004). Proponents of rail advanced a number or arguments for investing in rail. These included arguments that rail systems reduce the need for new highway construction, help contain suburban sprawl, carry more passengers at lower labor costs than buses, reduce air pollution, stimulate higher density development and redevelopment around rail stations and stops, and attract the affluent middle-class back to the cities (Altshuler & Luberoff, 2004; Ehrenhalt, 2012).

Decision-makers at all levels of government accepted the need for transit and investment in transit rose considerably. By 1999, capital expenditures (in 2002 dollars) for transit had increased from $1.4 billion to $9.2 billion (Altshuler & Luberoff, 2004), and increased investment in rail infrastructure have continued (Ganning, Beaudoin, Brewer, Kim, & Park, 2016). A substantial portion of these expenditures was devoted to heavy rail, light rail, and commuter rail. Between 1975 and 1999, ridership on heavy rail rose by 51%, on light rail by 133%, and on commuter rail by 44% (Altshuler & Luberoff, 2004).

As the rail ridership growth suggests, many of the 51 largest urban areas have devoted resources to the development or expansion of one or more of the three types of urban rail transit systems. Remarkably, 31 of these systems were either built or expanded since 1984. By 2012, 27 U.S. cities operated light rail or streetcar/trolley systems. Twenty-two of these light rail systems were built or expanded since 1980. Since 2012, several more have begun service or are under construction. The contribution of these urban transit systems to the growth of public transportation in the U.S. is clear. The American Public Transportation Association reports that "[F]rom 1995-2013, public transportation ridership grew 37.2 percent, almost double the amount of population growth at 20.3 percent" (American Public Transportation Association, nd).

A Return to the Center

Urban economic development policies have a mixed track record. Some, such as investment in building convention centers, have failed to generate the promised middle-class jobs and residents (Farmer, 2018). But investment in rail may have the desired effects as the number of young college-educated residents living near the city center of the largest 51 American metropolitan areas has grown significantly over recent years (Cortright, 2014; Kolko, 2016) and many private sector companies are moving downtown (Kneebone, 2013). Moreover, the emerging preference for residing close to the central business district is not regional, as 49 of the 51 largest metropolitan areas in the U.S. have experienced rising numbers of college-educated young adults (Cortright, 2014). Another indicator of the strength of this movement towards the city center is the finding that among the 27 cities with declining populations, 25 witnessed an increase in the number of college-educated residents, ages 25 to 34 years, residing within 3 miles of the central business district (Cortright, 2014). Kolko (2016) notes similar urban trends using public use micro-data samples from the 2000 Census and 2014 American Community Survey. Within the 25 to 49 year old age group, those with college education are not only 5% more likely to live in urban neighborhoods in 2014 compared to 2000, but also, 17% more likely to live in higher-density urban core neighborhoods.

In all likelihood, the growth in transit ridership and the percentage of the young college educated moving to the center has multiple sources. Clearly, immigration is producing demographic changes likely to increase demand for transit services, as immigrants are more likely to live in large cities, less likely to own automobiles, and more likely to rely on transit (Glaeser, 2011). This suggests that immigrants are attracted to the cities offering rail services.

The rise of college-educated residents moving close-in to the city center suggests another trend that can increase ridership, especially on rail, and spur reinvestment in the central city. Using data from the American Community Survey, Cortright (2014) found that increasingly large numbers of college-educated young people are moving into close-in neighborhoods in the nation's 51 largest metropolitan areas where in 2012 more than 170 million Americans lived.

A close-in neighborhood is defined as a neighborhood located within three miles of a metropolitan area's central business district. Cortright (2014) reported that, "In 2000, young adults with a four-year degree were 77 percent more likely to live in close-in neighborhoods than other metro residents. Now, these well-educated young adults are about 126 percent more likely to live in these close-in urban neighborhoods" (p. 2). Overall, the number of college-educated residents, ages 25 to 34, living in close-in neighborhoods rose 37.3 percent between 2000 and 2012—an increase almost twice the 19 percent rise of similar young people across the 51 metropolitan areas on the whole (Cortright, 2014). Kolko (2016) finds that the college-educated residents who live in dense urban neighborhoods are more likely to be young, white, affluent, and childless. Population growth in Washington, D.C. reflects the trend. The New York Times reports that approximately all of the District's population growth between 2000 and 2010 was in the 20 to 34 year old demographic group (Morello, Keating, & Hendrix, 2011; Weiland, 2017).

Not only are the young moving toward the city center, they are driving less. From 1995 to 2009, urban drivers, ages 25 to 34, reduced their vehicle miles of travel per day from 30.1 miles to 24.5 miles, while urban drivers, ages 35 to 44, reduced their daily mileage by a negligible amount from 30.3 to 30.1 miles (Santos, McGuckin, Nakamoto, Gray, & Liss, 2011). This suggests that the young, college-educated residents may be more willing to rely on transit for some or even all of their travel needs, in which case, they might be more likely to move to cities with transit options. Additionally, research has found that the more affluent residents of large cities are more likely than the average American to ride transit (Rashidi, Mohammadian, & Zhang, 2010).

No doubt many of the more educated will move to the suburbs when they have children. However, the rising age of first births for mothers, especially those who hold college degrees, suggests that the life course for many of the well-educated in their 20s and 30s will include a lengthy stay near the city center (Glaeser, 2011). Currently, those who move to suburbs are replaced each year by a new contingent of college graduates, an annual event that is expected to continue into the foreseeable future.

The Appeal of Rail

Commuter rail and urban rail transit connects suburbs and residential neighborhoods in the city to the central business district. However, as Kotkin (2010) notes, many commuters prefer traveling on urban rail transit to riding on city buses. Rail transit tends to be much faster than buses because buses share the street with automobiles and make more stops than rail, and in general, rail trips tend to be smoother and more comfortable. Middle-class Americans are more likely to ride heavy or light rail than buses, (Kotkin, 2010) and cities with these systems tend to have more affluent transit users.

Many of these urban rail transit systems are relatively new or newly expanded (Garrett & Taylor, 1999). The new and expanded rail transit systems may attract new ridership when they promote transit-oriented residential and commercial development around rail transit stops. By design, many of these developments are a quick ride from the downtown area, as developers appear to assume college-educated people find neighborhoods with access to rail transit to be desirable ("Transit Oriented Denver: Transit Oriented development Strategic Plan," 2014).

While this chapter's focus is on rail transit broadly, for analysis, it is useful to distinguish commuter rail from urban rail transit (light rail and heavy rail). Commuter rail systems bring suburban residents living far from the city center into the city. They tend to have fewer stops in the city. Urban rail transit systems, in contrast, are more likely to serve city residents with more stops, especially those living close-in.

Hypothesis One: Cities with light and/or heavy rail systems experienced a larger increase in the percent of young college graduates residing in close-in neighborhoods than cities with only commuter rail or bus service.

As shown in Table 1, there are two categories of commuter rail: legacy systems developed before 1975 and new start systems developed since 1975. Fourteen of the 20 commuter rail systems in the 51 largest metropolitan areas in the U.S. are new start systems and only six are legacy systems. New commuter rail systems are more likely to be built with redevelopment of the inner city and transit development in mind. The presence of these new start commuter rail systems could be associated with more college-educated residents living in close-in neighborhoods.

Hypothesis Two: Cities with recently built (new start) commuter rail systems experienced a larger increase in the percent of young college graduates moving close-in than cities with old (legacy) commuter rail systems.

DATA AND METHODS

This chapter's analysis uses a sample of the 51 largest metropolitan areas in the U.S in 2012 (see Table 2 for the list of metropolitan areas). The data include information from the Federal Transit Agency on the types of transit services available in each metropolitan area, including whether or not they are new start or legacy systems. Of the light rail systems operating across the U.S., all but three are operating in the 51 largest metropolitan areas used in this chapter.

In terms of population change, the data include the percentage change, between 2000 and 2012 in college-educated, 25 to 34 year olds, living within three miles of the city center, as reported by Cortright (2014) from American Community Survey data. Data on city population and demographics from the U.S. Census Bureau are also included. Note that the population of the central city is used rather than that of

Table 2. List of the 51 metropolitan areas (in 2012) in the study sample

Atlanta-Sandy Springs-Roswell	Minneapolis – St. Paul – Bloomington
Austin – Round Rock	Nashville – Davidson – Murfreesboro – Franklin
Baltimore – Columbia - Towson	New Orleans – Metairie
Birmingham – Hoover	New York – Newark – Jersey City
Boston – Cambridge – Newton	Oklahoma City
Buffalo – Cheektowaga – Niagara Falls	Orlando – Kissimmee – Sanford
Charlotte – Concord – Gastonia	Philadelphia – Camden – Wilmington
Chicago – Naperville – Elgin	Phoenix – Mesa – Scottsdale
Cincinnati	Pittsburgh
Cleveland – Elyria	Portland – Vancouver – Hillsboro
Columbus	Providence – Warwick
Washington – Arlington – Alexandria	Raleigh – Cary
Dallas – Fort Worth – Arlington	Richmond
Denver – Aurora – Lakewood	Riverside – San Bernardino – Ontario
Detroit – Warren –Dearborn	Rochester
Hartford – West Hartford – East Hartford	Sacramento – Roseville – Arden – Arcade
Houston – The Woodlands – Sugar Land	Salt Lake City
Indianapolis – Carmel – Anderson	San Antonio – New Braunfels
Jacksonville	San Diego – Carlsbad
Kansas City	San Francisco – Oakland – Hayward
Las Vegas – Henderson – Paradise	San Jose – Sunnyvale – Santa Clara
Los Angeles – Long Beach – Anaheim	Seattle – Tacoma – Bellevue
Louisville/Jefferson County	St. Louis
Memphis	Tampa – St. Petersburg – Clearwater
Miami – Fort Lauderdale – West Palm Beach	Virginia Beach – Norfolk – Newport News
Milwaukee – Waukesha – West Allis	

the metropolitan area, as transit is more heavily used by central city residents and used far more for commuting to work because of the high cost of downtown parking.

Some key characteristics of the sample are provided in Table 3, which shows the descriptive statistics. All the cities at the center of the 51 largest metropolitan areas operate a bus system with 18 cities (35.3%) offering only bus service. Thus, 33 cities (64.7%) provide some type of rail service. Twenty cities (39.2%) have commuter rail, 11 cities (21.2%) have heavy rail, and 26 cities (51%) have light rail.

Table 3. Descriptive statistics (N = 51)

	Mean	Std. Deviation	Minimum	Maximum
City Population	864,883.59	1,232,776.38	124,775	8,175,133
% Change in Young College Graduates Living Close-in (2000-2012)	46.73	29.85	-16.00	138.00
% White	71.27	9.66	47.90	87.50
% Black	16.15	10.84	1.60	54.00
% Hispanic	15.83	13.19	1.40	54.40
Light Rail	.51	.50	.00	1.00
Heavy Rail	.22	.42	.00	1.00
Light Rail or Heavy Rail	.59	.50	.00	1.00
Light Rail and Heavy Rail	.14	.35	.00	1.00
Commuter Rail	.40	.49	.00	1.00
Legacy Commuter Rail	.12	.33	.00	1.00
New Start Commuter Rail	.28	.45	.00	1.00
Buses Only	.35	.48	.00	1.00

Six central cities (11.8%) have legacy commuter rail systems, and 14 cities (27.5%) have new start commuter rail systems.

City size and racial composition vary greatly across the sample. Population size ranges from 124,775 to 8.2 million with a mean of 864,883 residents. The average city's population is 71.3% white, 16.2% black, and 15.8% Hispanic. Resident diversity by racial category also varies with the central cities having between 1.6% and 54.0% black residents and between 1.4% and 54.4% Hispanic residents.

In terms of the population, variable of interest – the percent change in young, college educated 25 to 34 year olds living within three miles of the central business district – there is also variation. The mean percent change was 47%, and the median was 40%. Two cities saw a movement of this group away from the central business district. Birmingham (AL) and Detroit (MI) experienced a -16% and -6% change in the population of young graduates living close-in, respectively. In contrast, St. Louis (MO) and Miami (FL) saw growth of 138% and 118%, respectively.

Building on the descriptive statistics, the analysis identifies and examines different transit system configurations and analyzes association patterns in terms of the different transit systems. Comparisons of means are used to analyze how the change in the percentage of young college graduates residing close-in differ across the transit system configurations featuring light and heavy rail, and new start and legacy commuter rail. Finally, correlation analysis is conducted to assess the two hypotheses and specifically, to determine how characteristics of commuter rail and urban rail transit systems covary with the movement of the young college educated into the inner city.

RESULTS

Transit System Configurations

All 51 metropolitan areas in the sample have transit systems that include bus routes, and 18 cities (35.3%) offered bus service to their citizens but no rail service. A total of 33 central cities in the sample complemented their bus service with some combination of the three possible types of rail service—light, heavy, and commuter. As shown in Table 4, urban transit systems can have seven possible configurations of bus and rail services. Combined with bus service, 11 cities (21.6%) offer only light rail. None of the cities offer the combination of only heavy rail and bus, but two cities (3.9%) provide a combination of light rail, heavy rail, and bus; while four cities (7.8%) offer the combination of commuter rail, heavy rail, and bus. Three central cities (5.9%) provide commuter rail and bus; eight cities (15.7%) offer light rail, commuter rail, and bus; and five cities (9.8%) offer all three types of rail service along with bus service.

As noted earlier, 20 cities (39.2%) have commuter rail, and of these 6 were legacy rail systems and 14 were new start commuter rail systems. 11 cities (21.2%) have heavy rail; and 26 cities (51%) have light rail. Thirty central cities (58.8%) have

Table 4. Distribution of transit system configuration in 51 largest cities in 2012

Transit System Configuration	Percent
Bus only	35.3%
Bus + Light Rail	21.6%
Bus + Commuter Rail	5.9%
Bus + Light Rail + Heavy Rail	3.9%
Bus + Commuter Rail + Light Rail	15.7%
Bus + Commuter Rail + Heavy Rail	7.8%
Bus + Commuter Rail + Light Rail + Heavy Rail	9.8%

Note: There is one additional possible combination of bus and heavy rail, but no city in the sample had this combination.

either heavy or light rail, and seven (13.7%) have both light and heavy rail. Table 5 shows the correlations between the commuter rail and urban rail transit variables.

Table 5 shows the correlations between the different commuter and urban rail transit systems. Commuter rail, which largely serves the suburbs, is also positively correlated with the presence of heavy rail transit systems. The correlations between commuter rail and any combination involving heavy rail transit were positive, ranging from a low of 0.306 (between legacy commuter rail and having either light rail or heavy rail) and a high of 0.574 (between new start commuter rail and heavy rail transit). The presence of a light rail transit system was not related to heavy rail transit or commuter rail (new start or legacy).

Assessment of Hypotheses

Table 6 summarizes the average percent change in college-educated, 25 to 34 year olds, living within three miles of the city center between 2000 and 2010 across the different transit system configurations. As a group, central cities with a combination of bus, commuter rail, and heavy rail had the highest increase in the percent of young college graduates living close-in. Cities with transit systems comprised of bus and light and heavy rail had, on average, a 60.5% increase in this demographic group. Cities with bus and light rail transit service experienced an average 57.1% increase. At the lower end of the growth spectrum, cities with only bus service saw a smaller increase in 25 to 34 year old college graduates living close-in of 35.3%, and cities with bus and commuter rail had an increase of 35.7%.

Table 5. Correlations between commuter rail and urban rail transit variables

	Light Rail	Heavy Rail	Light Rail or Heavy Rail	Light Rail and Heavy Rail	New Start Commuter Rail
Heavy Rail	0.133				
Light Rail or Heavy Rail	0.853	0.439*			
Light Rail and Heavy Rail	0.391**	0.761***	0.334*		
New Start Commuter Rail	0.211	0.574***	0.441***	0.363***	
Legacy Commuter Rail	0.115	0.548*	0.306*	0.385***	0.781***

Note: **** p<.0001 *** p<.001 ** p<.01 * p<.05

Table 6. Transit system configurations and average change in percent of young college graduates living within three miles of the city center

		Commuter Rail	Light Rail	Heavy Rail
Bus only	35.3%			
Bus +		35.7%	*57.1%*	
Bus + Commuter Rail +			*48.8%*	*65.8%*
Bus + Light Rail +				*60.5%*
Bus + Commuter Rail + Light Rail +				*37.3%*

Note: Boxes shaded in gray indicate transit configurations that include urban rail transit (light rail or heavy rail)

The results shown in Table 6 provide support for Hypothesis 1, as cities with some transit combination that included light and/or heavy rail systems (italics in Table 6) have experienced greater increases, on average, in the percent of young college graduates residing in close-in neighborhoods than cities with only bus service, only commuter rail, and bus service and commuter rail. The 18 cities offering only bus service saw an increase of 35.3%, while the three cities with only commuter rail and bus but neither light nor heavy rail had an increase of 35.7%. In contrast, the 30 cities with either light or heavy rail or both systems experienced an average increase in the percent of young college graduates living close-in of 54.1%. The differences between the two groups (bus only or bus and commuter rail vs. transit combinations that include urban rail transit) was statistically significant ($t = 2.012$, $p < .05$).

Hypothesis 2 expects that cities with new start commuter rail systems attracted more of the college-educated young to the city center than cities with legacy commuter rail systems. In a simple comparison of means, cities with new start commuter rail systems saw a larger increase in this demographic category than cities with legacy commuter rail systems (58.6% to 41.2%). However, the difference between the two groups, while in the expected direction, was not statistically significant.

In addition to the comparisons of the growth in the percent of young college graduates living close-in across the different groups (light and/or heavy rail vs. bus only or bus and commuter rail, and new start commuter rail vs. legacy commuter rail), the chapter also analyzed the correlation between the percent change in young college graduates living close-in, the presence of urban rail transit (light rail and heavy rail), and the presence of new start or legacy commuter rail systems. The correlation matrix is presented in Table 7.

The correlation analysis supports Hypothesis 1. The change in the percent of young college graduates living close-in was positively correlated with having light rail

Table 7. Correlations between population variables and commuter rail and urban rail transit

	% Change in Young College Graduates Living Close-in	Population	Population Density	% White	% Black
Light Rail	0.308*	-0.035	0.011	-0.154	-0.129
Heavy Rail	0.287*	0.412***	0.684****	-0.404*	0.126
Light Rail or Heavy Rail	0.416**	0.239	0.345*	-0.286*	-0.070
Light Rail and Heavy Rail	0.197	0.100	0.340*	-0.297*	0.064
New Start Commuter Rail	0.095	0.038	0.043	-0.032	-0.299*
Legacy Commuter Rail	-0.069	0.408***	0.645****	-0.258	-0.062

Note: **** p<.0001 *** p<.001 ** p<.01 * p<.05

(r=0.308, p=0.028) and heavy rail (r=0.287, p=0.041) transit systems. The change in the percent of college graduates ages 25 to 34 living in close-in neighborhoods are also positively correlated (r=0.416, p=0.002) with having either a light rail or heavy rail transit system. However, this population change was not correlated (r=0.197, p=0.167) with a central city having both light and heavy rail. The findings support that cities with light or heavy rail systems have experienced a larger increase in the percent of young college graduates residing in the urban center

The correlation between the percent of young college graduates living close-in and the presence of new start commuter rail was not statistically significant (r=0.095, p=0.506). The correlation with legacy commuter rail was similarly not statistically significant (r= –0.069, p=0.632). The correlation analysis does not provide support for Hypothesis 2.

Other population variables are related to some dimensions of transit. Population size is positively correlated with having a heavy rail transit system (r=0.412, p=0.003) and a legacy commuter rail system (r=0.408, p=0.003). Population density is similarly correlated with having a heavy rail system (r=0.684 p<.0001), either a heavy or light rail system (r=0.345, p=0.013) or both heavy and light rail systems (r=0.340, p=0.015). Population density is also positively correlated with having a legacy commuter system (r=0.645, p<.0001). These relationships suggest that more populated and denser central cities are more likely to have transit systems that include heavy rail and legacy commuter rail. Racial composition of the central city also covaries with urban rail transit. Cities with a larger percentage of its population that is white is also less likely to have a heavy rail system (r= –0.404, p=0.033) or a combination of light and heavy rail. Percent black is negatively correlated with having a new start commuter rail (r= –0.299, p=0.033).

DISCUSSION AND CONCLUSION

The hypothesis regarding the relationship between the presence of urban rail transit (light or heavy rail) and an increase in the percent of the young college-educated living close-in (within three miles of the urban center) was supported. The second hypothesis regarding new start and legacy commuter rail, however, was not supported. Cities with recently built (new start) commuter rail systems experienced a larger increase in the percent of young college graduates moving close-in than cities with old (legacy) commuter rail systems, but the differences were not statistically significant. Furthermore, the correlations between the percent of young college graduates living close-in and the presence of either new start commuter rail or legacy commuter rail were not statistically significant.

The results are consistent with the widely held belief of urban planners that urban rail transit systems will lure more affluent individuals such as college graduates into the inner city (Lucy & Phillips, 2006). This, of course, is one of the traditional goals of economic development near the city center. Affluent residents can afford to pay higher rents. Their landlords can afford to pay for basic residential maintenance or improvements. The presence of college-educated residents in a neighborhood draws commercial and residential investment, which is more profitable due to the increased spending power of the young newcomers. Over time, the cities with growing numbers of the affluent are likely to see property tax revenues go up significantly.

A caveat is called for—while the results suggest that the presence of rail transit can draw the young, college-educated to the city center, they do not establish a clear causal relationship for several reasons. This demographic group generally seeks professional work. No doubt the job structure of the city is a primary factor explaining its appeal (Polese, 2013). Moreover, these young, college-educated residents tend to have political influence (O'Connell, 2009), and they could be strongly represented in the city prior to the development of a new rail system. Their proportion of the population could then expand, leading to even more demand for rail and transit-oriented development. This scenario would constitute a reinforcing cycle of development.

Policy Implications

This chapter's findings suggest that it may be the case that government investment in rail connected to the downtown—heavy rail as well as light rail—can spur a virtuous cycle of inner-city development, including new residential multi-unit dwellings. Clearly, seeking to promote and shape the cityscape, government officials may have good reason to turn to investment in rail systems as a central part of a broader set of policies designed to enhance the economic, physical, and environmental sustainability of their cities.

Denver, Colorado, for instance, has been at the forefront of light rail expansion, increasing its mileage of rail transit from 5.3 miles in 1994 to 81 miles in 2016. The city views rail as vital to its growth and central to its strategy to promote the wellbeing of all its residents, wealthy and poor alike. In its plan for Transit Oriented Development (TOD), Denver actively encourages TOD in existing or planned communities near rail transit stations in order to create walkable, vibrant, mixed-use environments ("Transit Oriented Denver: Transit Oriented development Strategic Plan," 2014).

The overarching goal of Denver's plan is to provide an alternative to the automobile for everyday transportation. Residents living in TOD areas will use rail for the majority of their commutes as well as for some longer journeys to other neighborhoods and walking for trips to stores, restaurants, and most routine needs. Thus, dense walkable neighborhoods close to transit centers is the key to generating urban sustainability. The combination of less driving and more walking along with living in dense, often multi-story buildings, will reduce Denver's carbon footprint, the lane miles of new roadway it will need, the level of air pollution, and improve public health ("Transit Oriented Denver: Transit Oriented development Strategic Plan," 2014). Denver has had much success. Thousands of housing units have been built in walkable areas served by light rail, including Downtown, Union Station, and Lower Highlands. The city reports some indicators of positive results, as miles driven per capita have declined and the number of residents commuting by bike or on foot to work has gone up.

However, movement of the more affluent toward the city center during the early stages of their careers can also lead to increasing rents and the loss of housing for the low-income, but it might also provide the tax dollars needed for investment in low-income housing. It can also lead to new, high density housing around rail stations.

Could the popularity of light rail come to an end? Some argue that it is possible the future adoption of autonomous vehicles could undercut the appeal of rail transit (Anderson et al., 2014), and over time, the effect on transit ridership could be dramatic, as autonomous vehicles may have the capacity to drop people at their destinations and then move the vehicle to a remote parking facility or even pick up new riders who sequentially share use of the vehicle. This scenario, however, seems unlikely in the near term, as traffic congestion and an absence of inexpensive parking in dense urban centers will continue to deter automobile use and encourage rail transit.

This chapter looked at the 51 largest metropolitan areas. Clearly, more studies are needed, as research suggests that rail impacts location choices and related developments under some conditions but not others (Chatman, 2013). But the future conditions are favorable. People marry at a later age, crime is down, and many couples do not have children. Light rail makes it easier to move about without driving, cities are building walkable and bikeable facilities, and skilled as well as unskilled immigrants head to cities to find work and live in ethnic neighborhoods.

REFERENCES

Altshuler, A. A., & Luberoff, D. E. (2004). *Mega-projects: The changing politics of urban public investment*. Washington, DC: Brookings Institution Press.

American Public Transportation Association. (n.d.). *Public Transportation Use Is Growing—Here Are the Facts*. Retrieved from http://www.apta.com/mediacenter/ptbenefits/Pages/Public-Transportation-Use-is-Growing-.aspx

Anderson, J. M., Nidhi, K., Stanley, K. D., Sorensen, P., Samaras, C., & Oluwatola, O. A. (2014). *Autonomous Vehicle Technology: A Guide for Policymakers*. Santa Monica, CA: Rand Corporation. Retrieved from https://www.rand.org/pubs/research_reports/RR443-2.html

Brock, T. J., & Souleyrette, R. R. (2013). *An Overview of US Commuter Rail*. Lexington, KY: University of Kentucky, Kentucky Transportation Center.

Brock, T. J., & Souleyrette, R. R. (2014). Commuter Rail. In M. Garrett (Ed.), *The Encyclopedia of Transportation: Social Science and Policy* (pp. 392–394). Thousand Oaks, CA: Sage.

Chatman, D. G. (2013). Does TOD need the T? On the importance of factors other than rail access. *Journal of the American Planning Association*, *79*(1), 17–31. doi:10.1080/01944363.2013.791008

City and County of Denver. (2014). *Transit Oriented Denver: Transit Oriented development Strategic Plan*. Retrieved from www.denvergov.org/content/dam/denvergov/portals/193/documents/tod_plan/Tod_strategic_plan_final.pdf

Cortright, J. (2014, October). *City Report: The Young and Restless and the Nation's Cities*. Retrieved from http://cityobservatory.org/wp-content/uploads/2014/10/YNR-Report-Final.pdf

Ehrenhalt, A. (2012). *The great inversion and the future of the American city*. New York: Knopf Doubleday Publishing Group.

Farmer, L. (2018, May). How cities fell out of love with sports stadiums. *Governing*, 24–29.

Federal Highway Administration. (2011). *Summary of travel trends: 2009 National Household Travel Survey*. Washington, DC: US Department of Transportation.

Florida, R., & Adler, P. (2018). The patchwork metropolis: The morphology of the divided postindustrial city. *Journal of Urban Affairs*, *40*(5), 609–624. doi:10.1080/07352166.2017.1360743

Frank, L., Engelke, P., & Schmid, T. (2003). *Health and community design: The impact of the built environment on physical activity*. Washington, DC: Island Press.

Ganning, J., Beaudoin, M., Brewer, S., Kim, K., & Park, K. (2016, February 28). The Effects of Commuter Rail on Population Deconcentration and Commuting: A Salt Lake City Case Study. Portland, OR: National Institute for Transportation and Communities Report.

Garrett, M., & Taylor, B. (1999). Reconsidering social equity in public transit. *Berkeley Planning Journal*, *13*(1), 6–27.

Garvin, A. (1996). *The American city: What works, what doesn't*. New York: McGraw-Hill.

Gillham, O. (2002). *The limitless city: A primer on the urban sprawl debate*. Washington, DC: Island Press.

Glaeser, E. (2011). *Triumph of the City: How Our Greatest Invention Makes Us Richer, Smarter, Greener, Healthier, and Happier*. New York: Penguin Press.

Kneebone, E. (2013). *Job sprawl stalls: The Great Recession and metropolitan employment location*. Washington, DC: Metropolitan Policy Program at Brookings.

Kotkin, J. (2010, August 9). *Mass Transit: The Great Train Robbery*. Retrieved from https://www.forbes.com/2010/08/09/cities-transportation-class-opinions-columnists-joel-kotkin.html#a6aaac87208f

Lucy, W. H., & Phillips, D. L. (2006). Tomorrow's cities, tomorrow's suburbs. Chicago, IL: American Planning Association.

Morello, C., Keating, D., & Hendrix, S. (2011, May 5). Census: Young adults are responsible for most of D.C.'s growth in past decade. *Washington Post.* Retrieved from https://www.washingtonpost.com/local/census-young-adults-are-responsible-for-most-of-dcs-growth-in-past-decade/2011/05/04/AFJz5LtF_story.html

O'Connell, L. (2009). The Impact of Local Supporters on Smart Growth Policy Adoption. *Journal of the American Planning Association*, *75*(3), 281–291. doi:10.1080/01944360902885495

Polèse, M. (2013). Five principles of urban economics. *City Journal.* Retrieved from https://www.city-journal.org/html/five-principles-urban-economics-13531.html

Rashidi, T., Mohammadian, A., & Zhang, Y. (2010). Effect of Variation in Household Sociodemographics, Lifestyles, and Built Environment on Travel Behavior. *Transportation Research Record: Journal of the Transportation Research Board*, *2156*(1), 64–72. doi:10.3141/2156-08

U.S. Department of Housing and Urban Development. (2016). Urban Revival? Not For Most Americans. *Gentrification, 18*(3). Retrieved from http://jedkolko.com/2016/03/30/urban-revival-not-for-most-americans/

Weiland, N. (2017, November 27). Synagogue's Mix of Arts and Religion Helps Shape Jewish Life in Washington. *New York Times.* Retrieved from https://www.nytimes.com/2017/11/27/us/sixth-i-historic-synagogue-jewish-culture.html

KEY TERMS AND DEFINITIONS

Central Business District: The main business and commercial area of a town or city. In larger city, the central business district is often synonymous with the city's financial district. May also coincide with the city center or downtown, but this is not necessarily always the case.

Close-In Neighborhoods: A neighborhood located within three miles of a metropolitan area's central business district.

Commuter Rail: Passenger rail transport service that primarily connects the city center of a major city to the surrounding suburban communities or commuter towns. Also called regional or suburban rail.

Heavy Rail: A form of high-capacity urban rail transit that operates on a separated right-of-way, separate grade or on elevated railways. Also known as rapid transit or mass rapid transit.

Legacy Transit Systems: Transit systems built before 1975. Generally located in densely populated, older metropolitan areas. These systems generally face substantial issues in terms of overcrowding, aging infrastructure, and underfunding.

Light Rail Transit: A form of urban rail transit that operates tramcars or similar rolling stock on separated right-of-way, reserved corridors along highway medians or at-grade with street traffic.

New Start Transit Systems: Transit systems built after 1975. Generally built with redevelopment of the inner city and transit development in mind.

Urban Rail Transit: Encompasses various types of rail systems, including heavy rail (rapid transit) and light rail transit, that provide passenger service within and around urban or suburban areas.

Chapter 6
Support for and Behavioral Responses to Tolls:
Insights From Hampton Roads, Virginia

Juita-Elena (Wie) Yusuf
Old Dominion University, USA

Khairul Azfi Anuar
Old Dominion University, USA

Lenahan L. O'Connell
University of Kentucky, USA

David Chapman
Old Dominion University, USA

Donta Council
Old Dominion University, USA

Tancy Vandecar-Burdin
Old Dominion University, USA

Meagan M. Jordan
Old Dominion University, US

ABSTRACT

This chapter analyzes the experiences with tolling in the Hampton Roads region of Southeastern Virginia to better understand residents' and drivers' support for tolls and behavioral responses to tolls. The Hampton Roads region, with its population of 1.7 million and extensive network of highways, roads, bridges, and tunnels, has a long history of toll facilities that date back to the 1920s. The most recent tunnel tolls, associated with the Elizabeth River Crossing Project and introduced in February 2014, are the focus of this chapter. This chapter analyzes two sets of survey data to provide insights that have implications for policies regarding tolling: (1) The Life in Hampton Roads Surveys which includes questions about support for tolls and toll avoidance behaviors; and (2) two surveys (pre- and post- toll implementation) commissioned by the regional transportation planning organization.

DOI: 10.4018/978-1-5225-7396-8.ch006

INTRODUCTION

"In infrastructure, you get what you pay for and for decades we haven't been paying nearly enough" (American Society of Civil Engineers, 2017, p. 2). According to the 2017 Infrastructure Report Card, $4.59 trillion is the infrastructure investment need over a 10-year period for the nation's infrastructure to earn and maintain a grade of B (good, adequate for now). Every four years since 2001, the American Society of Civil Engineers (ASCE) releases its assessment of the nation's infrastructure, and once again, America's infrastructure received a grade of D+ (poor, at risk). According to the ASCE grading scale, a grade of D indicates that many elements of the infrastructure may be "approaching the end of their service life...exhibits significant deterioration...with strong risk of failure" (p.13). The nation's infrastructure has scored a D or D+ since the advent of the four-year periodic grading. Specifically, bridges have scored a C+ while the roads have scored a D. These bridge and road scores have been consistent during a time of increased use (American Society of Civil Engineers, 2017).

The ASCE's 2017 report explicitly emphasizes that federal and state funding of infrastructure is woefully inadequate for addressing deteriorating infrastructure. They conclude that "the U.S. has only been paying half of its infrastructure bill for some time" (p. 7). Reportedly, there is a $836 billion backlog of highway and bridge needs with about 50% of that needed to repair existing highways and 15% o needed for repairing bridges.

Given this inadequacy, funding mechanisms or revenue raising methods need to be reexamined. The Federal Highway Trust Fund is the primary source for federal highway funding. The federal motor fuels tax serves as the primary source of revenue for the Highway Trust Fund (Yusuf, 2014). However, the tax rate of the federal motor fuels tax has been stagnant at 18.4 cents per gallon of gasoline and 24.4 cents per gallon of diesel fuel since 1993. At the state level, the motor fuels tax has similarly been cited as a source of the crisis in highway finance (O'Connell & Yusuf, 2013; Yusuf & O'Connell, 2013) with roots in three primary issues: (1) the tax is levied on a per gallon basis that does not automatically adjust with inflation; (2) vehicle fuel efficiency has increased, reducing fuel consumption per mile traveled; and (3) the impact of inflation on construction costs. Local governments are not immune to the problem; as Yusuf, O'Connell, and Abutabenjeh (2011) point out, localities in the U.S.A. are also confronting a crisis in highway finance that forces local governments to look for new sources of funding.

The ASCE argues that infrastructure is the "backbone" of the economy and therefore infrastructure investment is an investment in the U.S. economy. In its 2016 economic impact study, Failure to Act: Closing the Infrastructure Investment

Gap for America's Economic Future (2016), the ASCE emphasizes the costs of deteriorating infrastructure to businesses and households. These costs for 2015 are estimated at $147 billion, which include higher operating and repair costs of vehicles, safety costs, environmental costs, and time costs. These costs are expected to increase substantially as the delay of sufficient funding continues. Furthermore, the deficient and deteriorating infrastructure negatively impacts productivity across job sectors. The ASCE estimates a loss of $3400 per year in household disposable income, millions of lost jobs, and a $4 trillion loss in gross domestic product by 2025 (American Society of Civil Engineers, 2016).

The ASCE (2017) recommends dedicated funding for federal, state, and local levels of government to support investment in infrastructure. Specifically, it recommends raising the federal motor fuels tax rates and indexing these rates to the consumer price index. Also, the ASCE argues that "infrastructure owners and operators must charge, and Americans must be willing to pay, rates and fees that reflect the true cost of using, maintaining, and improving infrastructure" (American Society of Civil Engineers, 2017, p. 9).

Essentially, the ASCE argues for funding that is sufficient to cover current needs and the flexibility to adjust to cover future needs. This emphasis on sufficiency and long-term adequacy are consistent with financially sustainable transportation. Leuenberger and Bartle (2009) approach transportation sustainability from the perspective of moving people and goods, economics, and also social, financial, and environmental perspectives. Similarly, Black (2010) defines sustainable transportation as a system that meets the needs of transportation and mobility in a way that is safe and efficient (reduced congestion), but with concerns for the environment (e.g., use of renewable fuels and reduction of emissions).

Many of these sustainability factors play into the fiscal sustainability approach of this chapter, consistent with Chen's (2014) argument that the financial component is a critical foundation for supporting other aspects of sustainability. This chapter focuses on the use of tolls as a means to generate revenues while simultaneously managing congestion. By providing an analysis of the acceptability of tolls and behavioral responses to tolls, the chapter also highlights implications for the long-term adequacy of tolls as a revenue source, and the effectiveness of tolls as congestion pricing mechanisms to change driver behavior.

Tolling is an alternative funding mechanism that seeks to address the lack of long-term sustainability of the primary sources of revenues for roads, highways and bridges – the motor fuels tax. Tolling is a user fee that is imposed on drivers for use of the tolled roadway (or facility). Tolls can also be used as a congestion pricing mechanism, where the price of the tolls can be adjusted to reflect demand for the tolled roadways and therefore be used to manage congestion (Brownstone,

Ghosh, Golob, Kazimi, & Van Amelsfort, 2003; Perez, Giordano, & Stamm, 2011; Schaller, 2010). In the longer term, tolls with congestion pricing built in also have the potential to address environmental sustainability by influencing land use patterns (Urban Land Institute, 2013).

BACKGROUND

"Tolling and road pricing have become part of contemporary transportation planning and policy making" (Zmud & Arce, 2008, p. 49), particularly in the current environment where governments face significant demands for transportation services and infrastructure, while experiencing increasingly limited resources to meet the demands. Many states and localities in the U.S. are experiencing transportation funding shortfalls coupled with growing needs to address congestion and increase mobility.

Tolls, defined as direct user fees charged for use of road capacity and services, have long been used in the U.S. as a supplemental source for transportation revenues. Toll roads of earlier centuries were largely owned by private toll or turnpike companies, while the toll facilities of the twentieth century have largely been associated with quasi-public authorities (Yusuf, O'Connell, & Anuar, 2014). In recent decades, tolling has become increasingly associated with public-private partnerships for specific transportation projects.

Over the years the traditional concept of tolls has changed very little, with the primary rationale for tolling being to obtain needed funding for specific projects, to shift the burden and costs to specific users, and to provide an immediate and direct source of revenue to service bonds and other financing obligations (Rusch, 1984). One additional function of tolling has recently been introduced, with tolls also taking on a congestion pricing function of managing demand (Yusuf et al., 2014).

Evidence in the popular media and in empirical research studies indicate that the public is not particularly supportive of tolls (Cook, 2014; Kimberlin, 2012; King, Manville, & Shoup, 2007; Laing, 2014; Odeck & Bråthen, 1998; Peirce, 2007; Podgorski & Kockelman, 2006; Rasmussen Reports, 2014; Schade & Schlag, 2003; Zmud & Arce, 2008). However, public support and response to tolls is important to consider when pursuing a tolling project (Santos & Fraser, 2006; Yusuf et al., 2014). Furthermore, the effectiveness of tolls as a congestion management tool also hinges on how the public (drivers, particularly) respond to the tolls.

Tolls and congestion pricing may prompt changes in drivers' use of a tolled roadway, as drivers may undertake specific behaviors to avoid tolls or reduce the financial impacts of tolls (Keuleers, Chow, Thorpe, Timmermans, & Wets, 2006). Examples of behavioral responses to tolls include changes in route, departure time, or destination; shift to using public transit; and telecommuting or working from home (Arentze, Hofman, & Timmermans, 2004; Francsics & Ingrey, 2000; Keuleers et al., 2006; Nielsen, 2004; Olszewski & Xie, 2005).

These behavioral responses to tolling exemplify desirable outcomes for transportation demand management (TDM). TDM represents a set of strategies with the goal of influencing travel behavior by providing alternative mobility options, including road pricing, ridesharing and HOV lanes (Meyer, 1997). By influencing drivers' use of a roadway or bridge, the implementation of tolls can play a role as TDM tools to manage and/or reduce congestion.

THE HAMPTON ROADS TOLLING EXPERIENCE

This chapter uses the recent tolling experience in Hampton Roads, Virginia to focus on two important aspects of tolling: perceptions of and support for tolls, and behavioral responses to tolls. The discussion begins with some background information about the transportation system in Hampton Roads and the region's experience with tolling, including the recent tolling experience – the Elizabeth River Crossing Project – that introduced tolls on two key tunnels serving the region. This is followed by an overview of the surveys that provide data for analysis. Survey results are analyzed and discussed to provide in-depth understanding of public support for tolls and behavioral responses to tolls. Suggestions for future research are offered and the Conclusion section summarizes key points and discusses implications for transportation finance.

Transportation and Tolling in Hampton Roads

The Hampton Roads Transportation System

Bridges and tunnels are distinctive characteristics of the Hampton Roads transportation system. Located on the southeastern coast of Virginia, the Hampton Roads region is one of the world's largest natural harbors. It encompasses several rivers, including the Elizabeth River and the James River, that empty into the Chesapeake Bay and then into the Atlantic Ocean.

Thus, the region's economy relies on the transport of goods and people over or under spans of waterways. The region is home to the Port of Virginia, which is ranked third highest in volume of containerized cargo on the East coast of the U.S. and sixth highest ranked in the country (Nichols, 2017). In 2016, a record setting 21 million tons of general cargo came through the port, of which 61% arrived or departed on the roadways. Additionally, Hampton Roads has well over a dozen military facilities, including the world's largest naval base and facilities for the Army, Marines, Air Force, and Coast Guard.

The region requires a multifaceted transportation system that supports and connects its many communities. Tunnels and bridges are the most widely used methods for transporting goods and people. There are five tunnels that submerge the waters and two drawbridges that allow for passage of maritime vessels during nonpeak periods. The most commonly used bridges are the Hampton Roads Bridge Tunnel and the Monitor-Merrimac Memorial Bridge Tunnel, followed by the Coleman Bridge, James River Bridge, and High-Rise Bridge. The two major sub-regions of Hampton Roads, the Southside (south of the Chesapeake Bay) and the Peninsula, are connected by bridges and tunnels. Figure 1 shows a map of the Southside of Hampton Roads and the major bridges and tunnels in the area.

After years of decreasing roadway usage that began around the Great Recession, roadway usage is on the rise as measured by vehicle-miles of travel and traffic volume (Nichols, 2017). Congestion tells a clearer story of the Hampton Roads travel experience, with a 1.23 travel time index, indicating that it takes 23% longer to travel during peak travel periods than during uncongested periods (Nichols, 2017). This travel time index places Hampton Roads in ninth place among the 35 large metropolitan areas with populations of 1 million to 3 million. Hampton Roads is ranked fifth among the 35 largest metropolitan areas, experiencing almost six hours of congested conditions during the work week in 2016 (Nichols, 2017).

Therefore, roadway conditions, especially the bridges and tunnels that are so vital for mobility around the region, are critical for Hampton Roads' viability. Yet, like much of the rest of the country, the region's infrastructure needs are vast. The aging infrastructure of the bridges and tunnels and increase in urbanization create unique challenges for the Hampton Roads area. The average age of bridges in Hampton Roads is 41 years old, while more than 400 bridges are over 50 years old (Nichols, 2017). As of 2017, 64 bridges were designated as structurally deficient and 248 bridges were designated as functionally obsolete (Nichols, 2017). Therefore, Hampton Roads is faced with the necessity to repair and/or expand its aging infrastructure in order to effectively and safely meet the persistent and growing demands of the users.

Tolling in Hampton Roads

Roadway infrastructure is a massive capital expense that requires specific capital planning and financing. The state of Virginia was one of the first in the modern U.S. highway era to use tolls to pay for roadway projects (Nichols & Belfield, 2016). The Norfolk-Portsmouth Bridge (later called the Jordan Bridge) first opened as a toll bridge in 1928. When traveling the southern branch of the Elizabeth River, this

Figure 1. Map of major bridges and tunnels in the southside of Hampton roads
Note: The Midtown Tunnel, Downton Tunnel and Jordan Bridge are tolled facilities. The Gilmerton and High Rise Bridges are not tolled.
Source: Created by authors from Google Maps

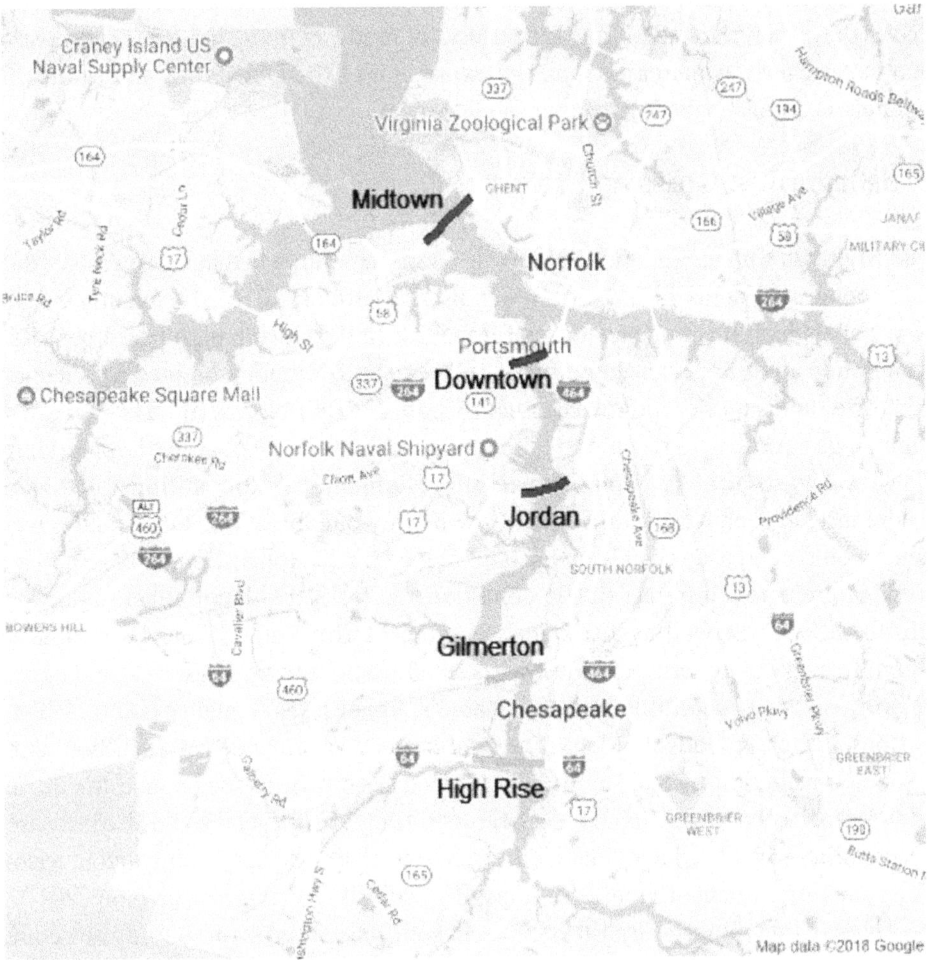

toll bridge was the first road between Norfolk and Portsmouth, creating the first continuous connection roadway from Richmond to the oceanfront (South Norfolk Jordan Bridge, n.d.). Also in 1928, the privately owned but state chartered James River Bridge created the first roadway connecting the Peninsula and the Southside and also created the system of roads and tolled bridges that approached the James River Bridge. The state bought the James River Bridge system in 1949 and implemented tolls from 1955 to 1975. Other tolled roads in the region included the Coleman Bridge and the Hampton Roads Bridge-Tunnel which became tolled facilities in the 1950s, and the Chesapeake Bay Bridge-Tunnel and the Norfolk-Virginia Beach Expressway which saw tolls introduced in the 1960s. The Coleman Bridge and the Chesapeake Bay Bridge-Tunnel remain tolled facilities today (Virginia Department of Transportation, 2018). Despite the varied popularity of tolls, Hampton Roads drivers have experienced tolling as an aspect of roadway financing for decades, and that experience has increased with the expansion of the Downtown and Midtown Tunnels via the Elizabeth River Crossing Project.

Elizabeth River Crossing Project

The Midtown Tunnel and the Downtown Tunnel are tunnels that serve among the most congested areas in Virginia (Nichols & Belfield, 2016). Frequently, these two tunnels had travel queues of more than four miles during high travel periods. Congestion and a drastic need for infrastructure improvement prompted a project to add travel tubes to the Midtown Tunnel (Nichols & Belfield, 2016). The Midtown Tunnel was constructed as a one tube tunnel with travel lanes in both directions. Travel safety would likely improve with the elimination of the bi-directional tube where instead each tube's traffic will flow in only one direction. Additional travel lanes would decrease travel back-ups and reduce traffic congestion.

This project to improve traffic conditions added a two-lane tunnel under the Elizabeth River as well as an extension to the Martin Luther King Extension to I-264. Because of government constraints, a public-private partnership would likely be adequate to successfully fund and enable a higher project quality (Daito, Chen, Gifford, Porter, & Gudgel, 2013). The total project was financed at $2 billion with a state contribution of $503 million combined with private sector investment. In addition to the benefits of improved traffic conditions, this project would also increase accessibility to jobs, educational facilities, medical services, shopping and tourism (Virginia Department of Transportation Office of Public-Private Partnership, 2014).

Through a public-private partnership, tolls were introduced to finance infrastructure and traffic condition improvements. According to Lee and Miller (2016), tolls have been an approach to manage rather than control highway congestion. Currently, both the Midtown Tunnel and Downtown Tunnel have implemented the use of all electronic tolling. With electronic tolling, the traditional toll collection plaza is replaced by an all-electronic video system that captures vehicle license plates and an E-ZPass system that relies on a transponder that allows drivers to pay for tolls automatically. Drivers that do not have an E-ZPass are mailed an invoice for their toll fees.

The public-private partnership project was financed and administered between the Virginia Department of Transportation and the Elizabeth River Crossings OpCo, LLC (Nichols & Belfield, 2016). Tolling began in 2014. As of this writing, tolls are generally unpopular (Pascale, 2017) and vary from $1.73 to $2.09 for E-ZPass passenger vehicles to as much as $3.81 for registered "pay by plate" passenger vehicles and $5.53 for unregistered "pay by plate" passenger vehicles (Elizabeth River Tunnels, 2018).

Data and Surveys for Tolling in Hampton Roads

This chapter uses two sets of survey data to understand residents' and drivers' support for tolls and behavioral responses to tolls in Hampton Roads. The first set of survey data comes from the Life in Hampton Roads (LIHR) Survey, conducted annually by the Old Dominion University Social Science Research Center. The analysis includes data from the 2012 survey (following the announcement of the tolls on the Downtown and Midtown Tunnels) through the 2016 survey and focuses on receptiveness and responses to tolls more broadly. The second set of data comes from the regional transportation planning organization, the Hampton Roads Transportation Planning Organization (HRTPO). The HRTPO commissioned two pre- and post-toll surveys that were conducted in early January 2014 (before the tolls went into effect) and in November 2014 (after toll implementation).

The Life in Hampton Roads Survey

The Social Science Research Center at Old Dominion University conducts an annual (every summer) Life in Hampton Roads (LIHR) survey, beginning in 2010. The principal goal of the survey is to gauge the quality of life in the Hampton Roads area. Residents from the cities of Chesapeake, Hampton, Newport News, Norfolk, Portsmouth, Suffolk, and Virginia Beach are surveyed about their perceptions of life in the region. Additional goals of the survey include determining the attitudes and perceptions of citizens regarding local topics such as transportation and traffic, local and state government, education, and crime, as well as providing a source of data to local decision and policy-makers. The analysis in this chapter utilizes data from the 2012 through 2016 surveys (Luetke, Gibbs, Pronier, Vandecar-Burdin, & Richman, 2012; Parker, Bush, Richman, & Vandecar-Burdin, 2014; Parker, Close, Gainey, & Vandecar-Burdin, 2015, 2016; Resnick et al., 2013).

The survey utilizes a computer-assisted telephone interviewing (CATI) system. Starting in 2011, both landline and cellphone numbers were utilized for the survey. A random digit dial telephone sample is utilized comprised of landline telephone numbers with Hampton Roads exchanges. A cellphone sample is also utilized based on switch points within the Hampton Roads area. The sample size for the LIHR surveys vary between 762 and 962 (see Table 1).

Table 1. Life in Hampton Roads survey sample sizes and margins of error

Year	Sample Size	Margin of Error
2012	762	5.3%
2013	812	5.3%
2014	853	3.7%
2015	883	3.9%
2016	962	3.6%
2017	908	3.3%

The Life in Hampton Roads survey also includes several questions useful for describing the travel characteristics of Hampton Roads residents. First, the average one-way commute to work or school, summarized in Table 2, ranges from about 22 minutes to 24 minutes across the years of analysis. Hampton Roads residents also indicated that congestion is a major concern for drivers in the region. For example, over time, there has been an increasing percentage of residents who avoided a business in a neighboring city due to traffic congestion from about 40% in 2012 to over 52% in 2017 (see Figure 2).

Life in Hampton Roads survey respondents also highlight the reliance of Hampton Roads residents on bridges and tunnels. As shown in Table 3, in 2012 just over ten percent of respondents traveled through a bridge or tunnel five to six times a month and almost 40% of respondents traveled through bridge or tunnel more than once a week. From 2014 to 2017, the survey asked respondents if they use a bridge or tunnel to commute to work or school (see Table 4). Between 13% and 18% of respondents indicated regular use of a bridge or tunnel for travel to work or school.

Table 2. Average one-way commute (in minutes) to work or school

2012	23.4
2013	23.7
2014	23.8
2015	23.6
2016	22.2
2017	24.0

Note: This includes people whose commute was greater than zero minutes or whose commute was zero minutes, but they were not retired or unemployed.

Figure 2. Percent of respondents who avoided visiting a business in a neighboring city due to concerns about traffic congestions

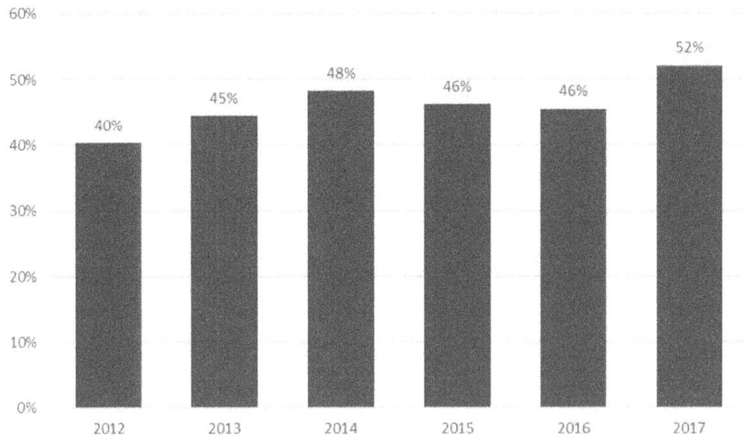

Table 3. Frequency of travel through a bridge or tunnel in Hampton Roads (2012)

Not at all	10%
Once or twice in the past month	28%
Three to four times in the past month	13%
Five to six times in the past month	10%
More than once a week	39%

Table 4. Percent of respondents who use a bridge or tunnel to commute to work or school

2014	13%
2015	15%
2016	18%
2017	17%

The South Hampton Roads Midtown and Downtown Tunnels Tolls Survey

This study also uses data from surveys obtained from the Hampton Roads Transportation Planning Organization (HRTPO). The HRTPO commissioned Christopher Newport University's Judy Ford Wason Center for Public Policy to conduct before and after telephone surveys of tolling in January and November 2014 (Christopher Newport University Judy Ford Wason Center for Public Policy, 2014a, 2014b). The January 2014 survey was a pre-toll survey conducted just before the Downton and Midtown Tunnels tolls went into effect, and the December 2014 survey was a post-toll survey conducted approximately nine months after tolls were implemented.

The goal of the January survey was to "assess the public's views and anticipated behavior in light of the implementation of tolling" (Christopher Newport University Judy Ford Wason Center for Public Policy, 2014a, p. 3) and the stated goal of the November 2014 was to "assess the public's views and behavior ten months out from the implementation of tolling on the Midtown and Downtown tunnels" (Christopher Newport University Judy Ford Wason Center for Public Policy, 2014b, p. 3). Both survey instruments included questions about the respondent's commuting experiences, opinions about tolls, E-ZPass usage, and behavioral changes in response to tolls.

Both HRTPO surveys utilized random digit dial telephone sample comprising of landline telephone and cellphone numbers of residents of five cities that make up the core of the Hampton Roads region: Norfolk, Portsmouth, Chesapeake, Suffolk, and Virginia Beach (Christopher Newport University Judy Ford Wason Center for Public Policy, 2014a, 2014b). The pre-toll survey was conducted between January 26 and 31, 2014 and the post-toll survey was conducted between November 7 and 16, 2014. The margin of error for both surveys was ±3.9% at the 95% level of confidence for the sample size of 601 and 629 for the January and November surveys, respectively. This margin of error is standard for samples of this size, as well as the population size; no corrections were otherwise made.

Analysis and Findings

In this section the findings from the HRTPO and LIHR surveys are discussed. Different survey questions are considered in the analysis to provide a broad understanding of perceptions of and support for tolls, and behavioral responses to tolls. The analysis and findings are divided into four different categories: (1) awareness of toll projects, (2) preferences for tolls, (3) use of tolled roadways and (4) toll avoidance behavior. This section also summarizes the results of a traffic analysis conducted by HRTPO transportation engineers that describe the quantitative impact of tolls.

Awareness of Toll Projects

Tolling has become a contentious issue in many of the communities where toll facilities have been implemented, including in Hampton Roads (Cook, 2014; Kimberlin, 2012; Laing, 2014; Rasmussen Reports, 2014; Zmud, 2008). As such, a key pre-cursor to public preference of and support for tolls is the level of public awareness of potential toll projects. In advance of the Elizabeth River Crossing Project, LIHR survey respondents were asked about their awareness of toll roads within the Hampton Roads region. The Elizabeth River Crossing Project was agreed upon in 2011, and in the 2012 and 2013 LIHR surveys, respondents were asked "Are you aware of any planned toll roads in the Hampton Roads area?" Survey results suggest that there was a high level of awareness among the region's residents of planned toll projects. In 2012, 68% of respondents indicated awareness and in 2013, 60% indicated they were aware of planned toll projects.

In the HRTPO January 2014 survey, respondents were asked "How much have you heard about the tolls that are coming to the Midtown and Downtown Tunnels?" A majority of respondents (76%) indicated they had heard 'a lot' about the future tolls. Eleven percent said they heard about coming tolls 'some' while only 9% indicated they had heard 'a little' and 4% had heard 'none.' Overall, the findings indicate that the majority of Hampton Roads residents were aware of the impending toll roads in the area.

Preferences and Support for Tolls

HRTPO and LIHR survey respondents were asked about their opinion for funding improvements to the Midtown and Downtown tunnels through tolling. The LIHR survey included questions about toll preferences in the years leading up to toll implementation. In 2012, respondents were asked what sources of funding they would support if additional funds were needed to maintain or expand the road, highway, and bridge systems in Hampton Roads. Among the top three responses was implementing tolls on highways (31.3%), in addition to increasing state vehicle registration (32.1%) and increasing the state tax on vehicle purchases (31.7%).

In 2013, the LIHR survey respondents were asked to think about the Downtown and Midtown Tunnels and how to pay for the expansion of existing bridges and tunnels. Less than one in four respondents supported contracting with a private company to set tolls at the downtown and midtown tunnels to pay for tunnel expansion (23.4%). Paying for the expansion in some other way was supported by just under half of the respondents (47.4%) while almost one in five opposed tunnel expansion (19.7%). Almost ten percent (9.5%) of respondents responded "don't know."

Respondents who indicated that they did not want private contractors to use tolls to expand the downtown and midtown tunnels were asked to specify another way to fund these projects. The most commonly recommended solution was to increase taxes to cover the cost of expansion. More specifically, respondents were willing to pay more in gas taxes, but a few also mentioned increasing property tax and/or sales tax. Another general suggestion was that the local and federal government should be responsible by using bonds and/or grants. Reworking the budget was also recommended by several individuals, although these respondents did not specify state, local, or federal budget. Some of these suggestions included borrowing money from the state lottery, ceasing to fund ineffective government-sponsored programs, and reducing the salaries of high paid government employees.

When looking at the results from the 2016 LIHR survey (about two years after tolls had been implemented), there appeared to be general support for the tolls when used to finance tunnel improvements. Respondents were told that the tolls on the Midtown and Downtown Tunnels were being collected to finance transportation improvements, including constructing a second tunnel for the Midtown Tunnel and increasing it from two to four lanes. The tolls would also be used to rehabilitate the existing Midtown and Downtown Tunnels. Respondents were asked if they generally support or oppose the tolls being used for these improvement purposes. As shown in Table 5, almost two-thirds of respondents reported they generally support the tolls being used for transportation improvements (61.9%), while less than a third of respondents reported they oppose the tolls being used to pay for transportation improvements (28.7%). Another 6.1% of respondents reported they had no opinion on the tolls being used to finance transportation improvements and 3.2% of respondents reported they did not know if they support or oppose the tolls.

The HRTPO surveys provide a comparison of acceptance for tolls before and after toll implementation. In both the January and November 2014 surveys, respondents were asked the question: "The tolls on the Midtown and Downtown tunnels are being collected to finance transportation improvements, including expanding the Midtown Tunnel from one tube to two tubes, resulting in a total of four lanes. Would you say that you generally support or generally oppose the toll for this purpose, or don't you have an opinion either way?" As shown in Table 6, support for tolls outweighed opposition to tolls in January 2014 (44% to 36%). However, the support for toll decreased only 34% of respondents expressing general support for tolls in November. In the short-term period following toll implementation, the HRTPO survey data show that support for tolls decreased and opposition increased.

Table 5. Preferences for Tolls from the LIHR 2016 survey

Generally support	62%
No view	6%
Generally oppose	29%
Don't know	3%

Note: Percentages may not add to 100% due to rounding.

Table 6. Preferences for tolls from HRTPO surveys

	January 2014	November 2014
Generally support	44%	34%
No view	18%	24%
Generally oppose	36%	42%
Don't know	2%	0%

Note: Percentages may not add to 100% due to rounding.

Similar to the LIHR 2013 survey, the HRTPO surveys also asked respondents to indicate support for different methods of paying for transportation improvements in Hampton Roads. The options included tolls, regional or state gas taxes, and regional or state sales taxes. Consistent with the LIHR 2013 survey results, 24% of the HRTPO survey respondents (in both January and November 2014) identified tolls as the funding option they support the most.

Use of Tolled Facility

Both the LIHR and HRTPO surveys question respondents about their use of toll bridges and tunnels (LIHR survey), and the Midtown and Downtown Tunnels specifically (HRTPO survey). Use of the tolled facility is critical to the success of tolling as a revenue raising mechanism. While a decline in usage would indeed accomplish one of the goals of relieving congestion, such decline in usage may also fail to raise sufficient funds for maintenance and improvements of the tunnels. Understanding drivers' anticipated and actual use of a tolled roadway can provide insight for decision makers into issues such as messaging and toll rates.

After the announcement of tolls on the Midtown and Downton Tunnels, the LIHR survey asked respondents about their likelihood of using a tolled roadway (see Figure 3 for a summary of responses). Specifically, the 2012 and 2013 LIHR surveys asked if respondents would be more or less likely to use a bridge tunnel when there is a toll on it or if the toll would not affect their decision. The response in 2012 was that about half of the respondents (49%) indicated no effect; however almost as many (45%) indicated that they were less likely to use a tolled bridge or tunnel. A year later, but still prior to the implementation of tolls, slightly more than half of the respondents (51%) indicate that they would use a tolled roadway less. The likelihood of no effect in usage of a tolled facility declined to 40%. Perhaps the additional year was enough time for more people to become aware of the new tolls, especially on facilities that were previously not tolled. Interestingly, there

Figure 3. Likelihood of using a tolled bridge or tunnel
Note: Does not include 'Don't Know' responses.

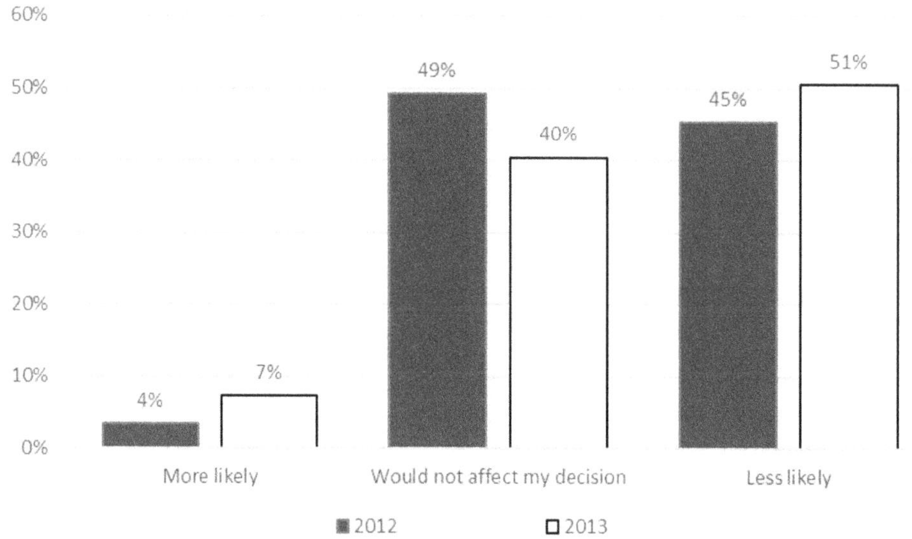

was an increase from 4% to 7% of respondents who stated that they have a greater likelihood of increasing use of a tolled facility, which may reflect an anticipation of decrease in congestion or reflects an increase in the number of respondents who will be affected by the new tolls.

In 2016, the LIHR survey further probed the weekly use of a tolled bridge or tunnel. All respondents who reported a commute time of greater than zero minutes were asked how many times in a typical week they use a toll bridge or tunnel (see Figure 4). While the majority of respondents reported using the toll facility very infrequently (less than once a week), 13% used a toll bridge or tunnel once or twice a week. Just over three percent of respondents used a toll bridge or tunnel more than ten times a week.

The HRTPO survey asked respondents how frequently they use the Midtown and Downtown Tunnels. The results are shown in Table 7. In January 2014, a month before implementation of the new tolls, 16% of respondents reported daily usage of the Midtown and Downtown Tunnels and 13% reported using the tunnels once a week. Following toll implementation, in November 2014, daily usage increased to 18% and once a week usage decreased to 11%. Interestingly, the largest change in reported usage frequency was in the never category, with reported 10% of respondents reporting never using the tunnels in January and 15% in November. Responses in the very seldom and almost never categories declined between January and November. This suggests that those who used the tunnels less than a few times a month may

Figure 4. Frequency of use of a toll bridge or tunnel in a typical week (2016)
Note: Percentages may not add to 100% due to rounding.

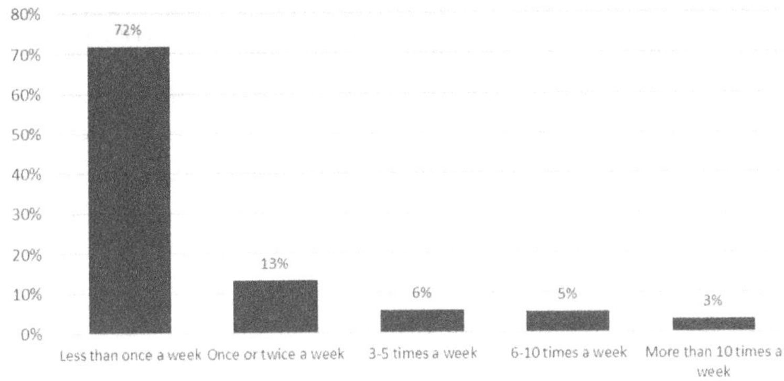

have resorted to never using the tunnels. Obviously, those who use the tunnels daily have a greater need to travel on a route via the tunnels than the least frequent users who may have simply stopped using the tunnels.

Toll Avoidance Behavior

Survey respondents were asked about avoidance of the tolled facilities. In the HRTPO survey, respondents of the pre-toll (January 2014) survey were given a list of potential ways to avoid the toll. In the post-toll (December 2014) survey, the respondents were asked if they had actually engaged in the toll avoidance behaviors. Table 8 reports the percent answering yes or no to each type of toll avoidance behavior. The results reveal a clear pattern of less involvement in actual avoidance behavior than predicted in the earlier survey. This could be due to the uncertainty of the effect

Table 7. Usage frequency of midtown and downtown tunnels

	HRTPO January 2014	HRTPO November 2014
Daily	16%	18%
Once a week	13%	11%
Few times a month	22%	22%
Very seldom	19%	16%
Almost never	18%	20%
Never	10%	15%

Note: Percentages may not add to 100% due to rounding.

Table 8. Anticipated (January 2014) and actual (November 2014) toll avoidance behaviors

	HRTPO January 2014	HRTPO November 2014
Change where you live or work to avoid tolls		
Yes	19%	12%
No	77%	86%
Change your commute to avoid the tolls		
Yes	57%	41%
No	38%	55%
Avoid traveling to destinations that require paying the tolls		
Yes	58%	45%
No	38%	53%
Telecommuting or working from home		
Yes	33%	15%
No	60%	80%
Ride the bus		
Yes	20%	7%
No	77%	92%
Ferry		
Yes	40%	12%
No	57%	87%
Carpool		
Yes	53%	24%
No	44%	75%

Note: Does not include 'don't know' responses

of tolling prior to the start of tolling. Once respondents understood the full impact of tolling on their travel times, the toll avoidance behavior may have become less attractive or more burdensome in time and effort than expected. Residents seem to have found the tolling less disruptive than they had imagined, as they generally reported avoiding tolls to a lesser degree (Christopher Newport University Judy Ford Wason Center for Public Policy, 2014b).

Still, the advent of tolling appears to have spurred some toll avoidance behaviors. In November 2014, 12% of respondents had moved or changed their place of employment, 41% changed their commute, and 45% avoided traveling to destinations that require paying tolls (Christopher Newport University Judy Ford Wason Center for Public Policy, 2014b).

Furthermore, approximately 20% of respondents indicated an improvement in travel time, which might be attributed to less peak time usage by casual drivers. This suggests that "regular commuters are using the tunnels a bit more while occasional commuters are using them a bit less" (Christopher Newport University Judy Ford Wason Center for Public Policy, 2014b, p. 15). Residents' bluffs may have been called, as the comparison of survey responses generally showed fewer than predicted instances of telecommuting or working from home, changing commute to avoid the tolls, avoiding traveling to destinations that require paying the tolls, or changing places of work or residence.

In the LIHR surveys (from 2014 through 2017), respondents were asked the question of whether they undertook specific actions to avoid a toll, including changing or intending to change job locations, reducing travel during peak periods, increasing use of buses or light rail, and carpooling with others. Responses across the years post-toll implementation are summarized in Table 9. Of those who indicated that they intentionally avoided tolls, taking a different route to work or school was the most common avoidance behavior (51% to 62% depending on the year). Reducing travel during peak periods became more popular in recent years (22% in 2014 compared to 49% in 2017).

In the LIHR 2014 and 2015 surveys, respondents were also asked: "If you take a different route now for your commute (to avoid tolls), how much more time does it take you?" As summarized in Figure 5, about half of respondents in each year indicated their alternative commute took 10 to 20 minutes longer. Only 5% percent of respondents indicated that their extra commute took more than 30 minutes in either year. Only 5% (2015) and 9% (2014) said that their alternative route did not take more time.

Table 9. Percentage of LIHR respondents reporting specific toll avoidance behaviors

	2014	2015	2016	2017
Changed or intend to change your job location	13%	11%	8%	13%
Changed or intend to change your home location	14%	6%	8%	9%
Carpooled with others	9%	13%	15%	17%
Taken a different route to get to work or school	62%	51%	59%	57%
Changed your work or school schedule	9%	6%	9%	11%
Increased your use of buses or light rail	6%	5%	6%	6%
Reduced your travel during peak periods	22%	37%	38%	49%

Note: Does not include 'Other' responses.

Figure 5. Increase in commute time by taking an alternate route to avoid tolls
Note: Percentages may not add to 100% due to rounding.

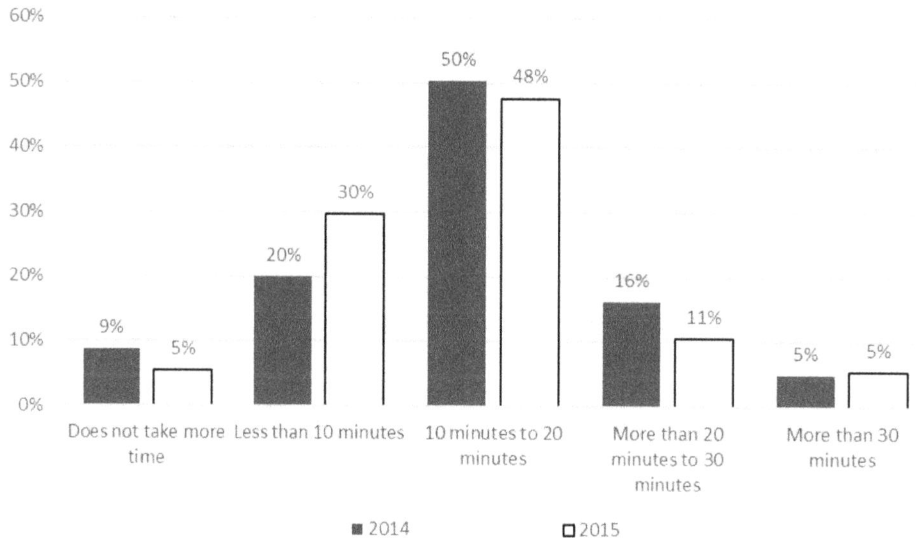

Quantitative Impact of Tolls

Nichols and Belfield (2016) quantified the impact of the Midtown and Downtown Tunnels tolls, given the heavy daily congestion before the tolls during commuting hours or other traffic disruptions (e.g., accidents, tunnel repairs). They compared different factors present before and after the implementation of the tolls on both tunnels (pre- and post- January 2014), including traffic volumes, travel time, and public transportation ridership. Data were collected from twelve locations, including the two tunnels.

The analysis found that there was a decrease in travel time at tolled locations and, conversely, travel time increased at non-tolled locations (Nichols & Belfield, 2016). Specifically, the study found that weekday peak period delays decreased by 53% at the tolled facilities, as compared to similar periods in 2013 and 2014. In contrast, two of the untolled bridge facilities showed a 16% increase during this period. However, it was noted that, over time, travel volume on the tolled facilities was slowly increasing, erasing some of the reductions in travel time that was observed in the period immediately following toll implementation. The research also found that public transportation ridership increased during the first month following the toll implementation but returned to normal a few months later.

FUTURE RESEARCH DIRECTIONS

As an area that has limited alternative travel routes between its cities, Hampton Roads is a natural laboratory for further research on tolling implementation and subsequent behaviors. The rivers, creeks, and bay provide environmental barriers to vehicular traffic. Given these physical characteristics, conditions in the landscape will not change significantly and allow researchers to observe changes in future driver behavior.

For example, will tolls be normalized and just become routine? Will tolling attitudes grow in resistance or will passive acceptance rule? The method of tolling, for those with E-ZPass transponders, is nearly invisible; a user account can be replenished at will via debit or credit card. This becomes a small bill for those with disposable income but may have a significant impact on lower wage earners. The question to be answered is whether sufficient numbers would consider this fee to be a significant part of daily life or just a price to be paid.

Additionally, moving behavior could be a factor of interest in future studies. As this chapter's results show, the surveyed response of threat of moving or job change (pre-toll) may not happen as a matter of reality (post-toll). Nevertheless, what will need to be examined is not only the net movement of people but changes in locational economic factors. The net change of population may be zero or low, but the changes in income and other financial factors may change if there are moves that achieve home-and-job convergence.

Finally, increased availability of alternative transportation options through transportation network companies (TNCs) such as Lyft and Uber may also need to be considered in terms of how it may affect behavioral responses to tolls. In 2015, the Virginia Department of Motor Vehicles passed legislation that authorized the operation of TNCs and required registration of partner (driver) vehicles. Vehicle registrations began in June of 2015, and by the end of September of 2015, there were over 50,000 registered vehicles. The number of registered vehicles in the state grew by over 185% by the end of September 2016 (Virginia Department of Motor Vehicles, 2016). In the first month of the requirement, Virginia Beach and Norfolk had the most registered TNC vehicles, outnumbering licensed taxis in those cities at a ratio of 2.5 to 1 (Forster, 2015). However, despite the growth in the availability of TNC vehicles as a transportation option for commuters, it is unclear how this option may support toll avoidance. As opposed to reducing toll costs by carpooling or using public transportation, toll costs are passed on to TNC passengers as part of the fare or surcharge.

High Occupancy Toll (HOT) Lanes in Hampton Roads

In January 2018, high occupancy toll (HOT) lanes were initiated by the Virginia Department of Transportation on interstate I-64 in southeastern Hampton Roads (Hafner, 2018). The intent of the HOT lanes was to increase the number of carpools and other transit use, while allowing drivers to pay a fee drive on less congested roadways (Pascale, 2017). The former HOV lanes had previously been open with free access during defined non-commuting hours and open only to HOV car, hybrid vehicle, or motorcycle traffic during specific commuting periods. The switch from HOV to HOT brought a dynamic pricing structure to the area during commute hours (Hafner & Pascale, 2018).

HOT lanes had been implemented in Northern Virginia with some mixed success (Pascale, 2017). While helping with traffic flow, the dynamic pricing model has been controversial due to a wide variance in the upper price range (Chesley, 2018). However, a large portion (80%) of the Northern Virginia HOT lane users used them in a range of one to five times in a month (Pascale, 2017). Approximately 5% used the Northern Virginia HOT lanes on a regular basis.

In Hampton Roads, the HOT lanes are only on the restricted portion of the highway, thereby creating a market-based choice (Pascale, 2017). The end-goal is to lessen congestion on the standard lanes and make better use of highway space compared to the prior HOV system in this particular corridor. The Virginia Department of Transportation believes that the HOT lanes will assist in reducing traffic in the standard lanes by up to 20%. Given the popularity of HOT lanes, research is needed to understand both public decision-making related to the use of HOT lanes and the effectiveness of HOT lanes for congestion management. Future research could use the Hampton Roads HOT experience as a basis for expanding knowledge about the behavioral responses beyond this chapter's focus on the introduction of tolls on previously untolled facilities (to expand and improve the facilities) to include scenarios where tolls are implemented on previously untolled HOV lanes.

CONCLUSION

Chen (2014) argues that financial sustainability is a critical foundation for overall transportation sustainability. This chapter used the case study of the recent tolling experience in Hampton Roads to assess, from the public's perception, tolls as a means to generate revenues while simultaneously managing congestion. This chapter utilizes survey data pre- and post-toll implementation to better understand not only the public's preferences and support for tolls, but also the expressed intent to avoid tolls and actual toll avoidance behaviors undertaken. Analysis of survey responses indicate a decline, over time, in opposition to toll projects and less reported toll avoidance behaviors.

The analysis shows that as a source of revenue for transportation infrastructure, between a fifth to a third of residents prefer tolls. The percentages of residents who prefer tolls are about on par with those who prefer increases to vehicle registration fees, vehicle sales taxes, gas taxes, or sales taxes. General support for tolls, even when the revenues are dedicated for specific transportation improvements, range between a low of 34% (November 2014) to a high of 62% (summer 2016). The analysis also shows that the tolls may not be as decisive a factor in deciding on whether to use a bridge or tunnel. In 2012, for example, 49% of LIHR survey respondents indicated the toll would not affect the decision to use a tolled bridge or tunnel, while 45% indicated the toll would make them less likely to use the bridge or tunnel. However, the analysis suggests that tolls may make a difference to infrequent users; those who use tunnels less frequently may decide not to use the tolled facility at all.

Not surprisingly, the findings indicate that people do respond to tolls by undertaking specific toll avoidance behaviors such as changing commutes or taking different routes to avoid tolls. However, analysis of pre- and post-toll survey data show that respondents reported fewer actual toll avoidance behaviors once tolling was implemented compared to toll avoidance behavior they anticipated using before tolling was implemented. Toll avoidance behaviors also change over time. Specific behaviors that reduce congestion, such as carpooling and reducing travel during peak periods, increased over time. Other behaviors that can be considered more drastic, such as changing job or home location and changing work or school schedules, on the other hand, remained fairly consistent or declined. This suggests that tolling with congestion pricing built in may be an effective tool for reducing congestion by impacting driver behavior. Simultaneously, the findings indicate that policy makers must remain conscious of social justice implications of tolls on under-resourced individuals. The financial costs of tolls may prompt some individuals to undertake toll avoidance behaviors such as taking alternative routes that can produce longer commutes. That more than half of LIHR survey respondents from 2014 through 2017 report taking a different route to school or work (to avoid tolls), which for about half of them add an extra 10 to 20 minutes to the commute, indicate that this is a concern that should not be taken lightly.

Finally, quantitative analysis using traffic data and public transportation ridership provide additional support for the findings from the survey data. Following the introduction of tolls, tolled facilities saw a decrease in travel time while some non-tolled facilities saw increases in traffic. This is consistent with the toll avoidance behavior reported in the LIHR and HRTPO surveys. The quantitative analysis also indicates that some of the effects of tolls, such as a decrease in tolled facility use and an increase in public transportation use, were only temporary.

This chapter's analysis of the acceptability of tolls and behavioral responses to tolls provide some insights into the long-term adequacy of tolls as a revenue source and the effectiveness of tolls as congestion pricing mechanisms to change driver behavior. Specifically, the findings suggest that opposition to tolling, while initially quite vocal, does appear to diminish over time. Tolls appear to become more acceptable in the longer term, assuming there are no incidents (e.g., massive toll rate hikes, infrastructure quality issues, etc.) that may generate opposition to tolls. This suggests that tolls may have longevity as a revenue source and may contribute to long-term financial sustainability of the transportation system.

The analysis presented in this chapter, while US-centric, has implications beyond tolling experiences in the U.S. The introduction of tolled facilities has increased not just in the U.S., but across the world, due in part to greater reliance on public-private partnerships (PPPs) to deliver infrastructure (Gurgun & Touran, 2013; Liyanage & Villalba-Romero, 2015; Willems et al., 2017). Much of the literature on transportation PPPs and tolling has focused on lessons learned in terms of management and governance (Dyble, 2011; Hodge, Boulot, Duffield, & Greve, 2017; Puentes & Istrate, 2011; Rouhani, Gao, & Geddes, 2015), managing risk (Chung, Hensher, & Rose, 2010; Lemp & Kockelman, 2009; Roumboutsos & Pantelias, 2015; Shan, Garvin, & Kumar, 2010; Wang, 2015), pricing and rate setting (Gross & Garvin, 2011; Jang, Song, Choi, & Kim, 2014; Light et al., 2015; Roumboutsos & Pantelias, 2015), and success factors (Hwang, Zhao, & Gay, 2013; Liyanage & Villalba-Romero, 2015; Osei-Kyei & Chan, 2015; Shi, Chong, Liu, & Ye, 2016; Willems et al., 2017). Fewer have examined public support for and behavioral responses to tolls (Gomez, Papanikolaou, & Vassallo, 2016; Jagers, Matti, & Nilsson, 2017; Yusuf et al., 2014). This chapter's analysis and findings fit well within this body of research and provide another in-depth case study of tolling and its implications

REFERENCES

American Society of Civil Engineers. (2016). *Failure to Act: Closing the Infrastructure Investment Gap for America's Economic Future*. Retrieved from https://www.infrastructurereportcard.org/wp-content/uploads/2016/05/2016-FTA-Report-Close-the-Gap.pdf

American Society of Civil Engineers. (2017). *2017 Infrastructure Report Card*. Retrieved from https://www.infrastructurereportcard.org/

Arentze, T., Hofman, F., & Timmermans, H. (2004). Predicting multi-faceted activity-travel adjustment strategies in response to possible congestion pricing scenarios using an Internet-based stated adaptation experiment. *Transport Policy, 11*(1), 31–41. doi:10.1016/S0967-070X(03)00016-7

Black, W. R. (2010). *Sustainable transportation: Problems and solutions*. Guilford Press.

Brownstone, D., Ghosh, A., Golob, T. F., Kazimi, C., & Van Amelsfort, D. (2003). Drivers' willingness-to-pay to reduce travel time: Evidence from the San Diego I-15 congestion pricing project. *Transportation Research Part A, Policy and Practice, 37*(4), 373–387. doi:10.1016/S0965-8564(02)00021-6

Chen, C. (2014). Measuring state highway sustainability: Taking the fiscal dimension into account. *Public Works Management & Policy, 19*(3), 255–276. doi:10.1177/1087724X14528475

Chesley, R. (2018, February 28). Settling-out process reduces howls about NOVA tolls -- and maybe on I-64, too. *The Virginian-Pilot.* Retrieved from https://pilotonline.com/news/local/columnist/roger-chesley/article_398fa1b9-1e64-5c3e-917f-dc12ab400039.html

Christopher Newport University Judy Ford Wason Center for Public Policy. (2014a). *South Hampton Roads Midtown and Downtown Tunnels Tolls Survey Part I: Pre-Tolling Report and Findings*. Retrieved from http://www.hrtpo.org/uploads/docs/South%20Hampton%20Roads%20Midtown%20and%20Downtown%20Tunnels%20Tolls%20Survey-Part%20II.pdf

Christopher Newport University Judy Ford Wason Center for Public Policy. (2014b). *South Hampton Roads Midtown and Downtown Tunnels Tolls Survey Part II: Post-Tolling Report and Findings*. Retrieved from http://www.hrtpo.org/uploads/docs/South%20Hampton%20Roads%20Midtown%20and%20Downtown%20Tunnels%20Tolls%20Survey-Part%20II.pdf

Chung, D., Hensher, D. A., & Rose, J. M. (2010). Toward the betterment of risk allocation: Investigating risk perceptions of Australian stakeholder groups to public–private-partnership tollroad projects. *Research in Transportation Economics, 30*(1), 43–58. doi:10.1016/j.retrec.2010.10.007

Cook, M. (2014). Toll Poll: Voters opposed to 241 extension. *Orange County Register.* Retrieved from https://www.ocregister.com/2014/09/20/toll-poll-voters-opposed-to-241-extension/

Daito, N., Chen, Z., Gifford, J. L., Porter, T., & Gudgel, J. E. (2013). Implementing public private partnerships during challenging economic times: Case study of the 495 Express Lanes on the Virginia portion of the Washington Capital Beltway Project (USA). *Case Studies on Transport Policy*, *1*(1), 35–45. doi:10.1016/j.cstp.2013.07.001

Dyble, L. N. (2011). Tolls and control: The Chicago Skyway and the Pennsylvania Turnpike. *Journal of Planning History*.

Elizabeth River Tunnels. (2018). *Toll Rates Information*. Retrieved from https://www.driveert.com/toll-info/toll-rates/

Forster, D. (2015, July 5). Drivers flock to Uber-style operations. *The Virginian-Pilot*. Retrieved from https://pilotonline.com/news/local/transportation/article_e5e2d0be-af03-5abc-abcd-5e42b3dd97d6.html

Francsics, J., & Ingrey, M. (2000). *The EUROTOLL Project: Road user responses to transport demand management*. Paper presented at the Tenth International Conference on Road Transport Information and Control. 10.1049/cp:20000117

Gomez, J., Papanikolaou, A., & Vassallo, J. M. (2016). Measuring regional differences in users' perceptions towards interurban toll roads. *Journal of Transport Geography*, *54*, 22–33. doi:10.1016/j.jtrangeo.2016.05.001

Gross, M., & Garvin, M. (2011). Structuring PPP toll-road contracts to achieve public pricing objectives. *Engineering Project Organization Journal*, *1*(2), 143–156. doi:10.1080/21573727.2011.572256

Gurgun, A. P., & Touran, A. (2013). Public-private partnership experience in the international arena: Case of Turkey. *Journal of Management Engineering*, *30*(6), 04014029. doi:10.1061/(ASCE)ME.1943-5479.0000213

Hafner, K. (2018, January 10). HOT lanes on I-64 open today. *The Virginian-Pilot*. Retrieved from https://pilotonline.com/news/local/transportation/traffic/article_c7570800-6a65-575e-ae0d-8c2d8111b258.html

Hafner, K., & Pascale, J. (2018, January 5). HOT lanes are coming to I-64 Wednesday. Here's what you should know. *The Virginian-Pilot*. Retrieved from https://pilotonline.com/news/local/transportation/article_00b922b9-94e0-51b3-b50e-48c3270122c7.html

Hodge, G., Boulot, E., Duffield, C., & Greve, C. (2017). After the ribbon cutting: Governing PPPs in the medium to long term. *Australian Journal of Public Administration*, *76*(3), 330–351. doi:10.1111/1467-8500.12239

Hwang, B.-G., Zhao, X., & Gay, M. J. S. (2013). Public private partnership projects in Singapore: Factors, critical risks and preferred risk allocation from the perspective of contractors. *International Journal of Project Management, 31*(3), 424–433. doi:10.1016/j.ijproman.2012.08.003

Jagers, S., Matti, S., & Nilsson, A. (2017). How exposure to policy tolls transforms the mechanisms behind public acceptability and acceptance - The case of the Gothenburg congestion tax. *International Journal of Sustainable Transportation, 11*(2), 108–119. doi:10.1080/15568318.2016.1197348

Jang, K., Song, M. K., Choi, K., & Kim, D.-K. (2014). A bi-level framework for pricing of High-Occupancy Toll lanes. *Transport, 29*(3), 317–325. doi:10.3846/16484142.2014.952248

Keuleers, B., Chow, V., Thorpe, N., Timmermans, H., & Wets, G. (2006). Behavioural change in activity-travel patterns in response to road user charging. *Journal of Transport Economics and Policy, 40*(1), 119–134.

Kimberlin, J. (2012, March 4). Tolls, tolls, tolls. How did we get into this mess? *The Virginian-Pilot*. Retrieved from http://pilotonline.com/news/local/transportation/tolls-tolls-tolls-how-did-we-get-into-this-mess/article_5bccaafd-87d5-5d68-b435-6eb9ccb74d30.html

King, D., Manville, M., & Shoup, D. (2007). The political calculus of congestion pricing. *Transport Policy, 14*(2), 111–123. doi:10.1016/j.tranpol.2006.11.002

Laing, K. (2014). Poll: 79 percent supports tolling to pay for highways. *The Hill*. Retrieved from http://thehill.com/policy/transportation/217476-poll-79-percent-support-tolling-to-pay-for-highways

Lee, C., & Miller, J. S. (2016). *Conditions contributing to the attitudes for toll facilities in the United States with a specific focus on Virginia*. Academic Press.

Lemp, J., & Kockelman, K. (2009). Understanding and accommodating risk and uncertainty in toll road projects: A review of the literature. *Transportation Research Record: Journal of the Transportation Research Board, 2132*(1), 106–112. doi:10.3141/2132-12

Leuenberger, D. Z., & Bartle, J. R. (2009). *Sustainable Development for Public Administration*. Armonk, NY: M.E. Sharpe.

Light, T., Patil, S., Erhardt, G. D., Tsang, F., Burge, P., Sorensen, P., & Zmud, M. (2015). *The Impact of Adopting Time-of-Day Tolling: Case Study of 183A in Austin, Texas*. Santa Monica, CA: RAND Corporation.

Liyanage, C., & Villalba-Romero, F. (2015). Measuring success of PPP transport projects: A cross-case analysis of toll roads. *Transport Reviews*, *35*(2), 140–161. doi:10.1080/01441647.2014.994583

Luetke, S. R., Gibbs, T., Pronier, C., Vandecar-Burdin, T., & Richman, J. (2012). *How is Life in Hampton Roads? 2012 Report for the 3rd Annual Life in Hampton Roads Survey*. Retrieved from http://images.bimedia.net/documents/odu_lifestyle_study.pdf

Meyer, M. (1997). *A toolbox for alleviating traffic congestion and enhancing mobility* (I. T. Engineers, Ed.). Washington, DC: Institute of Transportation Engineers.

Nichols, K. M. (2017). *The State of Transportation in Hampton Roads 2017*. Chesapeake, VA: Hampton Roads Transportation Planning Organization.

Nichols, K. M., & Belfield, S. S. (2016). Impact of New Tolls on Existing Roadway Facilities: Hampton Roads, Virginia, Experience. *Transportation Research Record: Journal of the Transportation Research Board*, *2597*(1), 28–36. doi:10.3141/2597-04

Nielsen, O. A. (2004). Behavioral Responses to Road Pricing Schemes: Description of the Danish AKTA Experiment. *Journal of Intelligent Transportation Systems: Technology, Planning, and Operations*, *8*(4), 233–251. doi:10.1080/15472450490495579

O'Connell, L., & Yusuf, J.-E. W. (2013). Improving Revenue Adequacy by Indexing the Gas Tax to Indicators of Need: A Simulation Analysis. *Public Works Management & Policy*, *18*(3), 229–243. doi:10.1177/1087724X12451575

Odeck, J., & Bråthen, S. (1998). The Planning of Toll Roads—Do Public Attitudes Matter? Case of the Oslo Toll Ring. *Transportation Research Record: Journal of the Transportation Research Board*, *1649*(1), 72–80. doi:10.3141/1649-09

Olszewski, P., & Xie, L. (2005). Modelling the effects of road pricing on traffic in Singapore. *Transportation Research Part A, Policy and Practice*, *39*(7), 755–772. doi:10.1016/j.tra.2005.02.015

Osei-Kyei, R., & Chan, A. P. (2015). Review of studies on the Critical Success Factors for Public–Private Partnership (PPP) projects from 1990 to 2013. *International Journal of Project Management*, *33*(6), 1335–1346. doi:10.1016/j.ijproman.2015.02.008

Parker, S., Bush, C., Richman, J., & Vandecar-Burdin, T. (2014). *Life in Hampton Roads Report The Fifth Annual Life in Hampton Roads Survey*. Academic Press.

Parker, S., Close, J., Gainey, R., & Vandecar-Burdin, T. (2015). *Life in Hampton Roads Report: The Sixth Annual Life in Hampton Roads Survey*. Retrieved from http://www.odu.edu/content/dam/odu/col-dept/ssrc/docs/lihr-report-final.pdf

Parker, S., Close, J., Gainey, R., & Vandecar-Burdin, T. (2016). *Life in Hampton Roads Report The Seventh Annual Life in Hampton Roads Survey*. Academic Press.

Pascale, J. (2017, April 12). FAQ: VDOT officials explain how High Occupancy Tolls lanes will work on I-64. *The Virginian-Pilot*. Retrieved from https://pilotonline.com/news/local/transportation/faq-vdot-officials-explain-how-high-occupancy-tolls-lanes-will/article_b80f01e5-a2eb-546b-a53b-e5e70136c058.html

Peirce, N. (2007). Selling Our Toll Roads: Good or Retrograde Idea? *Government Finance Review*, *23*(3), 81–82.

Perez, B. G., Giordano, R., & Stamm, H. (2011). *Evaluation and Performance Measurement of Congestion Pricing Projects*. Washington, DC: Transportation Research Board.

Podgorski, K. V., & Kockelman, K. M. (2006). Public Perceptions of Toll Roads: A Survey of the Texas Perspective. *Transportation Research Part A, Policy and Practice*, *40*(10), 888–902. doi:10.1016/j.tra.2006.03.002

Puentes, R., & Istrate, E. (2011). *Moving Forward on Pubilc Private Partnerships: U.S. and International Experience with PPP Unites*. Brookings-Rockefeller.

Rasmussen Reports. (2014, May 7). *65% Oppose Tolls on the Interstate Highways*. Retrieved from http://www.rasmussenreports.com/public_content/politics/general_politics/may_2014/65_oppose_tolls_on_interstate_highways

Resnick, S., Gibbs, T., Bush, C., Parker, S., Vandecar-Burdin, T., & Richman, J. (2013). *Life in Hampton Roads Report: The Fourth Annual Life in Hampton Roads Survey*. Retrieved from http://www.odu.edu/content/dam/odu/col-dept/al/docs/lihr-report2013withcomparisons.pdf

Rouhani, O. M., Gao, H. O., & Geddes, R. R. (2015). Policy lessons for regulating public–private partnership tolling schemes in urban environments. *Transport Policy*, *41*, 68–79. doi:10.1016/j.tranpol.2015.03.006

Roumboutsos, A., & Pantelias, A. (2015). Allocating revenue risk in transport infrastructure public private partnership projects: How it matters. *Transport Reviews*, *35*(2), 183–203. doi:10.1080/01441647.2014.988306

Rusch, W. A. (1984). *Toll Highway Financing*. Washington, DC: Transportation Research Board.

Santos, G., & Fraser, G. (2006). Road pricing: Lessons from London. *Economic Policy*, *21*(46), 263–310. doi:10.1111/j.1468-0327.2006.00159.x

Schade, J., & Schlag, B. (2003). Acceptability of Urban Transport Pricing Strategies. *Transportation Research Part F: Traffic Psychology and Behaviour*, *6*(1), 45–61. doi:10.1016/S1369-8478(02)00046-3

Schaller, B. (2010). New York City's congestion pricing experience and implications for road pricing acceptance in the United States. *Transport Policy*, *17*(4), 266–273. doi:10.1016/j.tranpol.2010.01.013

Shan, L., Garvin, M. J., & Kumar, R. (2010). Collar options to manage revenue risks in real toll public-private partnership transportation projects. *Construction Management and Economics*, *28*(10), 1057–1069. doi:10.1080/01446193.2010.506645

Shi, S., Chong, H.-Y., Liu, L., & Ye, X. (2016). Examining the Interrelationship among Critical Success Factors of Public Private Partnership Infrastructure Projects. *Sustainability*, *8*(12), 1313. doi:10.3390u8121313

South Norfolk Jordan Bridge. (n.d.). *Jordan Bridge History*. Retrieved from http://www.snjb.net/about-us/jordan-bridge-history

Urban Land Institute. (2013). *When the Road Price Is Right: Land Use, Tolls, and Congestion Pricing*. Retrieved from http://www.uli.org/wp-content/uploads/2013/03/When-the-Road-Price-is-Right_web_F.pdf

Virginia Department of Motor Vehicles. (2016, December). *Transportation Network Companies, 2016 Report*. Retrieved from https://rga.lis.virginia.gov/Published/2016/RD524/PDF

Virginia Department of Transportation. (2018, September 10). *Hampton Roads Tunnels and Bridges*. Retrieved from http://www.virginiadot.org/travel/hro-tunnel-default.asp

Virginia Department of Transportation Office of Public-Private Partnership. (2014). *Elizabeth River Tunnels: Increasing travel options and improving safety*. Retrieved from https://www.p3virginia.org/projects/elizabeth-river-tunnels/

Wang, Y. (2015). Evolution of public–private partnership models in American toll road development: Learning based on public institutions' risk management. *International Journal of Project Management*, *33*(3), 684–696. doi:10.1016/j.ijproman.2014.10.006

Willems, T., Verhoest, K., Voets, J., Coppens, T., Van Dooren, W., & Van den Hurk, M. (2017). Ten lessons from ten years PPP experience in Belgium. *Australian Journal of Public Administration*, *76*(3), 316–329. doi:10.1111/1467-8500.12207

Yusuf, J.-E. W. (2014). Highway Trust Fund. In M. Garrett (Ed.), *The Encyclopedia of Transportation: Social Science and Policy* (Vol. 2, pp. 745–747). Thousand Oaks, CA: Sage Reference.

Yusuf, J.-E. W., & O'Connell, L. (2013). The Crisis in State Highway Finances: Its Roots, Current Effects, and Some Possible Remedies. *Journal of Public Budgeting, Accounting & Financial Management*, *25*(3), 502–521.

Yusuf, J.-E. W., O'Connell, L., & Abutabenjeh, S. (2011). Paying for Locally-owned Roads: A Crisis in Local Government Highway Finance. *Public Works Management & Policy*, *16*(3), 250–269. doi:10.1177/1087724X11402357

Yusuf, J.-E. W., O'Connell, L., & Anuar, K. A. (2014). For whom the tunnel be tolled: A four-factor model for explaining willingness-to-pay tolls. *Transportation Research Part A, Policy and Practice*, *59*, 13–21. doi:10.1016/j.tra.2013.10.021

Zmud, J. (2008). The Public Supports Pricing If... A Synthesis of Public Opinion Studies on Tolling and Road Pricing. *Connecting People, Places and Ideas*, 29-39.

Zmud, J., & Arce, C. (2008). *Compilation of Public Opinion Data on Tolls and Road Pricing*. Washington, DC: Transportation Research Board of the National Academies. doi:10.17226/14151

KEY TERMS AND DEFINITIONS

Congestion Pricing: A surcharge imposed to manage the flow of traffic or curtail the overcrowding of roadway by regulating demand and thus making it possible to manage congestion without increasing demand. The objective of congestion pricing (also known as congestion charging) is to use price mechanisms to make drivers conscious of their costs during peak demand periods and encourage them to redistribute their demand in time and/or space.

E-Z Pass Transponder: A tool mounted on the vehicle that signals the equipment at the tolled facility to electronically charge the owner of the transponder for the use of the tolled facility without requiring the driver to stop.

Federal Highway Trust Fund: A fund that receives the proceeds of the federal fuel tax on gallons of gasoline and diesel fuel.

Hampton Roads: A region of southeastern Virginia bounded by the Atlantic Ocean, the Chesapeake Bay, James River, and the Elizabeth River.

High Occupancy Toll (HOT) Lanes: Lanes accessed by vehicles with multiple occupants and charged a fee for use (especially at high traffic times) to encourage carpooling and a less congested route for the drivers while also reducing congestion on alternative routes. A form of transportation demand management.

High Occupancy Vehicle (HOV) Lanes: Lanes accessed by vehicles with multiple occupants (especially at high traffic times) to encourage carpooling and a less congested route for the drivers while also reducing congestion on alternative routes. A form of transportation demand management.

Motor Fuel Tax: In the U.S.A. the motor fuel tax is a federal or state excise tax levied per gallon of gasoline or diesel fuel used by vehicles. The motor fuel tax is the primary source of revenue for transportation in the U.S.A.

Public-Private Partnership: A cooperative agreement between at least two public and private sector organizations for the delivery of goods or services to the public.

Road Pricing: A direct charge imposed for the use of roads to generate revenues and/or as a transportation demand management tool to manage the flow of traffic or curtail the overcrowding of roadway. Also known as road user charges, road pricing is a broad term that includes tolls, distance or time-based fees, and congestion charges.

Social Equity: A concept concerned with the fair and equitable provision, implementation, and impact of services, programs, and policies.

Toll Road: A road or highway that drivers must pay a fee or toll to use. In the U.S.A. toll roads are also known as a turnpike or tollway. The toll revenues are generally used to recoup the cost of road construction and maintenance.

Tolling: A form of road pricing in which a fee is assessed for use of the tolled facility.

Transportation Demand Management: A set of strategies and policies with the goal of influencing travel behavior to reduce travel demand or redistribute demand, thus managing or reducing congestion.

Transportation Network Company: "Ride-share" company that matches contracted, independent drivers and passengers via mobile apps as an alternative to public transportation and taxi cab service. Sometimes known as mobility service providers or ride-hailing services. TNC services include those provided by Uber and Grab.

Chapter 7
Financing Transportation Through Local Sales Taxes

Whitney B. Afonso
The University of North Carolina at Chapel Hill, USA

ABSTRACT

Fuel taxes have historically been the key revenue source for a great deal of transportation infrastructure, especially roads. For many reasons, such as reduced fuel tax receipts, governments at all levels have begun to explore additional financing options. This chapter explores an option that local governments have available to them in many states: local sales taxes earmarked for transportation projects. This chapter briefly discusses the literature on local sales taxes, the diversity of the laws regarding local sales taxes earmarked for transportation, potential consequences of increased reliance on local sales taxes earmarked for transportation, and briefly discusses a similar revenue source—the local fuel tax. This research is important to understanding the changing patterns of how public transportation is being financed in many states, and if the spread of non-earmarked local sales taxes are any indicator, how it is likely to be financed in many others moving forward.

INTRODUCTION

Transportation financing has been evolving in the United States for many decades. The federal government has devolved a great deal of responsibility for the maintenance of roads and highways to states and local governments, and states have devolved responsibility increasingly to local governments. These same local governments are often responsible for other areas of transportation expenditures such as public transit

DOI: 10.4018/978-1-5225-7396-8.ch007

and airports. In addition to the changing roles that different levels of government are playing, the way in which transportation is financed has been changing dramatically. This is in part due to the declining revenue capacity of what has historically been the primary source of revenue for transportation, the fuel tax. The revenue raising capacity of the fuel tax plateaued in the 1990s and has been steadily declining since then. While many states have increased their rates, it has not been sufficient to keep up with inflation and this reliance is not sustainable (Puentes and Prince 2003; Yusuf and O'Connell 2013; Bartle and Chen 2014; Zhao, Guo, and Coyle 2015). This chapter examines a revenue source that has been growing in importance over this same period that is available to many local governments, local sales taxes earmarked for transportation. As responsibility to fund transportation projects devolve to local governments it is critical to understand the ramifications of the ways in which local governments fund it and too little attention has been given to earmarked sales taxes. Local sales taxes earmarked for transportation are a revenue instrument available in many states and are an option to not only increase revenues but to link transportation expenditures to a sustainable revenue source. Other chapters in this book discuss financing concerns and opportunities from the perspective of user fees, tolls, public private partnerships, and this chapter expands upon those by adding local option taxes with an emphasis on sales taxes, but includes a brief discussion of local fuel taxes as well.

What is the state affairs in transportation finance? According to a report published by the National Conference of State Legislatures on transportation finance (Rall et al. 2011) transportation finance is not on a sustainable path. Federal funding, such as federal-aid highway and transit programs and congressional earmarks, make up approximately 20 percent of highway and transit spending. Two-thirds of federal revenue is generated by the fuel tax followed by general fund expenditures and the vehicle tax. States have taken on a more substantial role than what they have had historically. States are responsible for almost half of surface transportation funding and approximately 20 percent of transit funding. Approximately a third of state revenue for transportation comes from the state motor vehicle fuel tax, with a fifth being generated from the vehicle tax, another fifth from other forms of taxation, and the remaining coming from debt, tolls, and general funds. The laws governing how these funds can be used varies tremendously. Twenty-six states have either constitutional or statutory provisions that earmarks[1] state fuel taxes for highways and roads, for example. Local governments, which are the focus of this chapter, are responsible for approximately 30 percent of surface transportation funding and are responsible (or own) more than three-quarters of the nation's roadway miles. Local governments rely heavily on general funds, which make up almost half of the money spent on transportation projects, property and other taxes, debt, with very little revenue being generated by tolls and vehicle taxes (less than ten percent combined) (Rall et al. 2011, Oliff 2015). The current emphasis on fuel taxes is largely recognized as unsustainable due to the prevalence of lower fuel consumption vehicles and the lack of rate adjustment to incorporate inflation (Yusuf and O'Connell; O'Connell and Yusuf 2013; Zhao, Guo, and Coyle 2015). This snapshot of transportation finance suggests the need to better understand the specific revenue instruments being used by local government, for both roads and highways and transit for which they are responsible for approximately 60 percent of the expenditures. Furthermore, politics and citizen perceptions are also critical factors. The political acceptability of a local sales tax earmarked for transportation has made them a popular choice (Wachs 2003a, Afonso 2015a) and a lack of understanding of the nature of fuel taxes have made them less popular and feasible (Yusuf et al. 2011; O'Connell and Yusuf 2013).

These issues are not new. In a 2003(b) Brookings Institution report Martin Wachs highlights the decline in fuel taxes and the devolution of responsibility for funding transportation to local governments. Wachs calls for a policy change demonstrating that the incremental increases in fuel taxes are not an effective way of financing transportation given the trends and notes that local governments are increasingly responsible for financing roads and using local taxes to do so. He ultimately recommends increased state funding and a greater use of electronic tolls. However, the trend of increased local responsibility has continued (Yusuf et al. 2011) and not through the use of revenue sources like tolls.

Local governments are financing transportation in numerous ways throughout the United States. While fuel taxes have historically been the primary source of revenue for financing transportation, that is often not the case at the local level. This is due to the combination of fuel taxes being unavailable at the local level in many states, fewer intergovernmental transfers, and the fuel taxes' declining base. As a result there has been a notable shift in how local governments finance the transportation projects they are responsible for. Local governments are opting for financing approaches such as public–private partnerships and broader base taxes, such as local sales taxes earmarked for transportation (Taylor 1995; Adams et al., 2001; Goldman and Wachs 2003; Rodrigue 2013; Afonso 2015a). The potential role of local sales taxes in creating a more sustainable source of financing for transportation at the local level is a critical one. As previously discussed, many authors highlight that fuel taxes are not sustainable in their current form. There are numerous alternatives available such as the vehicle miles travelled tax, but it would be expensive to establish and administer and likely receive a great deal of resistance from the public (Forkenbrock 2006; Yusuf and O'Connell 2013; Zhao, Guo, and Coyle 2015). Whereas, a sales tax has a broad base, potential for a low rate, and is politically palatable—thus linking it to transportation presents a stable (though elastic) revenue source.

While these broad base taxes shift the burden away from financing sources that may satisfy the benefit principle of equity, they also pass the financing decision to voters in many cases, because of requirements for referendum. For example, there were 34 sales tax ballot measures earmarked for transportation at the local level in 2016. In contrast, there was only one fuel tax, 25 property taxes (primarily in Michigan and West Virginia), and seven bond referendums (Center for Transportation Excellence 2018). While it is apparent that local sales taxes are by no means the only way that local governments are financing transportation, they are a key element in many states and not as well understood as the property tax or even the fuel tax. The numbers in the referenda put forward echo the national trends. Whereas, approximately 90 percent of federal funding for transportation is generated from fuel taxes and vehicles fees, only four percent of local funding is. The majority of local funding is generated by general revenue or from earmarked sales and property taxes. In fact, approximately 30 percent of local transportation financing comes from special earmarked taxes like retail sales taxes (PRINCIPALS: Intergovernmental Forum on Transportation Finance 2008). A great deal of the local sales tax for transportation revenue goes to finance roadways and highways but depending on the state the revenue can be used for several modes of transportation including transit and pedestrian paths.

The focus of this chapter is on local sales taxes earmarked for transportation. The next section provides a brief overview of local sales taxes and the corresponding literature. The chapter proceeds with an overview of the variation of state laws surrounding local sales taxes with an emphasis on discretionary authority manifested in the ability to earmark revenue for transportation and examples of states that permit local governments to adopt local sales taxes earmarked for transportation and the restrictions on those sales tax instruments. From there, the discussion moves to the potential benefits and consequences of greater dependence on local sales taxes earmarked for transportation and how the revenue generated from earmarks may change the level of expenditures for the recipient programs. This broader conversation is briefly applied to another local option tax, the local fuel tax. The chapter concludes from there.

OVERVIEW OF LOCAL SALES TAXES

Local sales taxes are sales taxes that are adopted at the local level (i.e., county, municipal, and/or special districts). Local sales taxes are currently permitted in 39 states and vary in their structure and scope tremendously; e.g., Montana and

Pennsylvania have laws that restrict the adoption to an extremely limited number of local governments (Afonso 2017).[2] Thus, understanding local sales taxes requires not simply knowing which states permit local governments to adopt them, but also how those laws shape the implementation and adoption of them. An example of this is that many states allow local governments to adopt more than one type of local sales tax and often they come with different structures and requirements including how the revenue can be used and what types of jurisdictions can adopt them. Despite all of these differences and complicated structures, both the use of local sales taxes and the research on local sales taxes has continued to grow. In fact, local sales taxes have become the second largest source of own source revenue at the local level, second only to the property tax.

The research on local sales taxes has traditionally explored themes of volatility, expenditure levels, cross border shopping, and the timing of adoptions of local sales taxes. Fundamentally, the majority of this literature can fall into one of two themes: the relationship between sales taxes and property taxes and tax competition. Largely, the literature has found that greater reliance on local sales taxes both reduces property tax burdens and increases total expenditures (Jung 2001; Sjoquist et al. 2005; Afonso 2014a, 2015b). As the trend of reducing property tax reliance grows, the literature has also found that increased revenue diversification (not limited to local sales taxes) leads to greater stability (Jung 2001; Sjoquist et al. 2005; Zhao and Jung 2008; Afonso 2013). As for the second theme of tax competition the literature has largely found that urban areas and regional retail centers benefit from tax exportation from their neighbors and are more likely to be early adopters (Luna 2004, Rogers 2004, Zhao 2005, Burge and Rogers 2011, Afonso 2016a). However, there is also a closely related vein of literature that has found that there is less inequity between the revenue raising capacity of urban and rural jurisdictions than has been assumed (Zhao and Hou 2008, Afonso 2016b). While there may be regional competition, local sales taxes earmarked for transportation present the ability for regional cooperation in many states. For example, Virginia has the ability for localities to adopt a local sales tax earmarked for transportation for regional projects and California's transportation special districts can adopt them as well (Weinreich 2015, Afonso 2018). This is a feature of this tax instrument that is less common and presents greater opportunity for regional planning, cooperation, and collaboration in states where available.

As the use and reliance on local sales taxes has grown, so has the research attention it has received, and the scope of the questions being considered. For example, one of the more common features of local sales taxes across the country is the availability of earmarked local sales taxes, meaning that the revenue must be used for certain government purposes. One of the first regions that these earmarks have been examined is within the state of Georgia where the first local sales tax instrument's revenue must be used to reduce property tax burden. However, over the last decade, this research has expanded to include additional earmarks including those for transportation which will be discussed at greater length below.

Ultimately, what the research community has observed is that local sales taxes are growing in both importance in funding local government[3] and within the literature. There are many elements that must be considered when evaluating the existing research on local sales taxes and how they may impact policy decisions and outcomes and they will be described here as jurisdictional eligibility and discretionary authority from the framework laid out in Afonso (2017). These considerations must be integrated carefully into applying research from one state to another and understanding potential consequences of structure.

The Diversity of State Local Sales Tax Laws

The legal structures that define the scope of local sales taxes vary tremendously from state to state. This is one reason the majority of the literature is done at the single state level. Afonso (2017) lays out a framework for evaluating and understanding local sales tax laws composed of two dimensions: jurisdictional eligibility and discretionary authority. Jurisdictional eligibility is defined as "the specific set of criteria that identify which local governments in a given state are permitted to adopt local sales taxes" (28). Discretionary authority is defined as "the set of criteria that define how much autonomy local governments have over the implementation and use of local sales taxes" (28). While these two dimensions capture the majority of differences across the states, they do not capture all of them such as whether the jurisdiction levying the tax is required to share the revenue with other governments and what is included in the tax base. Afonso (2017) refers to these as the technical complexities of the law. Table 1 presents where states fall along these dimensions.

This chapter is most interested in the discretionary authority, though the application of the policies will clearly be constrained by jurisdictional eligibility as well. Specifically, within discretionary authority is an element of how the revenue generated from local sales taxes can be allocated by the levying government. The focus of this chapter is transportation and thus that dimension will be discussed most thoroughly. However, it is important to note that many states have general purpose local sales taxes, where the levying government can use the revenue generated by the local sales tax on any lawful government provision—including transportation. Additionally, many states also have earmarked local sales taxes where the revenue must be used for capital or infrastructure, which may include transportation projects. See Afonso (2017) for more details. While these revenues may go to fund transportation projects, this chapter focuses on those limited to transportation projects, though many

Table 1. State local sales tax laws by jurisdictional eligibility and discretionary authority

Jurisdictional Eligibility	High	California	Nevada	Alabama	Kansas	Oklahoma	Wyoming
		Florida	Ohio	Alaska	Louisiana	Tennessee	
		Illinois	West Virginia	Arizona	Missouri	Texas	
		Minnesota	Wisconsin	Arkansas	New York	Utah	
		Nebraska		Colorado	North Dakota	Washington	
	Low	Hawaii	South Carolina	Georgia			
		Iowa	South Dakota	Idaho			
		Mississippi	Vermont	Montana			
		New Jersey		New Mexico			
		North Carolina		Pennsylvania			
		Low		High			
		Discretionary Authority					

Note: States with no jurisdictional eligibility and no discretionary authority, i.e. those without local sales taxes are not included. Those states are: Connecticut, Delaware, Indiana, Kentucky, Maine, Maryland, Massachusetts, Michigan, New Hampshire, Oregon, and Rhode Island.

Taken from Afonso (2017).

of the same benefits and concerns will apply to those earmarked for transportation and those which are not.

Fifteen states have earmarked local sales tax instruments specified for transportation or transit projects. In addition to those fifteen, an additional seven states allow local governments to adopt earmarked local sales taxes for specific projects of their choosing. The states with the specified purposes (denoted as SP on Table 2) are states where the jurisdiction is able to present a set of projects to votes that will be funded if the local sales tax is adopted. An example of this is the state of Georgia where they are referred to as SPLOSTs (Jung 2002). The states where these options are available are presented in Table 2. Table 2 also includes the states that allow local fuel taxes to be levied, which will be discussed briefly later in this chapter. Despite the availability of these earmarks for transportation there is still great diversity in the structure (both jurisdictional eligibility and discretionary authority) of local sales taxes in these states.

Table 2. States that permit either local sales taxes earmarked for transportation or local fuel taxes

State	Local Sales Taxes Earmarked for Transportation	Local Fuel Taxes
Alabama		X
Arizona	X	
California	X	X
Florida	X	X
Georgia	X	
Hawaii	X	X
Illinois	X	X
Iowa	SP	
Kansas	SP	
Louisiana	SP	
Minnesota	X	
Mississippi	X	X
Montana	SP	X
Nebraska	SP*	
Nevada	X	X
New Mexico	X	X
North Carolina	X	
North Dakota		X
Ohio	SP	
Oregon		X
Tennessee		X
Texas	X	
Utah	X	
Virginia	X	
Washington	X	X
Wyoming	SP	

X=Available; SP=Specified Purpose

Many other states include earmarked local sales taxes for infrastructure and capital, which may include transportation expenditures.

*Specified purpose granted by legislature

Data taken from Afonso (2017) and American Association of State Highway and Transportation Officials (2016).

LOCAL SALES TAXES EARMARKED FOR TRANSPORTATION

Local sales taxes, as briefly discussed above, vary tremendously from state to state in how they are structured, which level of government (municipal, county, and/or special district) they are available to, and the level of discretion that the levying jurisdictions have. Local sales taxes earmarked for transportation are already more restrictive than many local sales taxes, because there are requirements on how the revenue generated must be used. However, there is still a great deal of diversity in the structure of just this type of local sales tax. Additionally, while many of the concerns that are present with sales taxes (local and otherwise) still apply, there are some who consider these earmarked local sales taxes unique and having a slightly different set of benefits and consequences (Brennan and Buchanan 1980, Dye and McGuire 1992, Crabbe et al. 2005, Crowley and Hoffer 2012). Lastly, it is critical to understand how the use of these earmarked sales taxes may impact program expenditures. All of these areas are examined in this section of the chapter.

Examples of Diversity of Laws Across the States

This section briefly describes a selection of state local sales tax earmarked for transportation laws across the United States and how they are being implemented.[4] Much like the variation that Afonso (2017) categorizes, local sales tax earmarked for transportation laws vary by jurisdictional eligibility and discretionary authority with regard to factors like tax rate and what the revenue is earmarked for. It is also important to note that there are many important laws that vary across states and that are relevant to this conversation. There are no *typical* states, however California, Minnesota, North Carolina, and Utah present a cross-section of many of the features, restrictions, and diversity in state laws. Those key distinguishing features are discussed. This chapter focuses on local sales taxes earmarked for transportation but there is also great diversity in other aspects of transportation finance across the states. For example, in Michigan local governments are responsible for 92 percent of roads whereas in North Carolina there are no county road departments and the state Department of Transportation builds and maintains all secondary roads (Rall et al. 2011).

California

California first permitted counties to adopt local sales taxes earmarked for transportation in 1976 and currently 24 of the 58 counties have adopted local sales taxes earmarked for transportation. These 24 counties cover 88 percent of the state's population. Sixteen counties have more than one local sales tax earmarked for transportation in place. While, the legislation changed in 1992 to require a super-majority to pass a local sales tax earmarked for transportation, they continue to be popular and feasible revenue instrument. Seventy-six measures have been put forward and 48 have passed (Albrecht, et al. 2017).

Local sales taxes earmarked for transportation in California annually generate approximately $4 billion dollars of revenue and the revenue generated is used to finance road construction and maintenance primarily (Afonso 2015a, Albrect, et al. 2017). Urban counties devote larger shares to finance public transit than suburban and rural counties that levy local sales tax earmarked for transportation revenue primarily to pay for roads. On average, in California, 34 percent of local sales tax earmarked for transportation revenue goes to local roads, 27 percent to highways, 31 percent to public transit, and 8 percent to transportation for the elderly and disabled and other smaller projects like bicycle and pedestrian facilities (Albrect et al. 2017).

Minnesota

The laws governing the use of local sales taxes earmarked for transportation in Minnesota closely echo the broader laws around local sales taxes nationally, where in many states only certain types of jurisdictions (e.g., high tourism, high population, home-rule) can adopt them. In Minnesota there are two types of local sales taxes earmarked for transportation: the Metropolitan Transportation Area Sales Tax and the Greater Minnesota Transportation Sales and Use Tax. The Metropolitan Transportation Area Sales Tax is a quarter cent tax that can only be adopted in the Twin Cities area and the revenue may only be used for transit projects in that region. The Greater Minnesota Transportation Sales and Use Tax is for any county outside of the Metropolitan Transportation Area and the revenue is used for transportation or transit projects, including operating and capital costs. It has possible rates of 0.5 percent and 1 percent (Dalton 2016).

North Carolina

The state of North Carolina began authorizing local sales taxes earmarked for transportation in 1997, but at the time only one county was able to adopt one, Mecklenburg County where the city of Charlotte is located. The law was revised in 2007 to permit any county to adopt a local sales tax earmarked for transportation, though urban counties are able to adopt higher rates than non-urban counties (0.5 percent versus 0.25 percent respectively). In North Carolina, local sales tax revenue earmarked for transportation must be used for public transit systems. The majority of local sales taxes in North Carolina do not exempt food, but local sales taxes earmarked for transportation do not tax food (North Carolina Center for County Research 2015).

Utah

Utah has a series of local sales taxes earmarked for transportation available to local governments. They are presented in Table 3. While there are taxes on grocery sales in Utah, it is worth noting that the local sales tax instruments presented in Table 3 do not include grocery sales in the tax base (Utah State Tax Commission 2017).[5] Utah, like some other states, has elected to permit numerous instruments with relatively narrow purposes rather than fewer broader instruments.

Table 3. Local sales tax instruments earmarked for transportation in Utah

Local Sales Tax Instrument	Type of Jurisdiction	Rate
County option transportation	County	0.25%
Public transit	County, city, or town	0.30%
Public transit, airport facility or state high projects	County, city, or town	0.25%
Mass transit fixed guideway	County	0.30%
County of the Second-Class Airport, Highway and Public Transit	County, city, or town	0.25%
County Option Highways and Public Transit	County	0.25%

Note: Taken from Utah State Tax Commission. (2017). Sales and Use Tax General Information. Report. Publication 25. Revised November, 2017.

Others

California, Minnesota, North Carolina, and Utah present a wide spectrum of the different ways in which local sales taxes earmarked for transportation are structured across the United States. In some states like North Carolina, there is a single instrument available with a limited set of projects that can be funded from the revenue (public transit). In other states, like Utah, there are multiple instruments available with different rates, jurisdictional eligibility requirements, and constraints on recipient programs. There is still even more diversity. This subsection briefly discusses three additional states that are more restrictive, by either the set of jurisdictions able to adopt a local sales tax or how the revenue must be budgeted for according to state law, than the examples previously discussed.

Illinois's local sales tax earmarked for transportation revenue goes directly to the Regional Transit Authority for the most part, with a portion of it going to collar counties to be used for transportation and public safety projects (Illinois Revenue 2018). There is also a Metro-East Mass Transit District that collects local sales taxes earmarked for transportation in two counties, and the revenue is used exclusively for mass transit (Illinois Revenue 2018). Arizona has two counties that have adopted a local sales tax earmarked for transportation, Gila and Pinal. The revenue is used to fund regional transportation projects (Arizona State Treasurer 2018). Similarly, Hawaii has a more modest local sales tax earmarked for transportation with only the counties of Oahu and Honolulu having a local sales tax earmarked for transportation in place whose revenue funds the mass transit system (Hawaii Department of Taxation 2018).

Important Considerations Regarding Local Sales Taxes Earmarked for Transportation

Local sales taxes earmarked for transportation expenditures have both benefits and potential drawbacks that should be considered by both academic and practitioner audiences. This section addresses many of these considerations.

Benefits

There are numerous potential benefits to local sales taxes earmarked for transportation. First, they have a broad tax base. Typically, local sales taxes have the same base as the state sales tax. That often means that the tax base does not include grocery sales, prescription drugs, and services amongst others. Despite these exclusions the sales tax is still considered to have a broad base. This broad tax base will produce a great deal of revenue at relatively low tax rates. This is of course affected by what is included in the tax base; for example, food, prescription drugs, and services are often excluded from the sales tax base. What is the impact of these exemptions? It will vary by state and sales tax laws, but we can examine the grocery exemption that is the most common. Seven states including Hawaii, South Dakota, and Alabama tax grocery sales at the state and local level, when applicable. However, they are outliers. In 32 states, the state sales tax does not apply to grocery sales. In six states, groceries are taxed at a reduced rate at the state level. North Carolina, South Carolina, Louisiana, and Georgia do not charge state sales tax for grocery sales but at least a portion of the local sales taxes do include groceries in the base (Dunn 2018). The exemption of food from the sales tax base represents a reduction in sales tax revenue of between 5 and 30 percent depending on the size of the sales tax base otherwise (Johnson and Lav 1998). Nonetheless, local sales taxes generate a great deal of revenue, which helps explain their popularity amongst governments. In fact, local sales taxes are estimated to generate $114 billion annually, and be mindful that many states either do not permit local governments to levy local sales taxes or are extremely restrictive in terms of jurisdictional eligibility and discretionary authority. For example, just using California's self-help counties, local sales taxes earmarked for transportation are estimated to generate over $4 billion in revenue for transportation projects annually and in fiscal year 2016 sales taxes accounted for 15 percent of statewide revenue for transit alone (Albrecht, et al. 2017).

Second, they are politically feasible. Property taxes, the most common revenue instrument for local governments in the United States are extremely unpopular and are often limited by state level tax and expenditure limitations. Sales taxes on the other hand are more tolerable, most likely this is in part due to the fact they are less visible than the property tax (Afonso 2014b). As discussed previously in this chapter, the passage of local sales taxes often require a referendum vote and in some cases like California the passage of a local sales tax earmarked for transportation requires a supermajority vote. While not all pass, many do which reinforces their popularity compared to many other forms of taxation (Hamideh et al. 2008). There is research on the acceptability of local sales taxes earmarked for transportation and it finds that factors such as self-interest, political leanings, perception of services, and income impact their popularity and support (Hannay and Wachs 2007, Green et al. 2013, Palm and Handy 2016).

A third important benefit of local sales taxes earmarked for transportation is that sales taxes capture revenue from non-residents and non-automobile users. Commuters, tourists, students, and others that do not reside in the jurisdiction but benefit from the services provided may not contribute to the tax base via property taxes. The use of local sales taxes earmarked for transportation allows for these populations to help finance the services they benefit from, including roads and transit. Sales taxes are also more difficult to avoid and have a larger tax base than the traditional mechanism for financing transportation, the fuel tax. While the fuel tax is a good proxy for use of the roadways, though less so with the introduction of electric and an increasingly fuel efficient vehicles—it does not capture those who benefit from non-automobile modes of public transportation infrastructure like transit, cyclists, and even pedestrians (Goldman and Wachs 2003; Albrecht et al. 2017). The ability to capture revenue from these populations, that may not be taxed through fuel taxes and vehicle fees, may be perceived as increasing an element of fairness—because a broader base of those who benefit from transportation expenditures will help in financing them.

Drawbacks

Two of the largest potential negative implications of relying more heavily on local sales taxes earmarked for transportation as a critical revenue source are concerns over equity and the stability of sales tax revenue.

Equity

Equity is a key concept in many disciplines, especially those surrounding government provision of services and goods like public administration, public policy, and planning. However, there is no universal and agreed upon definition of equity, or more simply put: *what is fair*. That does not mean that there are no ways by which people can evaluate the equity of revenues though; there are frameworks that are commonly applied. The two most dominant ones are ability to pay and the benefit principle both are discussed here.

The ability to pay framework assesses equity through the lens of horizontal and vertical equity. Horizontal equity dictates that a revenue instrument is equitable if people with the same ability to pay have the same tax burden. This of course seems intuitive and straightforward, but can quickly get complicated when deciding how to establish what goes into ability to pay. Clearly income does, but what about cost of living, capital gains income, number of children, school loans, etcetera? However, there are arguments that sales taxes are reasonably horizontally equitable because taxpayers with similar abilities to pay have similar expenditure patterns and thus similar sales tax burdens (Albrect et al. 2017).

In addition to horizontal equity, we can view equity through the lens of vertical equity with regard to ability to pay. Vertical equity assesses the tax burden by income group and ability to pay. A revenue instrument is considered regressive if a larger share of income (or wealth) is taxed for low income individuals, i.e., a larger burden in terms of effective tax rate is placed on lower income taxpayers. A proportional tax is one that all income groups pay the same effective tax rate. Lastly, a progressive tax is one where the larger burdens, in terms of effective tax rate, fall on higher income taxpayers. From the ability to pay framework, most would consider either proportional or progressive taxes (and fees) to be equitable. Sales taxes are regressive, though the extent to how regressive they are depends on what is included in the sales tax base. While exempting food reduces revenue considerably, it also makes sales taxes much less regressive. Once again, to evaluate these local sales taxes earmarked for transportation, it becomes critical to examine the three elements of the state laws presented by Afonso (2017). Using the earlier example of North Carolina, those who only examine what the state sales tax includes in its base would see that food is exempt. However, upon inspection of local sales taxes one would see that local sales taxes include food in their base. Though, there is an exception, local sales taxes earmarked for transportation do not tax food. Therefore, while other local sales tax instruments in North Carolina may be considered especially regressive, those earmarked for transportation are less so.

In contrast, the benefit principle considers the equity of revenue instruments as how closely aligned those who pay and those who benefit from the service are. Under the benefit principle, it is ideal for the share of revenue paid to be proportional to use. Fees are often assessed under the benefit principle; park entrance fees and utility fees are clear examples of revenues that satisfy the benefit principle. Often the revenue instruments that best satisfy the benefit principle are regressive and thus would not be considered equitable from the ability to pay perspective.

The traditional mechanisms for financing transportation, fuel taxes, tolls, and vehicle fees, are typically perceived to be equitable under the benefits principle. These revenue instruments are fair for a large share of transportation projects like roads and highways, but may not capture the breadth of what falls within this category now such as public transit and bike paths. Additionally, as technology changes some of the equity considerations may shift as well. For example, as vehicles become more fuel efficient, fuel taxes may be less proportional to use because some consumers of the roadways and fuel may be investing in more fuel-efficient vehicles. In a broader notion of the benefit principle, it could be argued that good public transportation infrastructure is a benefit to the entire community—though more directly to some than others. This is a common argument with regards to the link between property taxes and public education. Even if you believe that local sales taxes earmarked for transportation satisfy a broader benefit principle, the burden would still not be considered proportional to use, and thus not as equitable as other instruments available. For example, taxpayers may not consume roads, transit, and other transportation investments at the same rate making a finance structure like vehicle miles traveled taxes more equitable (Zhao, Guo, and Coyle 2015).

In addition to ability to pay and the benefit principal, that apply largely to the consideration of financing—the equity considerations around how the money will be spent can have important equity considerations. Decisions around how projects should be chosen, with considerations about whether proposed projects should be evaluated by merit, need, economic development potential, etcetera, may have important equity impacts. For example, the spatial considerations of where communities are choosing to make strategic investments, addressing needs, and attracting industry can shape the future of the respective neighborhoods. These considerations are present with any type of financing, but are especially true given the broad based sales tax. These considerations can be referred to as spatial equity concerns. Spatial equity concerns may be minimized if these regional projects are being financed through tolls, tax increment financing, or fares—where those benefitting most directly from the projects are those financing them.

There is an additional type of spatial equity concern with regard to local sales taxes, that of cross-border shopping and tax competition. To be clear, there are two types of geographic equity concerns—for example, within the county and between counties. The concern over cross-border shopping and tax competition are areas that have received a great deal of attention from the academic community. It has manifested itself in two primary areas of research interest; first, analysis of whether sales tax rates being lower in neighboring jurisdictions will cause consumers to shift their shopping out of their home jurisdiction to a neighboring county to avoid higher rates, and second, whether consumers prefer retail agglomerations (or numerous shopping options) and may shop outside their jurisdiction to find these agglomerations. Cross-border shopping to avoid higher sales tax rates is an important consideration for local sales taxes earmarked for transportation because levying this additional tax will increase the tax rate, and often above that of neighboring jurisdictions. This may result in both negative impacts on the businesses in the community as well as the revenue raising capacity of the tax. It may also make the local sales tax earmarked for transportation less politically feasible. As the e-commerce sector continues to grow this became of even greater concern because for many purchases on-line sales taxes were not collected and use taxes are seldom remitted (Manzi 2012). However, with the 2018 *South Dakota v. Wayfair* ruling this tax evasion will be less common as remote vendors will be more likely to collect state and local sales taxes.

Lastly, there is a concern over temporal equity. The concern of fairness over time in this case refers to the structure of how local sales taxes earmarked for transportation are adopted in many states. This is due to the requirement that the projects that will be financed using the revenue must be laid out for voters to consider.[6] This often means committing to a course of action for the next 20 years. While it is reasonable for voters and taxpayers to understand the planned use of these monies, it is also difficult to know that the best use of these funds that have been identified now will not change over time and will reflect the changing community. Potential changes that may shift priorities may involve changing demographics, technologies, and preferences.

How do these equity concerns play out in local sales taxes earmarked for transportation referendum? A recent study in California presents excellent analysis, and suggests a marginal impact. Albrecht et al. (2017) conducted a study of 37 ballot measures for local sales taxes earmarked for transportation and they find that 32 of the 37 ballot arguments mention equity. Temporal equity and income equity (what is referred to here as ability-to-pay) are mentioned least frequently and, excluding what they categorize as general equity, geographic and modal (the type of project being financed, ie., rail, roadways, transit) equity are mentioned most frequently. They find that the opponents of local sales taxes earmarked for transportation are much more likely to cite specific equity concerns. Interestingly, they find that most of the ballot arguments that mention income equity concerns discuss them from the perspective of the projects being financed will disproportionately benefit low-income residents and only 2 of the 14 mentioned that they will disproportionately burden low income residents. While they find that equity is frequently discussed in the ballot measures, they find that its discussion is not prominent and that most of the debate is whether the projects will actually produce what is promised, such as reduced traffic (Albrecht et al. 2017).

Stability

Stability is one of the many criteria that taxes and fees are typically evaluated by; others include equity and revenue raising capacity that have been previously discussed in this chapter. Stability of the revenue source is considered beneficial because it allows for better planning and financial management. This will be especially true when a revenue source is directly linked with a program area and is a primary source of finance for that program. One of the key ways that local sales taxes have been evaluated, in terms of stability, is in the revenue diversification literature. This is, in part, because local sales taxes are now the second largest source of own source revenue for local governments (Brunori 2007). When local governments are diversifying (i.e., moving away from dependence on the property tax), it often means relying more heavily on local sales taxes.

While the literature on local revenue diversification most frequently finds that increased diversification leads to greater stability in revenues (Jordan and Wagner

2008, Carroll 2009, Carroll and Stater 2009, Afonso 2013, Hendrick and Crawford 2014) that literature often looks beyond just property taxes and sales taxes to other tax instruments, fees, and intergovernmental revenues. It is understood that sales taxes are elastic, with an estimated elasticity of 1.43 for non-food purchases, and sensitive to changes in the economy (Sobel and Holcombe 1996). Thus, in periods of economic prosperity sales tax revenue receipts are expected to increase faster than changes in income and during economic downturns sales tax revenue receipts are expected to decline more quickly than income. Therefore, sales taxes are a less predictable and reliable source of revenue than property taxes, which are inelastic (Afonso 2013). The elasticity of sales taxes is greatly increased when deductions and exemptions are made to its base; the most common ones are for food and prescription drugs. While these exemptions may offset some of the equity concerns, they exacerbate the stability concerns. A lack of stability has ramifications for programs that are reliant on sales taxes for their funding, such as transportation projects that are funded through local sales taxes earmarked for transportation. It may make planning and service delivery more uncertain and will require careful forecasting and planning that is not as necessary when revenues are predictable.

The Effect of Revenue Generated from Local Sales Taxes Earmarked for Transportation on Spending

In addition to the potential benefits and consequences of local sales taxes earmarked for transportation there is an additional relevant literature on funding programs through earmarked revenues. The literature presents three frameworks on how to understand the effect of earmarked revenues on programmatic spending: the rational model, the Leviathan model, and the flypaper effect hypothesis. Each of these frameworks would anticipate different levels of program spending when earmarked revenue is introduced. The rational model suggests that revenue earmarks that only partially fund the program should not increase the overall level of spending any more than a simple increase in the general fund would. This is due to the fungible nature of revenues. Since governments can substitute the earmarked funds for dollars that would have been allocated through the general fund (or others), then the new earmarked revenue can be treated like unrestricted revenues in a practical sense. For example, Dye and McGuire (1992) find that an additional dollar of earmarked revenues for highways only increase expenditures on highways by $0.19.

The Leviathan model of government suggests that government will take advantage of the ability to generate additional revenue by introducing new earmarked revenue streams for popular programs—which will simply replace money that would have been allocated from the general fund. While on the face this predicted result seems similar to the rational model, it may result in no increase in support for the program that is the recipient of the earmarked revenues. Crowley and Hoffer (2012) find that neither fuel taxes nor vehicle registration fees earmarked for highway spending increase actual spending in this area. In a more dramatic case, Blackwell et al. (2006) find that expenditures on tourism promotion in South Carolina are lower than the revenue generated by an accommodations tax earmarked for tourism development.

Finally, the flypaper effect hypothesis suggests that revenue earmarked for a program will "stick" to that program. The majority of this research has examined intergovernmental transfers rather than own source revenue. Afonso (2015a) presents a modified flypaper effect hypothesis regarding own source revenue. Nesbit and Kreft (2009) find that for every dollar of earmarked revenue a recipient program receives, spending increases for that program by \$0.94. Similarly, Afonso (2015a) examines local sales taxes earmarked for transportation in California and finds that for every dollar of local sales tax earmarked for transportation revenue generated transportation expenditures increase by \$1.76, with a corresponding decrease in non-transportation expenditures of \$0.73.

Thus, there is no consensus in the literature on the effect of an earmarked revenue source on recipient program spending. However, there is evidence that earmarked revenues at the local level for capital expenditures may be difficult for policymakers to divert to other areas because of how visible they are (Jung 2002). That, in addition to the one study (Afonso 2015a) that examines local sales taxes earmarked for transportation specifically, in the context of California counties, may suggest that local sales taxes earmarked for transportation may increase transportation expenditures (especially for capital) in a way most consistent with the flypaper effect hypothesis. It is advisable to be mindful of these three different frameworks, especially because there is no consensus.

LOCAL FUEL TAXES

Local fuel taxes share many similarities with local sales taxes earmarked for transportation including that they have not been examined a great deal in the literature. As presented in Table 2, local fuel taxes are available in 13 states and in some cases represent an important funding source for local government transportation. While both local fuel taxes and local sales taxes earmarked for transportation are consumption based and earmarked for transportation expenditures, local fuel taxes are excise taxes on motor fuel and can thus be considered a benefit based tax. While local fuel taxes satisfy the benefit principle, they are regressive and are becoming increasingly so as premium electric and hybrid cars continue to enter the market. They do not have the same revenue raising capacity as local sales taxes earmarked for transportation. Local fuel taxes are subject to the same trends and concerns as state and federal fuel taxes, i.e., a declining tax base. What local fuel taxes do provide is a release valve for the growing pressures of local governments facing devolution of responsibilities for highways and often increasing demand for public transit. Not surprisingly, the early adopters of local fuel taxes tend to be more populous governments that are closer to highways and experiencing high levels of fiscal stress (Chen and Afonso 2017).

Additionally, like local sales taxes, local fuel taxes are able to capture revenue from non-residents and those who use and benefit from the roadways but may not pay that jurisdictions property taxes or vehicle fees. They do have a narrower base and are less likely to capture those who use public transit or greenways and non-road and highway transportation infrastructure and services though. Depending on the nature of the laws surrounding the adoption and implementation of local fuel taxes, they may suffer from the same spatial and temporal equity concerns as local sales taxes earmarked for transportation.

CONCLUSION

Finance for transportation is having to evolve in the United States. The historical approach of financing transportation projects through fuel taxes at primarily the federal and state level is no longer feasible given the declining size of the tax base. It has been recognized that alternative funding strategies are necessary and while much of this analysis is at the federal and state level, there is a clear need to assess revenue sources at the local level as well. The majority of the recommendations and conversation has revolved around increased use of tolls and other user fees. This chapter presents local sales taxes earmarked for transportation and briefly discusses local fuel taxes as well. Local sales taxes present a revenue instrument that can increase the long-term sustainability of transportation finance. While, these earmarked local sales taxes have largely been omitted from the conversation they are growing in both use and importance. Growing use of local sales taxes earmarked for transportation comes with benefits and some potential consequences that should be carefully considered by both the academic and public service community. While they have a broad base and a large revenue raising capacity that is supported by residents and non-residents alike, they are also regressive and do not satisfy the benefit principle of equity as transportation financing instruments typically have. In addition to that, the state laws may require local governments to lay out the long-term projects that will be funded through the local sales tax earmarked for transportation that may lead to inefficient and temporally inequitable transportation spending in the future.

There is still a great bit of uncertainty about the future of transportation finance and the role that local sales taxes earmarked for transportation will play in it. Many of the scholars that study these issues propose re-tethering transportation finance to a more benefit-based system like tolls, vehicle fees, and transit fares (Wachs 2003b, PRINCIPALS: Intergovernmental Forum on Transportation Finance 2008, Albrecht et al. 2017). Despite this advice, the trend of both the widespread adoption and use of local sales taxes, including those earmarked, suggests that they will remain an integral part of the financing puzzle.

REFERENCES

Afonso, W. (2013). Diversification toward Stability? The Effect of Local Sales Taxes on Own Source Revenue. *Journal of Public Budgeting, Accounting & Financial Management*, *25*(4), 649–674. doi:10.1108/JPBAFM-25-04-2013-B004

Afonso, W. (2014a). Local Sales Taxes as a Means of Increasing Revenues and Reducing Property Tax Burdens: An Analysis Using Propensity Score Matching. *Public Budgeting & Finance*, *34*(2), 24–43. doi:10.1111/pbaf.12039

Afonso, W. (2014b). "Fiscal Illusion in State and Local Finances: A Hindrance to Transparency." 2014. *State & Local Government Review*, *46*(3), 219–228. doi:10.1177/0160323X14550103

Afonso, W. (2017). State LST Laws: A Comprehensive Analysis of the Laws Governing Local Sales Taxes. *Public Budgeting & Finance*, *37*(4), 25–46. doi:10.1111/pbaf.12171

Afonso, W. (2018). *Modal Equity of Transportation Expenditures: The Impact of Collaborative Governance and the Use of Earmarked Local Sales Taxes*. Paper presented at the Symposium on implementing collaborative governance: Models, experiences and challenges to foster policy coordination, and to enhance sustainable community outcomes and public value generation conference, University of Palermo, Italy.

Afonso, W. B. (2015a). Leviathan or flypaper: Examining the fungibility of earmarked local sales taxes for transportation. *Public Budgeting & Finance*, *35*(3), 1–23. doi:10.1111/pbaf.12072

Afonso, W. B. (2015b). LOST and Found Tax Dollars: Local Option Sales Taxes, Property Taxes, and Own Source Revenue. *Journal of Public Budgeting, Accounting and Financial Management, 27*(3), 318–351. doi:10.1108/JPBAFM-27-03-2015-B002

Afonso, W. B. (2016a). Time to Adoption of Local Option Sales Taxes: An Examination of Texas Municipalities. *Public Finance Review*. doi:10.1177/1091142116673147

Afonso, W. B. (2016b). The Equity of Local Sales Tax Distributions in Urban, Suburban, Rural, and Tourism-Rich Counties in North Carolina. *Public Finance Review, 44*(6), 691–721. doi:10.1177/1091142115588976

Albrecht, M., Brown, A., Lederman, J., Taylor, B., & Wachs, M. (2017). *The Equity Challenges and Outcomes of California County Transportation Sales Taxes. Report*. Center of Economic Competitiveness, University of California.

American Association of State Highway and Transportation Officials. (2016). *Transportation Finance Governance and Finance*. Retrieved from www.financingtransportation.org/pdf/50_state_review_nov16.pdf

Arizona State Treasurer. (2018). *Revenue Distributions List*. Retrieved from https://aztreasury.gov/local-govt/revenue-distributions/revenue-distributions-list/

Bartle, J. R., & Chen, C. (2014). Future Issues in State Transportation Finance. In Sustaining the States: The Fiscal Viability of American State Governments (pp. 211-233). New York: Taylor & Francis Group Press. doi:10.1201/b17267-12

Blackwell, C., Crotts, J. C., Litvin, S. W., & Styles, A. K. (2006). Local Government Compliance with Earmarked Tax Regulation. *Public Finance Review, 34*(2), 212–228. doi:10.1177/1091142105284213

Brennan, G., & Buchanan, J. (1980). *The Power to Tax: Analytical Foundations of a Fiscal Constitution*. Cambridge, UK: Cambridge University Press.

Brunori, D. (2007). *Local Tax Policy: A Federalist Perspective* (2nd ed.). Washington, DC: Urban Institute Press.

Burge, G., & Rogers, C. (2011). Local option sales taxes and consumer spending patterns: Fiscal interdependence under multi-tiered local taxation. *Regional Science and Urban Economics, 41*(1), 46–58. doi:10.1016/j.regsciurbeco.2010.08.001

Carroll, D. A. (2009). Diversifying Municipal Government Revenue Structures: Fiscal Illusion or Instability? *Public Budgeting & Finance, 29*(1), 27–48. doi:10.1111/j.1540-5850.2009.00922.x

Carroll, D. A., & Stater, K. (2009). Revenue Diversification in Nonprofit Organizations: Does It Lead to Financial Stability? *Journal of Public Administration: Research and Theory, 19*(4), 947–966. doi:10.1093/jopart/mun025

Center for Transportation Excellence. (2018). *Transportation Ballot Measures.* Retrieved from http://www.cfte.org/elections

Chen, C., & Afonso, W. (2017). *A First Look at the Time to Adoption of Local Option Fuel Taxes: Evidence from Florida Counties.* Paper presented at the Association of Budgeting and Financial Management conference, Washington, DC.

Crabbe, A., Hiatt, R., Poliwka, S., & Wachs, M. (2005). Local Transportation Sales Taxes: California's Experiment in Transportation Finance. *Public Budgeting & Finance, 25*(3), 91–121. doi:10.1111/j.1540-5850.2005.00369.x

Crowley, G., & Hoffer, A. (2012). Dedicating Tax Revenue: Constraining Government or Masking Its Growth? Working Paper. Mercatus Center at George Mason University, Arlington, VA.

Dalton, P. (2016). *Local Sales Taxes in Minnesota.* Information brief. Research Department, Minnesota House of Representatives.

Dunn, J. (2018). Sales Tax By State: Are Grocery Items Taxable? *TaxJar*. Retrieved from https://blog.taxjar.com/states-grocery-items-tax-exempt/

Dye, R., & McGuire, T. (1992). The Effect of Earmarked Revenues on the Level and Composition of Expenditures. *Public Finance Review, 20*(4), 543–556. doi:10.1177/109114219202000410

Forkenbrock, D. J. (2006). Financing local roads: Current problems and new paradigms. *Transportation Research Record: Journal of the Transportation Research Board*, (1960): 8–14.

Goldman, T., & Wachs, M. (2003). *A quiet revolution in transportation finance: The rise of local option transportation taxes.* University of California Transportation Center. Retrieved from https://escholarship.org/uc/item/2gp4m4xq

Green, A. D., Neiman, M., Bockman, S., & Sirotnik, B. (2013). Public Support for Transportation Sales Taxes in California: A Two County Assessment. *California Journal of Politics and Policy, 5*(4), 645–670. doi:10.5070/P2X886

Hamideh, A., Oh, J. E., Labi, S., & Mannering, F. (2008). Public acceptance of local government transportation sales taxes: A statistical assessment. *State & Local Government Review, 40*(3), 150–159. doi:10.1177/0160323X0804000302

Hannay, R., & Wachs, M. (2007). Factors influencing support for local transportation sales tax measures. *Transportation*, *34*(1), 17–35. doi:10.100711116-006-0006-4

Hawaii Department of Taxation. (2018). *County Surcharge FAQs*. Retrieved from http://tax.hawaii.gov/geninfo/a2_b2_7csurchg_faq/

Hendrick, R., & Crawford, J. (2014). Municipal Fiscal Policy Space and Fiscal Structure: Tools for Managing Spending Volatility. *Public Budgeting & Finance*, *34*(3), 24–50. doi:10.1111/pbaf.12042

Illinois Revenue. (2018). *Mass Transit District Taxes*. Retrieved from http://www.revenue.state.il.us/localgovernment/Overview/HowDisbursed/masstransit.htm

Johnson, N., & Lav, I. (1998). *Should States Tax Food? Examining the Policy Issues and Options*. Center on Budget and Policy Priorities. Retrieved from https://www.cbpp.org/archiveSite/stfdtax98.pdf

Jordan, M., & Wagner, G. (2008). Revenue Diversification in Arkansas Cities: The Budgetary and Tax Effort Impacts. *Public Budgeting & Finance*, *28*(2), 68–82. doi:10.1111/j.1540-5850.2008.00911.x

Jung, C. (2001). Does the Local-Option Sales Tax Provide Property Tax Relief? The Georgia Case. *Public Budgeting & Finance*, *21*(1), 73–86. doi:10.1111/0275-1100.00037

Jung, C. (2002). The Effect of Local Earmarking on Capital Spending in Georgia Counties. *State & Local Government Review*, *34*(1), 29–37. doi:10.1177/0160323X0203400103

Leuenberger, D. Z., Bartle, J. R., & Chen, C. (2014). Sustainability and Transportation. *Public Works Management & Policy*, *19*(4), 316–321. doi:10.1177/1087724X14545540

Luna, L. (2004). Local sales tax competition and the effect on county governments' tax rates and tax bases. *The Journal of the American Taxation Association*, *26*(1), 43–61. doi:10.2308/jata.2004.26.1.43

Mercier, J. (2009). Equity, Social Justice, and Sustainable Urban Transportation in the Twenty-First Century. *Administrative Theory & Praxis*, *31*(2), 145–163. doi:10.2753/ATP1084-1806310201

Nesbit, T., & Kreft, S. (2009). Federal Grants, Earmarked Revenues, and Budget Crowd-Out: State Highway Funding. *Public Budgeting & Finance*, *29*(2), 94–110. doi:10.1111/j.1540-5850.2009.00930.x

North Carolina Center for County Research. (2015). *Basics of North Carolina Local Option Sales Taxes*. Report. Author.

O'Connell, L., & Yusuf, J.-E. W. (2013). Improving Revenue Adequacy by Indexing the Gas Tax to Indicators of Need: A Simulation Analysis. *Public Works Management & Policy*, *18*(3), 229–243. doi:10.1177/1087724X12451575

Oliff, P. (2015). *Funding Challenges in Highway and Transit: A federal-state-local analysis*. Fiscal Federalism Initiative, The PEW Charitable Trusts. Retrieved from http://www.pewtrusts.org/en/research-and-analysis/analysis/2015/02/24/funding-challenges-in-highway-and-transit-a-federal-state-local-analysis

Palm, M., & Handy, S. (2016). Sustainable transportation at the ballot box: A disaggregate analysis of the relative importance of user travel mode, attitudes and self-interest. *Transportation*, 1–21.

Principals, Intergovernmental Forum on Transportation Finance. (2008). *Financing Transportation in the 21st Century: An Intergovernmental Perspective*. National Academy of Public Administration. Retrieved from https://www.napawash.org/uploads/Academy_Studies/08-16.pdf

Puentes, R., & Prince, R. (2003). Fueling Transportation Finance: A Primer on the Gas Tax. *The Brookings Institution*. Retrieved from https://www.brookings.edu/research/fueling-transportation-finance-a-primer-on-the-gas-tax/

Rall, J., Wheet, A., Farber, N., & Reed, J. (2011). *Transportation Governance and Finance*. National Conference of State Legislatures. Retrieved from http://www.ncsl.org/documents/transportation/FULL-REPORT.pdf

Rodrigue, J. (2013). The Financing of Transportation Infrastructure. In *The Geography of Transport Systems*. New York: Routledge. doi:10.4324/9780203371183

Rogers, C. (2004). Local option sales tax (LOST) policy on the urban fringe. *The Journal of Regional Analysis & Policy*, *34*(1), 25–50.

Sjoquist, D. L., Walker, M. B., & Wallace, S. (2005). Estimating Differential Responses to Local Fiscal Conditions: A Mixture Model Analysis. *Public Finance Review*, *33*(1), 36–61. doi:10.1177/1091142104270656

Sobel, R. S., & Holcombe, R. J. (1996). Measuring the Growth and Variability of Tax Bases Over the Business Cycle. *National Tax Journal*, *49*, 535–552.

Taylor, B. (1995). Public Perceptions, Fiscal Realities, and Freeway Planning: The California Case. *Journal of the American Planning Association, 61*(1), 43–56. doi:10.1080/01944369508975618

Utah State Tax Commission. (2017). *Sales and Use Tax General Information.* Report. Publication 25. Author.

Wachs, M. (2003a). Local Option Transportation Taxes: Devolution as Revolution. *Transportation Quarterly*, (22): 9–15.

Wachs, M. (2003b). *Improving efficiency and equity in transportation finance.* Brookings Institution, Center on Urban and Metropolitan Policy.

Weinreich, D. (2015). Thinking Big When Funding Is Local Assessing the Potential of Local Option Transportation Funding in a Multi-Jurisdictional Context. *Public Works Management & Policy.*

Yusuf, J.-E. W., & O'Connell, L. (2013). The Crisis in State Highway Finances: Its Roots, Current Effects, and Some Possible Remedies. *Journal of Public Budgeting, Accounting & Financial Management, 25*(3), 502–521. doi:10.1108/JPBAFM-25-03-2013-B007

Yusuf, J.-E. W., O'Connell, L., & Abutabenjeh, S. (2011). Paying for Locally-owned Roads: A Crisis in Local Government Highway Finance. *Public Works Management & Policy, 16*(3), 250–269. doi:10.1177/1087724X11402357

Zhao, Z., Guo, H., Coyle, D., Robinson, F., & Munnich, L. (2015). Revisiting the Fuel Tax–Based Transportation Funding System in the United States. *Public Works Management & Policy, 20*(2), 105–126. doi:10.1177/1087724X14539139

Zhao, Z. J., & Hou, Y. (2008). Local option sales taxes and fiscal disparity: The case of Georgia counties. *Public Budgeting & Finance, 28*(1), 39–57. doi:10.1111/j.1540-5850.2008.00896.x

Zhao, Z. J., & Jung, C. (2008). Does Earmarked Revenue Provide Property Tax Relief? Long-Term Budgetary Effects of Georgia's Local Option Sales Tax. *Public Budgeting & Finance, 28*(4), 52–70. doi:10.1111/j.1540-5850.2008.00916.x

ADDITIONAL READING

Afonso, W. (2017). State LST Laws: A Comprehensive Analysis of the Laws Governing Local Sales Taxes. *Public Budgeting & Finance*, *37*(4), 25–46. doi:10.1111/pbaf.12171

Afonso, W. B. (2015). Leviathan or flypaper: Examining the fungibility of earmarked local sales taxes for transportation. *Public Budgeting & Finance*, *35*(3), 1–23. doi:10.1111/pbaf.12072

Albrecht, M., Brown, A., Lederman, J., Taylor, B., & Wachs, M. (2017). *The Equity Challenges and Outcomes of California County Transportation Sales Taxes. Report*. Center of Economic Competitiveness, University of California.

American Association of State Highway and Transportation Officials. (2016). "Transportation Finance Governance and Finance." Retrieved from www.financingtransportation.org/pdf/50_state_review_nov16.pdf

Rall, J., Wheet, A., Farber, N., & Reed, J. (2011). "Transportation Governance and Finance." National Conference of State Legislatures, Washington, DC. Retrieved from http://www.ncsl.org/documents/transportation/FULL-REPORT.pdf

Ross, J. M. (2014). *A Primer on State and Local Tax Policy: Trade-Offs among Tax Instruments*. Arlington, VA: Mercatus Research, Mercatus Center at George Mason University.

Wachs, M. (2003). *Improving efficiency and equity in transportation finance*. Brookings Institution, Center on Urban and Metropolitan Policy.

KEY TERMS AND DEFINITIONS

Ability to Pay Principle: An equity principle that states that fairness should be assessed by vertical and horizontal equity. Where vertical equity dictates that taxpayers with different abilities-to-pay should have different tax burdens (i.e., the idea of progressive versus regressive taxation). Whereas, horizontal equity dictates that taxpayers with the same ability-to-pay be subject to the same tax burden.

Benefit Principle: An equity principle that suggests that those who use the service or program should bear the burden of its cost, and that use and burden should be proportional.

Discretional Authority: The criteria that determines the level of autonomy that governments have over the levying and use of revenue of tax instruments. Developed with regard to local sales taxes by Afonso.

Flypaper Hypothesis: A hypothesis that suggests that earmarked revenue, often intergovernmental transfers, tends to "stick" to the program that it is earmarked for. The outcome, if the flypaper hypothesis is accurate, is that overall expenditures will increase when new earmarked revenue streams are introduced.

Jurisdictional Eligibility: The criteria that establishes which governments are permitted to levy the revenue instrument. Developed with regard to local sales taxes by Afonso.

Leviathan Model of Expenditures: A model that suggests earmarked revenue will be dedicated to popular programs and will simply replace existing revenue streams and due to the fungible nature of money, the ultimate outcome may be that the recipient program does not have any additional monies allocated once the earmarked revenue is introduced.

Rational Model of Expenditures: A model that suggests earmarked revenue that only partially funds the recipient program will simply be treated as an increase in general fund revenues. Thus, the likely outcome is that the recipient program will only receive an increase in allocation that is a fraction of the earmarked revenue due to the fungible nature of money.

Tax Earmarks: Tax instruments where the revenue generated must be used on specific areas of expenditure.

ENDNOTES

1. In this chapter, the term earmark refers to a revenue from a financing instrument that is legally required to be appropriated to specific projects or areas of expenditure.
2. Kentucky has a state statute that appears to permit local sales taxes, but it is written in a way that no local government has been able to adopt a local sales tax. It is not included in the 39 states referenced.
3. Using county-area estimates from the US Census of Governments the percentage of general own-source revenue coming from the general sales tax almost doubled between the 1972 estimates and 2002, from 3 percent to over 7 percent respectively. These estimates are countywide and include states and governments where the local sales tax is not permitted.

4 Every state that authorizes local governments to adopt local sales taxes, with the exception of Kentucky, has jurisdictions that have adopted one.

5 Utah has capped the combined state and local sales tax on grocery sales at 3%.

6 Please note that it is a common feature of most local sales taxes with earmarks and other earmarked revenues.

Chapter 8
Reducing Risk in Public-Private Partnership Contracts:
Two Examples From Highway Tolling Projects

Martin Mayer
Old Dominion University, USA

Juita-Elena (Wie) Yusuf
Old Dominion University, USA

Lenahan L. O'Connell
University of Kentucky, USA

ABSTRACT

In an effort to address financial constraints and environmental concerns states have increasingly turned to a combination of un-tolled (HOV) and tolled (HOT) lanes. Public-private partnerships (3Ps) are a popular mechanism for this more sustainable approach to highway infrastructure that couples environmental sustainability (efficient utilization of existing lanes, less congestion) with financial sustainability (private investment). This chapter offers an approach to 3P contract writing for HOV/HOT facilities that is structured by a stakeholder analysis of actors in the project accountability environment. By analyzing two Virginia 3P highway projects, the chapter shows it is possible to build into a contract a set of terms and conditions to enhance the likelihood of meeting the goals of multiple stakeholders. By necessity, such contracts cannot specify precise monetary returns and other stakeholder benefits, but they can be written to include trade-offs to minimize losses to one party at the expense of another.

DOI: 10.4018/978-1-5225-7396-8.ch008

INTRODUCTION

Financial constraints and environmental concerns are promoting the development of tolled highway projects that encourage less driving with a combination of (un-tolled) high occupancy vehicle (HOV) lanes and high occupancy toll (HOT) lanes. HOV/HOT projects are financially sustainable when the private sector is a partner who invests money in the project. The projects are environmentally sustainable when they add fewer lane miles of pavement than a traditional roadway expansion, consume less land, and offer drivers a choice between un-tolled lanes and a tolled lane that is faster and less congested.

Public-private partnerships (3Ps) are a popular mechanism for this more sustainable approach to new highway infrastructure. However, many 3Ps are risk-prone due to their complexity and the unpredictable nature of the revenue streams they frequently create (Urban Land Institute, 2013). The risks and uncertainties challenge accountability. This is especially the case when public infrastructure investments assume adequate returns to the parties over an extended time horizon (Hodge, 2004). Given the ambiguity of future events, it is impossible to specify all the desired results in the contract. Thus, it is difficult for the parties to hold each accountable for any failure to deliver. In this regard, 3Ps to build tolled facilities can subvert the conventional approach to accountability and fail to generate the expected financial and environmental benefits.

Public and private sector actors thus may face the prospect of a failed project. Farmer (2018) provides a classic example of a local government suffering a substantial loss from a poorly designed contract with the private sector. The City of St. Louis lost money when its NFL team, the Rams, moved to Los Angeles. The Rams owners used a loophole in their contract with the St. Louis Regional Convention and Sports Complex Authority to avoid paying the rest of the Rams' share of the $259 million, 30-year bond used to finance the construction of a football stadium (Farmer, 2018).

The Rams were able to leave the taxpayers on the hook for the remaining financial obligation because in its leasing agreement with the team, St. Louis officials agreed that the new stadium would remain rated in the top 25 percent of all NFL stadiums. When the financially-strapped city refused to make the stadium upgrades to keep it in the top 25 percent, the team's lease was not renewed. St. Louis officials made a simple error—they did not stipulate in the contract that the team must continue to lease the stadium so long as any of the debt remained outstanding.

This chapter offers an approach to 3P contract writing for HOV/HOT facilities that is structured by a stakeholder analysis of the interests of all the actors in the accountability environment surrounding a project's field. Rather than specifying precise outcomes of the highway projects, the contracts analyzed were strategically designed to reduce risk by setting up a range of potential outcomes to increase the likelihood that all stakeholders in the project's field benefit. This can be viewed as a type of emergent accountability in that the contract contains mechanisms that adjust outcomes to minimize the prospect of extreme gains and losses by the interested parties.

The objective of this chapter is to show that it is possible to build into a 3P contract a set of terms and conditions to enhance the likelihood of meeting the goals of multiple stakeholders. By necessity, such contracts cannot specify precise monetary returns and other stakeholder benefits. They can, however, be written to include trade-offs to minimize losses to one party at the expense of another. This can be done, in part, by taking into account conditions that could lead to a one-sided distribution of outcomes.

The chapter begins by discussing advantages of a stakeholder approach to devising contracts for multi-sectoral public-private partnerships. This is followed by a description of the frequent elements of two common accountability environments. The authors then describe two Virginia highway projects, highlighting the risks involved. The contracts are then analyzed with a focus on the techniques deployed to foster accountability to all stakeholders when the ultimate distribution of gains and losses from tolling is unpredictable.

BACKGROUND

State Transportation Agencies, Privatization, and Public-Private Partnerships

As Yusuf and O'Connell (2014) note, local and state governments are increasingly forming partnerships with the private sector to provide transportation infrastructure in an effort to address expanding needs under the constraint of limited public resources. In the transportation arena, growing privatization has many roots, including shortages of government personnel and the resulting lack of expertise in the public sector, expansion of demand for transportation services and infrastructure, and pressures to reduce costs and improve quality (Gen & Kingsley, 2007; Ponomariov & Kingsley, 2008; Warne, 2003; Yusuf & O'Connell, 2014).

As of 2016, 34 states, the District of Columbia and Puerto Rico had enabling legislation allowing them to enter into 3Ps (Slone, 2016). From the late 1980s through 2013, the number of transportation-related 3Ps has steadily grown (Z. Chen, Daito, & Gifford, 2014). The popularity of 3Ps can be attributed to many factors, including innovation and new technology emanating from the private sector, the need for private sector expertise, and especially the potential for private funding for transportation infrastructure and services.

While the opportunity for successful public-private partnerships is great, it is certainly not without risk (Urban Land Institute, 2013). Public officials are accountable to many stakeholders and 3Ps can exacerbate accountability challenges (Yusuf & Leavitt, 2014). Moreover, the accountability challenge is multidimensional, especially when the contract extends over many years (Hodge 2004; Grimsey and Lewis 2001). Roumboutsos and Pantelias (2015) note that 3Ps "are not actual 'partnerships', but strict legal transactions where all parties are contractually bound into delivering their own part in the project under detailed specifications and requirements" (p. 186). As such, contracts for long-term 3Ps must address a variety of potential risks, among them engineering and design failures, poor construction, expensive delays, maintenance failures, revenue shortfalls, funding risks, environmental issues, changes in government policies, and possible project default (Grimsey and Lewis 2001).

As contracts lengthen unforeseen challenges to project success can mount (Hodge 2004). This places a premium on the creation of contract terms that build in flexibility over the course of the contract and a willingness to negotiate changes in partner expectations for results. We view the willingness to negotiate these changes as an emergent form of accountability in which outcomes for partners can evolve with changing circumstances. In this regard, contracts assume good faith collaboration among partners who cooperate with each other to create mutually acceptable solutions to the problems that arise (Willems, Verhoest, Voets, Copperns, Van Dooren, and Van den Hurk 2016). Contracts can be structured in ways to encourage creative responses to potential contingencies. A degree of flexibility can be built into contracts, for instance, some contracts call for a resort to third parties to provide dispute resolution to enable adjustments in fees and revenue allocation. (Shrestha, Chan, Aibinu and Chen 2017). Awareness of the need to respond to changing conditions has led to the elimination from many contracts of the non-competition clauses that prevented governments from building new facilities in the same geographic area in subsequent years (Siemiatycki 2009).

For successful 3Ps, Little (2011) contends, the various project risks should be "transparently identified, equitably allocated, and costed appropriately" (p. 246). Political risks, which could include cancellation of the concession or the introduction of regulatory policies, should be borne by the state. The private partner should be expected to control construction risks (e.g., incorrect or inappropriate design, project delays, etc.) and operation and maintenance risks (e.g., physical condition of the facility, poor construction quality, etc.). If income risk or any risks that would reduce use or profitability is to be held by the private contractor, careful consideration of usage and revenue projections should be addressed in the contract (Little, 2011).

We focus on the terms in the contracts that facilitate negotiation and mutual adjustment. We show that the parties had to build terms into the HOV/HOT contracts that facilitated working together to reach the twin goals of maintaining a smooth traffic flow while providing a profitable revenue stream. The greatest risk for the state was a possible failure to maintain the desired level of traffic flow; for the private sector firm a failure to maintain a profitable income stream.

Congestion and HOV/HOT Projects

With rising vehicle ownership in the U.S., traffic congestion has become a major threat to the quality of urban life. According to the American Society of Civil Engineers (2017), more than 40% of the country's major urban highways are congested, costing, in 2014, an estimated $160 billion in wasted time and fuel, 6.9 billion hours in traffic delays (42 hours per driver) and 3.1 billion gallons of gasoline. From 2013 to 2014, 95% of the 100 largest metropolitan areas in the U.S., had increased traffic congestion.

More lane miles are needed to reduce congestion but budgetary shortfalls persist. In response, states and localities have begun implementing creative strategies aimed at reducing congestion while simultaneously raising roadway funds. Policy makers are increasingly turning over the responsibilities for roadway management and upkeep to private companies even though this has produced mixed results (Boarnet & Dimento, 2004). Another common approach has been to increase the use of variable rate tolling on specifically designated HOV/HOT lanes (Copeland & Overberg, 2012). HOV/HOT lanes give drivers a choice. Drivers can car pool or pay a toll to use the less congested toll lane or they can avoid the toll by driving in the un-tolled lane(s).

HOV lanes are designed to encourage carpooling and, in turn, ease congestion and roadway stress, but the effectiveness of HOV lanes at addressing congestion has been limited. Research finds that HOV lanes are typically underused with little ameliorative impact on either congestion or the environment (Poole Jr & Orski, 2000; Safirova, Gillingham, Harrington, & Nelson, 2003).

As the questionable impact of HOV lanes has become increasingly clear, many states have started converting existing HOV lanes to HOT lanes (Poole Jr & Orski, 2000). HOT lanes can charge a set fee or charge variable fees based on congestion at a given time throughout the day. Variable fees maintain the smooth flow of traffic in the toll lanes by raising fees when traffic volumes increase and, conversely, lowering fees when traffic volumes decrease.(Copeland & Overberg, 2012). The basic premise of variable rate fees is that some people are willing to pay higher rates to avoid peak-time congestion.

In addition to controlling and limiting congestion, HOT lanes also generate higher toll revenues and often eliminate the need for new or additional roadway construction (Transcore, 2009). Perhaps, more importantly to drivers, HOT lanes typically offer the added peace of mind of predictable travel times (Harlow, 2013).

While many HOV lane structures have begun full conversions to HOT lanes, a significant number of federal roadways have begun incorporating both the HOT and HOV lanes together in unison (Perez, Batac, & Vovsha, 2012). One of the reasons for the combination is the federal government's regulations regarding tolling on federal roadways. By combining HOT lanes with existing HOV lanes, the government is able to retain the integrity of the HOV system, while generating additional revenue through additional HOT lane use by toll paying single occupant drivers (Perez et al., 2012). The benefits to this approach are many. In addition to the much needed revenue generated by HOT lane users, the typically underutilized HOV lanes receive increased traffic and usage, which in turn decreases overall congestion for all roadway users (Safirova et al., 2003).

The Stakeholder Approach to Public-Private Partnerships and Accountability

The perspectives of different stakeholders are important in considering 3Ps, since 3Ps are generally long-term partnership between two stakeholder groups - the public agency and the private firm – with implications for a third category of stakeholders – users and the general public (Forrer, Kee, Newcomer, & Boyer, 2010; Grossi & Thomasson, 2015; Liyanage & Villalba-Romero, 2015).

For transportation agencies, accountability to the public is a critical aspect of 3Ps (Forrer et al., 2010; Wu, Liu, Jin, & Sing, 2016). The delegation of authority to nongovernmental entities (i.e., private firms) can lead to the potential loss of legitimacy, since the former government function is now accomplished indirectly and at arms' length. However, while government agencies can transfer power to the private sector, they cannot transfer legitimacy and accountability in the same way. 3Ps change the venue within which transportation infrastructure and services are delivered; but they do not eliminate the transportation agency's responsibility to the public. 3Ps can only work well if the government agency manages the partnership effectively and ensures continued private partner accountability (Hodge & Greve, 2017; Milward & Provan, 2003; Milward & Provan, 2000; Skelcher, 2010).

Kearns (1996) coined the term 'accountability environment' to capture the complex nature of the public's search for accountable governance. Accountability frequently emerges from the interplay of the multiple actors who can be said to comprise the organizational environment or 'field.' Besides the focal organization, the field includes the external organizations and interested parties with a stake in the focal organization's activities. In the transportation arena, state transportation agencies (i.e., state Departments of Transportation, DOTs) must manage complex stakeholder relations (O'Connell, Yusuf, & Hackbart, 2009; Yusuf, O'Connell, Hackbart, & Wallace, 2008) that have increasingly been marked by contracting and partnerships (Lockwood, 1998; Warne, 2003)

Accountability is a multidimensional product of the stakeholders operating in the accountability environment (Hill & Hupe, 2002; Hult & Walcott, 1990; O'Connell, 2005, 2006), and, therefore, as O'Connell et al. suggest, accountability in multiparty situations results "from negotiations between the parties that share powers" (2009, p. 410). When forging a contract for a 3P transportation project the stakeholders include, among others, the state DOT, the road builder, environmentalists, drivers, local residents, taxpayers, and elected officials.

A public-private partnership is a collaboration involving actors and/or funding from the public sector on the one hand and the private sector on the other. It is a contractual agreement between the public sector and a private partner "wherein the private sector, in exchange for compensation, agrees to deliver facilities and/ or services that have been or could be provided by the public sector" (Little, 2011, p. 243). The private sector contributes resources and expertise, and has a fiscal incentive to maintain efficient management of the roadway (Krol, 2016; van Den Hurk & Verhoest, 2017). As a jointly developed endeavor, costs, risks, rewards, and resources can be shared in a variety of ways as stipulated in the contract that underpins the partnership. Table 1 compares the typical accountability environment under which public-private partnerships are constructed to that under which simple contracts are constructed. Accountability under a 3P is not necessarily the product of a hierarchical relationship in which the government agency can define the project results in advance and sanction the contractor (private partner) for any failure to reach the desired outcomes. When multiple actors with manifold ties to each other and different goals are in play, the degree and extent of accountability is best described as emerging from their multi-stranded interactions.

Simple contracting out works best when there is little inherent project or task uncertainty. The contract specifies the outcomes and obligations of each party. If a contractor fails to provide the agreed upon service in the stipulated manner, the government can sanction the contractor, who is accountable to the government agency and assumes all risk. The relationship between contractor and government is an exchange of fee for service, frequently formed after the submission of competitive bids with only a minimal amount of negotiation of the terms of the contract. Just as there is little or no sharing of risk, there is little or no sharing of property rights. Table 1 lists an example of a simple contract—a state government paying a firm to

Table 1. A comparison of the accountability environment for simple contracting out and tolling public-private partnership

Accountability Environment for Contract Development	Simple Contracting Out	Tolling Public-Private Partnership
Degree of project certainty	Higher	Lower
Specificity of contractor outcomes and obligations	Higher	Combination of specified and unspecified
Relationship formation	Competitive bidding with minimal discussion	Discussion and negotiation prior to and post contract finalization
Ownership of property and/or income rights	Government	Shared property and/or income rights
Burden of risk for project failure	Contractor assumes majority of risks	Risk is shared with a range of acceptable outcomes for multiple parties
Type of accountability	Hierarchical	Emergent
Typical project	Contract for 2-inch overlay on state highway	Contract for toll road with partner rights to toll for specified number of years

lay a 2-inch overlay of asphalt on a highway. Employees of the state DOT routinely inspect the work to ensure compliance with the terms of the contract. If the contractor fails to make a profit or performs poorly, the contractor suffers the consequences. The contractor can rarely re-negotiate the terms or ask for more payment.

Public-private partnerships, in contrast, tend to involve work that is less certain, producing effects that are unpredictable and sometimes undesirable, such as shortfalls in projected revenue streams. The loss of revenue can ultimately doom a project, as was the case in the recent bankruptcy of the State Highway 130 project in Texas. The project, noteworthy for its 85 mph speed limit, lasted just four years into a fifty-year pact before the concessionaire filed for bankruptcy citing traffic revenues and usage drastically less than initial projections (Wilson, 2017).

Such outcomes can produce losses for multiple stakeholders. When the government agency and the contractor enter a relationship, mutual trust, negotiation and discussion are necessary. In many cases, the relationship exists prior to the signing of the contract, as the objectives of the project and means used to attain them are jointly discussed. Often, there is sharing of property rights and income streams between the parties. Risk too is shared and the government, as in the case of the city of St. Louis and the runaway football Rams, can suffer financial losses.

3Ps often involve interdependent stakeholders from multiple sectors with diverse interests and goals (Gray, 1989; Liyanage & Villalba-Romero, 2015). Some goals are shared (Frey, Lohmeier, Lee, & Tollefson, 2006), but others diverge, which can increase the risks. Decision making tends to be shared between the parties and the distribution of outcomes is a product of negotiation. Therefore, accountability in public-private partnerships tends to emerge over the duration of the contract as the parties respond to unforeseen circumstances. Consequently, it is necessary to build into the contract carefully crafted but flexible incentive structures—specifications that provide ample room for negotiation (Dewatripont & Legros, 2005; Engel, Fischer, & Galetovic, 2013).

Accountability tends to emerge through the give and take of negotiation or through the construction of more elaborate terms in the contract, which can enumerate different outcomes based on contingencies that may arise over time. Some toll road contracts, for instance, may reduce contractor risk by allowing the contractor to raise tolls in the future but with a limit on the absolute size of the toll or the contractor's profit. Even with such provisions, forecasting accurate and realistic traffic projections remain critical to project success. Few cases illustrate the importance better than the failed 75-year agreement between the State of Indiana and the Indiana Toll Road Concession Company. With a nearly $4 billion upfront cost, it was estimated that the concessionaire would need roughly 11 million toll paying vehicles to travel the turnpike between Chicago and Ohio. Once complete, nearly half as many did so. The resulting bankruptcy and debt reconsolidation have only compounded the problems for all involved (Puentes, 2014).

This chapter's analysis of accountability builds on Gray's (1989) observation that collaboration between parties to a contract is needed when the stakeholders are interdependent and decision-making involves joint ownership, collective responsibility for future outcomes, and the ability to view accountability as emerging from negotiation in a constructive work environment. The analysis documents that the Virginia DOT devised 3P contracts that built in disclosure, feedback, and processes for collective voice that ensure each stakeholder is able to play a meaningful role in the lengthy life-cycle of the contract. Of course, the involvement of specific stakeholders can vary over time (El-Gohary, Osman, & El-Diraby, 2006).

One key to 3P success is the presence of a shared goal, which in this case of the two 3P projects analyzed in this chapter was the construction of toll lanes that Virginia drivers would pay to use in future years. Another key to successful 3Ps is contractual flexibility to ensure that incentives and risks remain balanced over the life of the contract. The explicit sharing of risk is a key aspect of 3P projects (Hodge & Greve, 2017), and flexibility in the contract helps to curb financial risk for both public and private stakeholders, while maximizing public benefit (Schank, 2011). The flexibility to undo a poorly written contract, "should such undoing become necessary," is vitally important for all involved (Boarnet & Dimento, 2004, p. 30).

Prior to construction the parties established a number of agreed upon metrics which were designed to limit loss to the stakeholders under defined circumstances

(Dewatripont & Legros, 2005). This creates an ongoing evaluation process that reassures partners that they can negotiate corrections to handle emergent contingencies (Ahadzi & Bowles, 2004). Indeed, the two 3P contracts analyzed were very detailed, containing numerous deadlines, oversight mechanisms, construction and maintenance standards, reporting requirements, and sanctions for failure to perform along with dispute resolution procedures including conditions for contract termination or re-structuring.

TWO PUBLIC-PRIVATE PARTNERSHIP CASE STUDIES

The core of this chapter is two 3P case studies of HOV/HOT projects in Northern Virginia, outside of Washington, D.C. A summary of the two projects is provided in Table 2.

The Virginia DOT entered into two long-term contracts for the construction of toll lanes: one with Capital Beltway Express LLC for the I-495 HOT lanes project and the other with I-95 Express Lanes LLC for the I-95 HOV/HOT lanes project. The toll roads were managed over the term of the contracts by Fluor-Transurban Corporation, which paid for the bulk of the construction and was allowed to set and collect tolls over the long life of the contract—80 years for I-495 and 76 years for I-95. The private sector partner, referred to in the contracts as the concessionaire, agreed to design, build, finance, operate, and maintain the facility (a DBFOM contract) and do so according to the conditions laid out in the contracts. The public partner, VDOT, for its part, paid 20% of the construction cost for the I-495 project and 11% for the I-95 project. VDOT was not financially responsible for operation and maintenance after construction.

Table 2. Summary of the 3P case studies

	I-495 HOV/HOT Project	I-95 HOV/HOT Project
Fiscal year approved	2008	2012
Projected/total cost	$2.07 billion	$948 million
Percent of public investment	20%	11%
Miles covered	14	29.4
Contract duration	80 years	76 years
Competing proposals	0	1

Source(s): Amended And Restated Comprehensive Agreement Relating To The Route 495 Hot Lanes in Virginia Project, 2007; Comprehensive Agreement Relating To The I-95 HOV/HOT Lanes Project, 2012.

The contracts obligate the concessionaire to operate and maintain the facilities over the life of the contracts according to a set of standards. In each contract, there is an agreed upon performance point system that applies five years after the service commencement date. The Virginia DOT monitors performance and "If the Department determines any breach or failure...has occurred, the Department shall...deliver to the Concessionaire written notice thereof describing the breach or failure in reasonable detail" (section 8.16(iii)). The concessionaire must 'cure' the breach or failure within a specified period of time to remove points and/or avoid incurring new ones. The contract stipulates that the concessionaire can object to the determination of a breach and the awarding of points.

The non-compliance performance point system is used as an accountability tool, albeit, one that may provide a range of divergent outcomes. In either agreement, if the number of performance points continues to accumulate and/or breaches are not being cured in a timely manner, several actions may then be taken, starting with increased departmental monitoring and proceeding to the mandatory development of a remedial plan. Failure to meet the goals of the remedial plan can result in contract termination. But the non-compliance performance point system affords flexibility to each sector and recognizes the inevitability of breaches. Increased monitoring only occurs after the assessment of 135 performance points during any 365 day cycle and a remedial plan is only demanded after the assessment of 200 or more points during any 365 day cycle.

To discourage improper resort to contract termination, damages can be sought by both the concessionaire and the department depending on the circumstances. The power to seek damages prevents either party from using the non-compliance performance point system to void the contract without valid reason.

The I-495 Project

The contract for the I-495 project -- *Amended and restated comprehensive agreement relating to the route I-495 HOT lanes in Virginia project* (2007) -- was agreed to on December 19, 2007 between the Virginia Department of Transportation and Capital Beltway Express LLC, an entity formed by Fluor and Transurban to design, finance, construct, and ultimately operate new high-occupancy tolling lanes, designed to moderate and ease congestion around the nation's capital.

The project called for the construction of two HOT express lanes in each direction along 14 miles of Virginia I-495. High Occupancy Vehicles would not pay a toll when using the HOT lanes. Single occupant vehicles in the toll lanes would pay a toll that varied by time of day and level of congestion. Thus, toll paying vehicles with only a driver and high occupancy vehicles with a driver and passenger(s) could use the same lanes. This can reduce overall highway congestion as well as congestion for those who use the un-tolled lanes (Safirova et al., 2003).

Sustainability Goals

This chapter's focus is on the aspects of the contract that introduce an element of uncertainty and potential failure that the contract attempts to minimize. The foremost goal of the I-495 3P project was to provide drivers with a choice of toll lanes that would provide a free flow operating standard for travel on I-495. The un-tolled lanes did not have an operating standard and could be congested with bumper-to-bumper traffic. Free flow would be accomplished with electronic tolling (with video equipment, transponders, or other technology) and an absence of toll booths and plazas.

Achieving the project's goal would also require variable toll fees to closely regulate driver use of the toll lane so as to meet the free flow standard. When the highway is congested and free flow is constrained, the toll rate would rise to a level high enough to divert drivers to the un-tolled lanes. The means to accomplish this are clearly stated in the contract: The Concessionaire can "establish, impose, charge, collect, use and enforce the collection and payment of Tolls" (Section 4.01). The ability to regulate the traffic speed is equally explicit in the contract: "The Concessionaire shall impose congestion pricing on the HOT Lanes, which may include dynamic tolling with potential toll rate changes at frequent intervals with a view to maintaining free flow conditions of traffic, and there shall be no restrictions on toll rates" (Section 4.04(a)).

Stakeholder Risk Reduction

In addition to HOVs, several other types of vehicles cannot be charged a toll for use of the HOT lanes. These include mass transit and commuter buses, school buses, and motorcycles. To reduce the Concessionaire's revenue risk, the contract contains this provision regarding the volume of un-tolled HOVs in the HOT lane: "The department agrees to pay the Concessionaire… amounts equal to 70% of the Average Toll applicable to vehicles paying tolls for the number of High Occupancy Vehicles exceeding a threshold of 24% of total flow of all Permitted Vehicles that are using such toll section going in the same direction for the first 30 consecutive minutes during any day and any additional 15 consecutive minute periods" (Section 13.05(b)). The complex wording reflects the difficulty of ensuring both free flow (a government goal) and profitability (a contractor goal).

The ability to set rates does not assure a net profit for the concessionaire, as it must maintain a free flow condition and accept a large number of un-tolled vehicles. If toll prices are too low, it may not be able to maintain free flow. If it raises tolls too high to obtain free flow, too many drivers may avoid the toll lanes.

The performance point system discussed above ensures a proper balance of interest between the Concessionaire and the Department. The point system can be likened to a scorecard, providing an accountability mechanism if the Concessionaire "breaches or fails to perform its obligations under the agreement", the Department has the ability to assess performance points (Section 8.16). The point system measures Concessionaire performance across the agreed upon metrics of the contract; the accumulation of non-compliance performance points by the Concessionaire may further trigger remedies and provisions set forth including but not limited to greater oversight, additional monitoring, and financial penalty.

Profitability is constrained by two other requirements in the contract. First, maintenance standards were specified in the contract, which clearly states that the I-495 HOT lane cannot become a 'federally degraded facility.' Second, Fluor/Transurban had to pay for all aspects of electronic tolling, including administration and enforcement.

The I-95 Project

The contract for the I-95 HOV/HOT project -- *Comprehensive agreement relating to the I-95 HOV/HOT lanes project* (2012) -- was agreed to on July 31, 2012. The project called for the expansion and conversion of 14 miles of HOV lanes into three lanes of HOV/HOT lanes. It also extended 9 miles of existing HOV lanes, which would be tolled. However, as noted with the I-495 project, high occupancy vehicles would not pay a toll when using the HOT lanes. Single occupant vehicles in the toll lanes would pay a toll that varied by time of day and level of congestion. Thus, toll paying vehicles with only a driver and high occupancy vehicles with a driver and passenger(s) would use the same lanes.

Sustainability Goals

The foremost goal of the project was to provide drivers with a choice of lanes with an operating speed performance standard (OSPS) of 55 miles per hour. The un-tolled lanes did not have an operating standard and could be congested with slow-moving or bumper-to-bumper traffic. This would be accomplished with electronic tolling.

The project would also require variable toll fees to closely regulate driver use of the toll lane so as to maintain the 55 mph OSPS. When speeds fall below the

OSPS, the toll rate would rise to a level high enough to divert drivers to the un-tolled lanes. The means to accomplish this are explicitly stated in the contract: "From and after the Service Commencement Date and continuing during the Term, the Concessionaire will have the exclusive right to establish, impose, charge, collect, use and enforce the collection and payment of the Toll revenues in accordance with the terms of this Agreement" (Section 5.01(a)). The ability to regulate the traffic speed is equally clear: "The Concessionaire will impose congestion pricing on the HOT Lanes, which may include dynamic tolling with potential toll rate changes at frequent intervals and there will be no restrictions on toll rates" (Section502(a)). The agreement specifies that the pricing methodology must "be designed to assure that the Project will meet the OSPS" (Section 5.02 (a)(iii)).

Stakeholder Risk Reduction

While the Concessionaire is empowered to set tolls, it is not guaranteed a profit. Section 5.07(a) of the contract states: "The Department [of Transportation] will not have any risk or liability related to actual traffic volume and revenue, including but not limited to the risk that actual traffic volume is less than the traffic volume projected in the Base Case Financial Model."

The Concessionaire's revenue risk, however, is reduced by a specific contract provision regarding the volume of un-tolled HOVs in the toll lanes. The contract reads: "From the period beginning on the second anniversary of the Service Commencement Date to December 31, 2030 (the 'First Measurement Period"), the department will pay the Concessionaire amounts equal to 70% of the Average Toll for the number of High Occupancy Vehicles exceeding a threshold of 35% of the total flow of all Permitted Vehicles in two consecutive Toll Sections that are then using such Toll Sections going in the same direction for any period of 15 consecutive minutes." (Section5.07(a)(i)). Similar provisions apply to the other two measurement periods that complete the 76-year term of the contract. How the 3P contracts analyzed in this chapter address other risks are discussed next.

In the highly congested Washington, D.C. area, the 55 mph OSPS may not be possible even with dynamic tolling with variable fee rates. The contract contains a process to address failures to meet the OSPS. Records of travel speeds are kept and a monthly report is mandated. The contract states: "the Concessionaire will notify the Department if the Concessionaire's scheduled monthly report identifies any instance of the Project's failure to meet the OSPS…The notice will describe such failure in reasonable detail. The Department will notify the Concessionaire within 30 days of its receipt of the Concessionaire's report whether or not it requires an OSPS Improvement Plan" (Section 5.08(a)). The contract further makes clear that the plan "will be required to propose a strategy to address the specific reasons which the Concessionaire reasonably believes caused such failure as described in the Concessionaire's report" Section 5.08(b). The state DOT then reviews the submitted plan and the Concessionaire is required to promptly implement the elements of the plan that are deemed to be within its control.

The contract does not automatically terminate if the OSPS is not met, but this may result in the accumulation of non-compliance points should the Department find the Concessionaire to be in breach of contract. In the event of a breach, the language calls for a good faith effort to reconcile or cure it; but recognizes the possibility of failure.

The first OSPS covers the initial 10 years of operation. Every 10 years a new one is negotiated; this includes the non-compliance performance point system. However, the contract stipulates that in a subsequent OSPS the 55 mph traffic flow requirement cannot be raised.

DISCUSSION

Provisions to Counter Risk

3Ps are inherently risky, primarily because of the many and varied interests of stakeholders. As Roumboutsos and Pantelias (2015) aptly summarize, the contract is the nexus that merges the multiple interests, and "Paramount in the determination of this balance of interests is the management of risks and their subsequent allocation among the various stakeholders" (p. 186).

Summing up, a number of stakeholder interests were addressed in the contracts but the exact outcomes are far from certain. Government officials and taxpayers benefit from the greatly reduced cost of construction, maintenance and operation. The expected reduction in congestion could save drivers time; it could also reduce air pollution and other environmental costs. Obviously, the drivers who paid the tolls or rode in high occupancy vehicles expect faster travel and less congestion. But, even those who choose not to pay the toll may experience some reduction in congestion when drivers switch to the toll lane. Of course, drivers can avoid the toll altogether and accept a slower drive. Alternatively, they can use local streets. Whatever their decision, all drivers benefit in their role as taxpayers, and they can choose to pay for the toll lane on occasion when time-pressured. The investors stand to make a reasonable profit so long as a sufficient number of drivers pay the requisite amount of toll.

While the taxpayers will save money on construction, the contractor may fail to adequately maintain the facility over the long durations of the contracts—80 years for the I-495 project and 76 years for the I-95 project. Moreover, the added toll lanes may fail to significantly reduce congestion. One potential reason for such a failure is the fact that the contractor is allowed to set the fee for the toll. Excessively high fees would impose a high cost for toll lane access and could shunt an excessive number of drivers into the un-tolled lanes. With greater congestion in the tolled lanes, there will be little or no net improvement in congestion and therefore, few, if any, environmental benefits. Allowing single occupancy vehicles to utilize HOV/HOT lane might reduce the number of HOVs, if drivers abandon carpools and drive their own vehicles, thereby increasing the number of vehicles on the highway.

As with many 3P projects, the two HOV/HOT lane projects presented significant revenue-related risks to the private sector. These projects rely on user-based remuneration (i.e., tolls) which are subject to demand risk (Roumboutsos & Pantelias, 2015). Demand risk reflects the uncertainty in predicted demand, which in the two cases are associated with traffic volumes and free flow conditions. Addressing the revenue risk, such as by allowing the private sector flexibility in setting toll rates, is an important challenge to achieving optimal risk allocation, which is considered the cornerstone of any successful PPP arrangement (Roumboutsos & Pantelias, 2015).

Table 3 lists some of the provisions in the contracts that reduce risks to stakeholders arising from private sector's right to set toll fees and other concerns. While taxpayers benefit from the substantial private sector investment and obligation to operate and maintain the toll lanes, poor construction and maintenance are a possibility along with little improvement in traffic flow. The contracts addressed these concerns with clear construction and maintenance standards and a system to address failures to create free-flow conditions on I-495 and a 55 mph OSPS on I-95. The contract also contains a very detailed, yet flexible, non-compliance performance point system. It is designed to ensure timely efforts to cure breaches and failures on the part of the concessionaire.

The 3P projects offered the environmental benefit of consuming less land and encouraging free HOV and bus use. Faster travel and less congestion is more likely with the use of variable rate tolling, which allows for more effective traffic regulation. The absence of mandatory tolling benefits low income drivers. However, it is possible that private sector investors would not realize an adequate profit due to toll lanes failing to attract a sufficient number of toll-paying drivers. Several features of the contract offer some protection for the private partner—the foremost being the right to set toll fees and to vary them in response to traffic flow conditions. In addition, when HOV traffic reaches specified thresholds in the toll lanes, the government reimburses the concessionaire for HOV use.

Table 3. Examples of potential stakeholder benefits and losses and contract provisions to reduce loss

Stakeholders	Some Potential Benefits	Some Potential Losses/ Drawbacks	Examples of Contract Provision to Reduce Loss
Taxpayers	Public money saved on construction and maintenance	Little change in congestion and poor construction and maintenance	Construction and maintenance standards; OSPS improvement system
Environmentalists	Fewer miles driven/ reduced pollution; less land use	No measurable change in environmental impacts	Fewer lanes added; traffic free-flow; OSPS 55 mph; electronic tolling
Drivers who use HOV lane or pay to use HOT lane	Faster travel; less congestion	High toll charges; fewer cars with multiple riders	Traffic free-flow; OSPS 55 mph; variable fees
Taxpayers who don't use tolled facility	Money saved on construction and maintenance/ management	More traffic on local streets from drivers avoiding toll road	No reduction in number of un-tolled lanes; diversion of drivers to toll lanes
Low income drivers	No mandatory tolling	Increased congestion in toll free lanes; large toll fees	No mandatory tolling; diversion of drivers to toll lanes
Private investors	Adequate long-term profit	Financial loss or low long-term profit; construction of competing facility by DOT	Payments for heavy HOV usage of toll lanes; no limit on toll fees; ability to compete for competing facility contract
Government officials	Money saved in construction costs and infrastructure maintenance/ management	Low savings; angry drivers and voters; little improvement in congestion; poor maintenance	Driver choice of un-tolled lanes; free-flow and OSPS 55; maintenance standards; non-compliance performance point system; ability to terminate for failure to meet standards

Protection against demand risk also come in the form of restrictions to competition such as when the 3P project can be considered a natural monopoly or in cases where a quasi- or temporary monopoly is created through the contractual arrangement (Roumboutsos & Pantelias, 2015). However, this risk to the private sector can come at a cost to public stakeholders who are forced into fewer options. Furthermore, the public sector seeks to minimize the likelihood of opportunistic behavior of the private partner (van Den Hurk & Verhoest, 2017).

To protect the public interest, the state of Virginia retains the right to build a competing facility to handle traffic congestion in the region. In recognition of the potential impact of this on the concessionaire's revenue, the contract states that the concessionaire can bid for the new highway project. This provision, which calls for fair treatment, is not unusual, as non-compete clauses have been phased out of recent 3P contracts due to their tendency to handcuff the public sector for the long term of the contract (Holeywell, 2013).

CONCLUSION

Public-private partnerships offer a way for resource-strapped government agencies to meet the needs of multiple stakeholders (Estache, Juan, & Trujillo, 2007). In this chapter, we focused on contract provisions related to partner risks and incentives, two aspects of 3P contracts identified by Dewatripont and Legros (2005) as most critical to 3P success. In optimal 3P agreements, the contract contains provisions to properly balance the risks and rewards for the public and private sectors, while attending (to the extent possible) to those of other stakeholders (Ahadzi & Bowles, 2004). With increasing frequency, 3Ps have been used to fund large-scale infrastructure projects (Bennett & Iossa, 2006; Grout, 1997). Frequently, 3Ps save money by 'bundling' service provision (Hart, 2003). Virginia DOT did so by tying construction, operation, and maintenance into each contract. With bundling, the public sector is able to limit budgetary strain (Engel et al., 2013).

However, electronic tolling technologies make it possible to pursue environmental benefits as well. Along with improvements in traffic flow, sustainable transportation projects can solve some of the social, financial, and ecological problems of urban community life. In this respect, sustainable transportation often focuses less on city growth and more on integrating transportation into current urban development, thereby consuming less land (Leuenberger, Bartle, & Chen, 2014). The 3Ps discussed in this chapter illustrate a more sustainable approach to congestion management by introducing HOV/HOT lanes and incorporating congestion pricing. The Virginia DOT avoided resorting to the traditional solution for traffic congestion—expanding traffic capacity through construction of many more additional lanes. In pursuit of sustainability, Virginia constructed fewer lanes, spent far less taxpayer money, and continued to encourage carpooling and bus transit.

Due to the array of stakeholder interests addressed in the 3P contracts and the unpredictability of long term returns from the tolling lanes, the exact outcome for each Virginia stakeholder is difficult to predict. To foster accountability, the contractual provisions promote discussion and negotiation over the course of the partnership. The Virginia 3Ps demonstrate that it is possible to build constraints into the contract to reduce losses and promote a range of acceptable outcomes. The non-compliance performance point system, for instance, attempts to provide recourse for the public in the event of a private sector breach of contract or failure to maintain desired performances. But it is accommodating, allowing for much discussion and creative negotiation within stated parameters, the hallmarks of emergent accountability

Of course, many of the devices used in these contracts apply to simple contracts as well as to the more complex conditions associated with public-private partnerships for tolling. For example, the contract contained detailed specifications for the new toll lanes and technologies along with an oversight role for the DOT to ensure proper materials and desired construction quality. But many of the provisions laid out in the contract were less specific and were meant to balance the needs of various stakeholders in ways that implicitly accept a range of outcomes over the long term. Still, despite the uncertainties involved over the many years of the contracts, Virginia's stakeholders appear to like this type of contract. Since the I-495 and I-95 HOV/HOT projects were begun, four additional HOT/HOV projects have been introduced, albeit not all are 3Ps.

Despite the positives of bundling service provision, leveraging private assets, and utilizing technology in a manner that promotes environmental sustainability, there are several reasons for caution. Given the scope of this research, examining two case studies within the same state, it is impossible to draw firm generalizable conclusions that may apply across the state and country. Each contract is unique, the needs of the locality and state vary from jurisdiction to jurisdiction; yet what the cases do reveal is how nuanced each situation and agreement is. The need for sustainable infrastructure, the contractual nuance that creates and maintains accountability, and the importance of shared risk all illustrate key provisions that can be studied and evaluated across contracts and jurisdictions.

Given their numerous benefits, we expect to see more HOV/HOT tolling projects in the future. In addition, to reducing reliance on state and local governments for the funding of new infrastructure, these projects add another layer of financial sustainability to transportation development by reducing reliance on the motor fuel tax, which in this era of fuel efficient and electric vehicles, is producing increasingly inadequate revenue for road building and maintenance (Chen, (2014). And they reduce the overall environmental harm of fossil fuel consumption by increasing the cost of driving, resulting in more use of transit and fewer miles driven in single occupancy vehicles (Bartle & Devan, 2006; Black, 2010; Leuenberger & Bartle, 2009; O'Connell & Yusuf, 2013; Wachs, 2003; Yusuf, O'Connell, & Abutabenjeh, 2011).

REFERENCES

Ahadzi, M., & Bowles, G. (2004). Public–private partnerships and contract negotiations: An empirical study. *Construction Management and Economics*, *22*(9), 967–978. doi:10.1080/0144619042000241471

Amended And Restated Comprehensive Agreement Relating To The Route 495 Hot Lanes in Virginia Project, Dated as of December 19, 2007 by and among Virginia Department of Transportation, an Agency of the Commonwealth of Virginia and Capital Beltway Expressway LLC, a Delaware limited liability company. (2007). Retrieved from http://www.p3virginia.org/wp-content/uploads/2017/02/1.-ARCA-Execution-Copy.pdf

American Society of Civil Engineers. (2017). *2017 Infrastructure Report Card: Roads*. Retrieved from https://www.infrastructurereportcard.org/wp-content/uploads/2017/01/Roads-Final.pdf

Bartle, J. R., & Devan, J. (2006). Sustainable highways: Destination or mirage? *Public Works Management & Policy*, *10*(3), 225–234. doi:10.1177/1087724X06287493

Bennett, J., & Iossa, E. (2006). Building and managing facilities for public services. *Journal of Public Economics*, *90*(10–11), 2143–2160. doi:10.1016/j.jpubeco.2006.04.001

Black, W. R. (2010). *Sustainable transportation: problems and solutions*. Guilford Press.

Boarnet, M. G., & Dimento, J. F. (2004). The private sector's role in highway finance: Lessons from SR 91. *Access*, *25*, 26–31.

Chen, C. (2014). Measuring state highway sustainability: Taking the fiscal dimension into account. *Public Works Management & Policy*, *19*(3), 255–276. doi:10.1177/1087724X14528475

Chen, Z., Daito, N., & Gifford, J. L. (2014). Do State Fiscal Constraints Affect Implementation of Highway Public-Private Partnerships? A Panel Fixed Logit Assessment. *Journal of the Transportation Research Forum*, *53*(2), 111–128.

Comprehensive Agreement Relating To The I-95 HOV/HOT Lanes Project, Dated as of July 31, 2012 by and between Virginia Department of Transportation, an Agency of The Commonwealth of Virginia and 95 Express Lanes LLC, a Delaware limited liability company. (2012). Retrieved from http://www.virginiadot.org/projects/resources/NorthernVirginia/Express_Lanes_Comprehensive_Agreement.pdf

Copeland, L., & Overberg, P. (2012, December 4). *Controversial HOT lanes spread nationally*. Retrieved from https://www.usatoday.com/story/news/nation/2012/12/04/controversial-hot-lanes-spread-nationally/1747319/

Dewatripont, M., & Legros, P. (2005). Public-private partnerships: Contract design and risk transfer. *EIB Papers, 10*(1), 120-145.

El-Gohary, N. M., Osman, H., & El-Diraby, T. E. (2006). Stakeholder management for public private partnerships. *International Journal of Project Management, 24*(7), 595–604. doi:10.1016/j.ijproman.2006.07.009

Engel, E., Fischer, R., & Galetovic, A. (2013). The basic public finance of public–private partnerships. *Journal of the European Economic Association, 11*(1), 83–111. doi:10.1111/j.1542-4774.2012.01105.x

Estache, A., Juan, E., & Trujillo, L. (2007). *Public-private partnerships in transport*. Policy Research Working Paper 4436. World Bank. Retrieved from https://openknowledge.worldbank.org/bitstream/handle/10986/7602/wps4436.pdf

Farmer, L. (2018, May). How cities fell out of love with sports stadiums. *Governing*, 24–29.

Forrer, J., Kee, J. E., Newcomer, K. E., & Boyer, E. (2010). Public-Private Partnerships and the Public Accountability Question. *Public Administration Review, 70*(3), 475–484. doi:10.1111/j.1540-6210.2010.02161.x

Frey, B. B., Lohmeier, J. H., Lee, S. W., & Tollefson, N. (2006). Measuring collaboration among grant partners. *The American Journal of Evaluation, 27*(3), 383–392. doi:10.1177/1098214006290356

Gen, S., & Kingsley, G. (2007). Effects of Contracting Out Engineering Services Over Time in a State Department of Transportation. *Public Works Management & Policy, 12*(1), 331–343. doi:10.1177/1087724X07302585

Gray, B. (1989). *Collaborating: Finding common ground for multiparty problems*. San Francisco: Jossey-Bass Publishers.

Grossi, G., & Thomasson, A. (2015). Bridging the accountability gap in hybrid organizations: The case of Copenhagen Malmö Port. *International Review of Administrative Sciences, 81*(3), 604–620. doi:10.1177/0020852314548151

Grout, P. A. (1997). The economics of the private finance initiative. *Oxford Review of Economic Policy, 13*(4), 53–66. doi:10.1093/oxrep/13.4.53

Hart, O. (2003). Incomplete Contracts and Public Ownership: Remarks, and an Application to Public-Private Partnerships. *Economic Journal (London)*, *113*(486), C69–C76. doi:10.1111/1468-0297.00119

Hill, M., & Hupe, P. (Eds.). (2002). *Implementing Public Policy*. London: Sage Publications.

Hodge, G. A. (2004). The risky business of public–private partnerships. *Australian Journal of Public Administration*, *63*(4), 37–49. doi:10.1111/j.1467-8500.2004.00400.x

Hodge, G. A., & Greve, C. (2017). On public–private partnership performance: A contemporary review. *Public Works Management & Policy*, *22*(1), 55–78. doi:10.1177/1087724X16657830

Holeywell, R. (2013). Public. Private. Practical? *Governing*, *27*(2), 34–41.

Hult, K. M., & Walcott, C. (1990). *Governing Public Organizations: Politics Structures, and Institutional Design*. Pacific Grove, CA: Brooks/Cole.

Kearns, K. P. (1996). *Managing for Accountability: Preserving the Public Trust in Public and Nonprofit Organizations*. San Francisco: Jossey-Bass.

Krol, R. (2016). *Tolling the Freeway: Congestion Pricing and the Economics of Managing Traffic*. Academic Press.

Leuenberger, D. Z., & Bartle, J. R. (2009). *Sustainable Development for Public Administration*. Armonk, NY: M.E. Sharpe.

Leuenberger, D. Z., Bartle, J. R., & Chen, C. (2014). Sustainability and Transportation. *Public Works Management & Policy*, *19*(4), 316–321. doi:10.1177/1087724X14545540

Little, R. G. (2011). The Emerging Role of Public-Private Partnerships in Megaproject Delivery. *Public Works Management & Policy*, *16*(3), 240–249. doi:10.1177/1087724X11409244

Liyanage, C., & Villalba-Romero, F. (2015). Measuring success of PPP transport projects: A cross-case analysis of toll roads. *Transport Reviews*, *35*(2), 140–161. doi:10.1080/01441647.2014.994583

Lockwood, S. (1998). *The Changing State DOT*. Washington, DC: American Association of State Highway and Transportation Officials.

Milward, H. B., & Provan, K. (2003). Managing the hollow state: Collaboration and contracting. *Public Management Review*, *5*(1), 1–18. doi:10.1080/1461667022000028834

Milward, H. B., & Provan, K. G. (2000). Governing the Hollow State. *Journal of Public Administration: Research and Theory*, *10*(2), 359–380. doi:10.1093/oxfordjournals.jpart.a024273

O'Connell, L. (2005). Program accountability as an emergent property: The role of stakeholders in a program's field. *Public Administration Review*, *65*(1), 85–93. doi:10.1111/j.1540-6210.2005.00433.x

O'Connell, L. (2006). Emergent accountability in state-local relations: Some lessons from solid waste policy in Kentucky. *Administration & Society*, *38*(4), 500–513. doi:10.1177/0095399706290635

O'Connell, L., Yusuf, J.-E., & Hackbart, M. (2009). Transportation Commissions as Accountability Structures A Review of their Statutory Roles and Other Attributes. *American Review of Public Administration*, *39*(4), 409–424. doi:10.1177/0275074008321892

O'Connell, L., & Yusuf, J.-E. W. (2013). Improving Revenue Adequacy by Indexing the Gas Tax to Indicators of Need: A Simulation Analysis. *Public Works Management & Policy*, *18*(3), 229–243. doi:10.1177/1087724X12451575

Perez, B. G., Batac, T., & Vovsha, P. (2012). *Assessing Highway Tolling and Pricing Options and Impacts*. Retrieved from https://www.nap.edu/download/22701

Ponomariov, B., & Kingsley, G. (2008). Applicability of the Normative Model of Outsourcing in the Public Sector: The Case of a State Transportation Agency. *Public Organization Review*, *8*(3), 253–272. doi:10.100711115-008-0059-2

Poole, R. W. Jr, & Orski, C. K. (2000). HOT lanes: A better way to attack urban highway congestion. *Regulation*, *23*(1), 15–20.

Puentes, R. (2014). The Indiana toll road. How did a good deal go bad? *Forbes*. Retrieved from https://www.forbes.com/sites/realspin/2014/10/03/the-indiana-toll-road-how-did-a-good-deal-go-bad/#4a33c8b62087

Roumboutsos, A., & Pantelias, A. (2015). Allocating revenue risk in transport infrastructure public private partnership projects: How it matters. *Transport Reviews*, *35*(2), 183–203. doi:10.1080/01441647.2014.988306

Safirova, E., Gillingham, K., Harrington, W., & Nelson, P. (2003). *Are HOT lanes a hot deal? the potential consequences of converting HOV to HOT lanes in northern virginia*. RFF Issue Brief.

Schank, J. (2011). *Public-Private Partnerships: Understanding the Tradeoffs*. EnoBrief.

Skelcher, C. (2010). Governance of public-private partnerships. In G. Hodge, C. Greve, & A. Boardman (Eds.), *International Handbook on Public-private Partnerships* (pp. 292–304). Cheltenham, UK: Edward Elgar. doi:10.4337/9781849804691.00021

Slone, S. (2016). *Evolution of the public-private partnership pipeline*. Retrieved from http://knowledgecenter.csg.org/kc/content/evolution-public-private-partnership-pipeline

Urban Land Institute. (2013). *When the Road Price Is Right: Land Use, Tolls, and Congestion Pricing*. Retrieved from http://www.uli.org/wp-content/uploads/2013/03/When-the-Road-Price-is-Right_web_F.pdf

van Den Hurk, M., & Verhoest, K. (2017). On the fast track? Using standard contracts in public-private partnerships for sports facilities: A case study. *Sports Management Review, 20*, 226-238.

Wachs, M. (2003). A dozen reasons for raising gasoline taxes. *Public Works Management & Policy*, *7*(4), 235–242. doi:10.1177/1087724X03253152

Warne, T. R. (2003). *State DOT Outsourcing and Private-Sector Utilization: A Synthesis of Highway Practice*. Retrieved from http://onlinepubs.trb.org/onlinepubs/nchrp/nchrp_syn_313.pdf

Wilson, K. (2017). Operator of Texas toll road with 85 mph speeed limit emerges from bankruptcy. *Texas Tribune*. Retrieved from https://www.texastribune.org/2017/06/28/sh-130-toll-operator-announces-new-post-bankruptcy-ownership-structure/

Wu, J., Liu, J., Jin, X., & Sing, M. C. (2016). Government accountability within infrastructure public–private partnerships. *International Journal of Project Management*, *34*(8), 1471–1478. doi:10.1016/j.ijproman.2016.08.003

Yusuf, J.-E. W., & Leavitt, W. M. (2014). The Case for Performance Management in Public Works and Infrastructure. *Public Works Management & Policy*.

Yusuf, J.-E. W., & O'Connell, L. (2014). Outsourcing Expert Services by State Transportation Departments: A Look at Effects on Cost, Quality, and Changing Employment Levels. *American Review of Public Administration*, *44*(4), 477–492. doi:10.1177/0275074012469460

Yusuf, J.-E. W., O'Connell, L., & Abutabenjeh, S. (2011). Paying for Locally-owned Roads: A Crisis in Local Government Highway Finance. *Public Works Management & Policy*, *16*(3), 250–269. doi:10.1177/1087724X11402357

Yusuf, J.-E. W., O'Connell, L., Hackbart, M., & Wallace, C. (2008). An Empirical Examination of the Statutory Characteristics and Effects of Highway and Transportation Commissions on DOT Capital Management Capacity. *Public Works Management & Policy*, *12*(3), 533–543. doi:10.1177/1087724X07311260

KEY TERMS AND DEFINITIONS

Capital Beltway Express LLC: An entity formed by Fluor and Transurban to design, finance, construct, and ultimately operate the I-495 high-occupancy tolling lanes, designed to moderate and ease congestion around the nation's capital.

Contracting Out: Is a means of delivering public services and/or performing public functions where the government provides compensation to an outside party in exchange for a defined set of services or functions. Also known as outsourcing.

Emergent Accountability: The flexible accountability mechanisms built into the contractual agreements to provide recourse and contingencies that ensure a fair and equitable process for all involved parties.

Express Lanes: Express Lanes are specially-designated highway lanes that allow drivers to choose to pay a toll to use the lanes and that are free to carpools, motorcycles, vanpools and other eligible vehicles during the designated hours of operation. Express lanes operate under the premise of reducing congestion by incenting and rewarding desired behaviors and/or those willing to pay a toll.

Fluor-Transurban: A principal parties in the I-495 and I-95 roadway projects. Fluor Corporation provides services on a global basis in the fields of engineering, procurement, construction, operations, maintenance and project management. Transurban Group is an international toll road investor and manager with more than 10 years of experience developing and operating complex toll road infrastructure.

High Occupancy Toll (HOT) Lanes: Lanes accessed by vehicles with multiple occupants and charged a fee for use (especially at high traffic times) to encourage carpooling and a less congested route for the drivers while also reducing congestion on alternative routes. A form of transportation demand management that reduces congestion through dynamic tolling practices.

High Occupancy Vehicle (HOV) Lanes: HOV lanes refer to high occupancy vehicles. HOV lanes reduce congestion by promoting carpooling and ridesharing practices. Lanes accessed by vehicles with multiple occupants (especially at high traffic times) to encourage carpooling and ridesharing, and to reduce congestion. A form of transportation demand management.

I-95 Express Lanes LLC: An entity formed by Fluor and Transurban to design, finance, construct, and ultimately operate the I-95 high-occupancy tolling lanes, designed to moderate and ease congestion around the nation's capital.

Public-Private Partnership (3P): A partnership between a government agency and the private sector in the delivery of goods or services to the public. P3s have been widely implemented in the U.S.A. and across the world for services and infrastructures related to transportation, social services, and waste disposal.

Tolling: A form of road pricing in which a fee is assessed for use of the tolled facility.

Virginia Department of Transportation (VDOT): The Virginia Department of Transportation is the state agency for the Commonwealth of Virginia (USA) with the primary responsibility for building, maintaining, and operating the roads, bridges, and tunnels in the state. For transportation-related P3s in the state, VDOT represents the interests of the Commonwealth.

Chapter 9
Cost and Cost Politics in Airport Takeovers:
Two Cases From North Carolina

Zachary Mohr
University of North Carolina at Charlotte, USA

Tonderai E. C. Mushipe
University of North Carolina at Charlotte, USA

ABSTRACT

While it is well known that cost is a politically salient issue, much less is known about the role of cost development and how costs get elevated into the political discussion. This chapter looks at the role of cost accounting and cost development in the recent takeover attempts of two airports by the state of North Carolina. The Charlotte-Douglas airport takeover was a failed attempt, and the Asheville Regional airport takeover succeeded. The chapter makes three important contributions to the study of sustainable transportation financial management. First, it shows that costs are important to the political discussion of the takeovers. Second, it provides a discussion of the relevant costs in each case. Third, it provides research on the political dimensions of cost in the United States in the management of transportation assets, which may limit sustainable transportation policy.

DOI: 10.4018/978-1-5225-7396-8.ch009

INTRODUCTION

Airport governance changes, such as converting municipally run airports to regionally governed airport authorities[1], are often thought to increase financial and resource efficiency (Oum, Yan, & Yu, 2008; Craig et al 2012). Globally, ownership by the private sector of airports may increase efficiency through better management. In the United States commercial service airports are owned by the public sector, primarily by municipal governments or by local authorities (Bacot & Christine, 2006). By being more efficient with regional airport infrastructure and potentially providing a lower cost of management, governance changes may make airport infrastructure more sustainable. However, little is known about airport financial management in the scholarly literature and discussions of how costs are perceived provides some indication of the possibilities and problems associated with this potential sustainability strategy.

While efficiency and sustainability are ostensible reasons for governance changes, there are often historical and political reasons for an airport's governance structure. A city or a county may have originally built the airport, which means that this entity often controls it. The reasons for a change of governance to these large and important assets is likely to be at least as political as it is based upon economic and accounting reasons. When governance changes are forced on a city by its state, we call this a takeover of the airport. Especially in non-home-rule or weak home-rule states, the state can take over a municipal airport, which it could then run itself or it could give it to a local authority. In the literature, it is unclear what financial considerations are given to these governance changes. To address this issue we ask the following research question: what costs and financial considerations are discussed in airport takeovers? Specifically, are the indirect costs associated with airport management like airport security, emergency response, and other administrative indirect costs a factor in the governance decision?

To address these research questions, this chapter looks at the costs discussed and the cost politics associated with a government airport takeover attempt by the State of North Carolina of Charlotte-Douglas Airport (CLT). It then applies the findings to the case of the Asheville Regional Airport (AVL) to see whether costs were important in that takeover attempt. Both airport takeovers were attempted by the North Carolina General Assembly between 2012 and 2014. CLT is the ninth largest airport in the United States and a hub airport for American Airlines, currently the largest airline in the world in terms of passengers (FlightGlobal 2017). While the General Assembly did pass legislation to take the airport from the City of Charlotte, this attempt was blocked in court. AVL is a regional airport, and the General Assembly was successful in changing the governance structure in this case. The chapter examines the statements made in the North Carolina media and in technical documents concerning the costs of the airports. This case represents a pathway case study (Gerring, 2007) of the politicization of costs in the relatively technical area of accounting and financial management of airports.

The chapter makes three important contributions to the study of government financial management and transportation. First, it shows that costs were important to the discussion of both takeover attempts. Second, it discusses the costs that were relevant in each case. Third, it provides more research on cost accounting and indirect costs in the United States, which is an area that is severely underdeveloped and in need of scholarship (Rivenbark, 2005; Mohr, 2017). Beyond merely technical issues, the process of how costs are discussed in the public is a fascinating part of the policy process and deserves more attention in scholarship.

BACKGROUND

At this point there is a significant literature on the relationship between ownership of airports and their costs (Albalate, Bel, and Fageda 2013; Assaf and Gillen 2012; Adler and Libert 2014; Carney and Mew 2003; Kutlu and McCarthy 2016). Generally, the global literature finds that there is an effect from ownership on costs, but primarily lower costs are associated with private airports, which may have different input factors such as lower capital investment and staff levels. However, in the United States commercial service airports are owned by public organizations either through municipal governments or through independent authorities (Bacot and Christine 2006; Craig et al 2012).[2] One study that looks at large US airport governance structures indicates that airport authorities do have lower costs, but this is due almost exclusively from municipalities paying higher wages and prices for goods (Craig et al 2012). Specifically, this study notes that while the technical efficiency of authorities may be as high as 41%, once the differences in prices is accounted for the efficiency drops below 5%. They suggest that their results indicate that the difference may be due to differences in objectives between cities and authorities.

One area where cities and authorities may have significant differences in objectives is in the area of indirect or supporting costs. For example, airports have to have security arrangements and be able to respond to emergencies. Where a city may look to use its police force for security, an airport authority may want to use a lower cost security option. Likewise, the city may want to use or share some of its fire response equipment between the city and the airport. The airport authority would prefer to have its own and not need to rely upon or justify to the U.S. Federal Aviation

Administration (FAA) the use of shared resources even if it can be technically done with adequate cost accounting.[3]

While most of the studies that have looked at organizational form and cost may account for some costs, they do not model indirect costs specifically and are significantly constrained by the available data (Kutlu and McCarthy 2016). There is little information available in the typical FAA datasources, and even the analysis by Craig et al (2012) would not include consistent costs reported for security and emergency response services because different governments may provide different levels of service from other funds. These services from other funds may be considered indirect costs or overhead service costs and may be significant. As noted in the annual report of Charlotte-Douglas the Airport Services Facility Expense, the overhead expenses of the airport represents 24% of the operating costs of the airport (CLT 2016). While the financial report of CLT discloses this amount, other municipally owned airports may not because the overhead and security costs may be reported in different funds and may have different cost allocation procedures on the necessary and reasonable costs (see Eger and McDonald 2017 from a grant standpoint; and see Pope and Mohr 2017 from an enterprise fund standpoint). From a researcher's perspective, this means that the costs of airports may differ because of the indirect cost category. More importantly for this chapter, we are not sure if this is a cost that is generally discussed when choosing a governance form.

Indirect and overhead costs can differ as Craig et al (2012) suggests. For example, cities may have a preference for more safety and security measures at the airport. If people can sneak into the airport or airport emergencies spread beyond the airport, the local area is the entity that will bear this cost. Therefore, the cities are likely to have a preference for greater security and emergency response that will be paid for by airline passengers. This is especially the case if a significant portion of the passengers comes from outside of the region and the cost can be effectively exported. In contrast, the airport authorities may be able to rely upon emergency response services of the city or county around it and thus minimize their costs. We think that this may be another way that municipal airports may have different preferences and objectives than airport authorities. By examining the costs associated with takeover attempts, we can ascertain whether these costs are relevant or not and whether there is a politics that is associated with costs as it concerns airports.

METHOD OF ANALYSIS

The Charlotte-Douglass International Airport (CLT) is the ninth largest airport in terms of take offs and landings and is the lowest cost per passenger hub airport in the country (Whymann 2013). It was a hub airport in the US Airways system and is now a hub airport in the American Airlines system, currently the largest airline in the world. The low cost of the airport is widely regarded as a reason that Charlotte has such a large airport. This efficient, hub airport is seen as a lynchpin of the economic strategy of the region. With the large number of banks and other financial institutions, including Bank of America, Wells Fargo, TIAA CREF and other major national financial institutions the ability to fly out of Charlotte to anywhere in the world is seen as an important competitive advantage. For these reasons, this is an unlikely case and even pathway case for understanding how costs would become politicized in an organization that prides itself on efficiency.

While the airport takeover is seen as an unlikely case study in terms of a need for reform (George & Bennett, 2005), the high visibility and importance of the case, especially to the middle and upper class segments of the community, means that the case would significantly be played out in the media. These attributes make it a good pathway case study for understanding the likely development of similar issues (Gerring, 2007). This unlikely but highly visible episode shows in a relatively brief period the interesting dynamics of the politics of cost at the local level.

To analyze the case, we downloaded all of the articles on CLT from the archives of the local newspaper, The Charlotte Observer, for the period January 1, 2012 to December 31, 2014. The total sample of articles consisted of 154 separate articles. These articles were coded for the term cost and then coded based upon the theme of the cost discussion. So, if cost (or costs) were being discussed in relationship to the cost of bond financing, this was coded as part of debt. If it was a general statement about cost at the airport, this was coded as a general statement and so on for all of the major themes related to cost. These 154 articles gave us 42 articles where cost was explicitly discussed and 125 separate mentions of the term cost, once mentions of costs that were not specifically related to CLT were taken out.

In contrast, Asheville Regional Airport (AVL) is a small hub airport that served over 400,000 passengers per year in 2016. While this is much smaller than CLT (21.5 million), this number of passengers makes it a hub airport and is similar to Springfield, Missouri and Midland, Texas. Primarily, the airport serves a multi-state region that includes North Carolina, Tennessee, and South Carolina. Also, it serves as the airport for the tourists that visit the Great Smokey Mountains.

The case of AVL was somewhat different than CLT. While CLT was very visible and played out significantly in the media, the City of Asheville was resigned to giving the AVL to the state. Therefore, the discussion of costs was much less played out in the media; however, there were still important discussions of costs in this case. Like the CLT case, we collected and coded articles in the local newspaper, The Asheville Citizen Times, for articles that mention the airport authority or takeover from January 1, 2011 to December 31, 2012. The total number of articles was six and there were 15 mentions of the term cost. As we suspected, the AVL case was less about costs, but cost and efficiency consideration were still relevant in that decision as well.

THE CASE OF THE CHARLOTTE-DOUGLAS ATTEMPTED TAKEOVER

The case of the Charlotte-Douglas takeover grabbed attention because it was a major shake-up in a city that is known for its business acumen and market principles (Bacot 2008). In fact the cost per enplaned passenger is the lowest in the country with a cost of $2.29 (Whymann 2013). The next lowest is Salt Lake City (SLC) at $3.91, and it ranges all the way up to $50.95 per enplaned passenger at New York JFK. In short, CLT is a very efficient airport and it seemed unlikely that the proponents of an airport authority would justify their decision based on cost grounds. However, cost was frequently being mentioned as a reason for the takeover and we wanted to know why.

Background on CLT and the Charlotte-Douglas Airport Authority

On its surface, the CLT takeover is an unlikely case for a discussion of costs becoming political because it is the lowest cost, large hub airport in the country. Functionally, the airport is a department of the City of Charlotte (the City). Because Charlotte is in North Carolina, which is a Dillon Rule state, the City can be controlled extensively by the state. The story of CLT and the fight to make CLT a separate Airport Authority with a legally separate Board is, thus, an intergovernmental contest between the City that fought to keep CLT as a department of the City and the coalition of state legislators and local leaders that fought to take the airport away from the City and form an Airport Authority. While there are important arguments for an Airport Authority, such as greater regionalization and sustainability, the question is how much concerns about costs played into the decision to try and switch the governance of the airport?

On the City's side, the important players were Mayor Anthony Foxx, who went on to become the Secretary of Transportation, city officials such as the City Manager and Finance Director, and Brent Cagle, who is eventually appointed the Aviation Director of CLT once the longtime Aviation Director Jerry Orr stepped down and was subsequently relieved of his position. On the side of creating an Airport Authority, Senator Rucho is the Airport Authority's primary sponsor in the State General Assembly, various business leaders in the community and eventually Jerry Orr becomes one of the main proponents of the Airport Authority. Shawn Dorsch is the chairman of the City's Airport Advisory Committee, but it comes out that he was one of the actors at the local level that was pushing for the regional airport authority. Jerry Orr is the long serving Aviation Director of CLT that is effectively terminated once the legislation is passed that removes CLT from the control of the City. While he does not figure prominently in the discussion either before or for a month after the legislation, he becomes a key figure and even the personification of cost for the proponents of the Airport Authority.

The timeline that can be ascertained precisely is that in 2013 there was a bill introduced in the Senate by Senator Rucho in early February. The Airport Authority bill is passed on July 18 (SB 81). This bill transfers authority of CLT from the city to the Airport Authority. Importantly, the bill specifies that the new Executive Director of the Airport Authority will be the former Aviation Director of CLT (Mr. Orr). When the bill is passed on July 18th, Mr. Orr began referring to himself as the Executive Director of the Airport Authority. Because the City has not conceded that this is a legal transfer of authority, the City claims that he has resigned as Aviation Director. Also, in July, Mayor Foxx took the position of Transportation Secretary. This puts the former Mayor in charge of the Federal Aviation Administration (FAA), which regulates commercial service airports for the federal government. This is important because on July 30th, the Superior Court made the FAA the final arbiter of the dispute. Although the FAA and the Superior Court go back and forth on who should be responsible for deciding the matter, the Superior Court ruled in October 2014 that the City will continue to run the airport until the FAA rules. Since the Superior Court did not address the issue that the FAA was seeking clarification, the FAA has effectively tabled the motion indefinitely.

While the facts of the case are fairly well known to anyone that watched the case, the question of why the Airport Authority reform was created remains. Particularly, the role of costs in the case has not been extensively explored. However, there seems to be clear evidence in the accounts that played out in the media that costs were a significant issue. The following analysis seeks to answer the deceptively simple question of what role costs played in the CLT takeover attempt and what costs were important to both sides.

Table 1. Costs noted in the analysis of the CLT airport takeover

	City	Authority Proponents
Before authority	• Bond and interest costs • Legal costs	• Mayor Meddling • Police and Security Costs
After authority created and during legal fight	• Contract and Purchasing Costs • Lax oversight and audits	• Support costs • Pay and incentive structure

Costs of the Attempted CLT Takeover

In our analysis, the cost of the airport was a significant concern from both the City of Charlotte and the proponents of the Airport Authority. The City of Charlotte was concerned about additional legal costs, and bond and interest costs that they anticipated coming from the separation of CLT from the City. On the side of the Airport Authority, their concerns noted wanting to keep CLT the lowest cost airport in the country and the perception by proponents was that the City was raising costs at the Airport. This was a very important issue to authority proponents. In sum, the fight over the perception of cost is a fascinating part of the fight over control of the airport.[4]

The City of Charlotte was originally concerned about what impact the airport takeover would have on their bond and interest costs (see table 1 for a summary of costs). The issue revolves around the fact that the City had borrowed money for the airport and other facilities attached to the airport such as parking decks. Even if the airport were to continue paying the bonds after the governance switched to the Airport Authority, it would be a technical default because the city would no longer be the ones paying the bonds. Discussing the options for an authority paying the outstanding bonds, the City Treasurer, Greg Gaskins, said "That's extra costs that you wouldn't have to pay if you weren't doing that. That would hurt the airport" (Roberts & Morrill, 2013). Also, early in the fight over the airport, the City raised concerns about the legal costs associated with the takeover. Later in the fight over the airport, the City charged that Jerry Orr had lax financial oversight that was actually leading to higher contract and construction costs for services like parking at the airport (Portillo, 2014b, 2014c).

On the side of the Airport Authority, the proponents originally claimed that they wanted an independent airport authority because of meddling by Mayor Anthony Foxx and that additional costs for police and security were driving up costs. After the Airport Authority was created and the city and state were legally fighting over the airport, the concern for cost shifted to personifying Jerry Orr as the manager that could keep costs low. At this point, Orr was no longer the aviation director of the airport, but he clearly wanted to be. To get back in the game, he began publicly discussing how he always wanted to keep costs low at the airport. The proponents of the airport authority imply that Orr is the only person that can keep the costs so low. Also in the later periods of the fight over the airport, proponents also point to a study of CLT governance structure that noted that an airport authority would have greater flexibility in its pay and incentive structure and would not need to make centralized services "contributions" (Portillo & Harrison, 2013; Whymann, 2013).

Clearly, the indirect costs were significant to this case, particularly from the side of the airport authority proponents. The proponents of the airport authority cited the City's level of police and security as an early reason for switching to an airport authority, and later in the legal fight they discussed the importance of central service contributions by referencing the Whymann study. The issue of the police and additional security at the airport is particularly interesting because it involves a case that attracted national attention. The reason that the City forced the airport to increase security was in response to Delvonte Tisdale, who was a teenager that snuck into the airport and climbed into the wheel well of an airplane bound for Boston. How Tisdale died in the wheel well is not known, but he fell out of the plane when it went to land in Boston. The case was horrifying and played out in the national news media. Furthermore, it raised fears that terrorists might be able to breach airport security in a similar manner (Zezima, 2010). The City's response to this was to put the Charlotte-Mecklenburg Police Department in charge of security and greatly raise the level of security at the airport. The cost of airport security went from $2.6 million dollars per year to $5.5 million dollars (Portillo, 2013).

This large increase seemed to indicate to proponents of the authority that the City's level of security and the cost for security at CLT was too high. When the Whymann study came back saying that other support costs could be avoided, this further indicated to proponents that indirect costs were too high.[5] However, the City documented its costs for the airport in a transparent way through a valid cost accounting plan (see also Pope & Mohr, 2017). So, it appears that the costs were reasonable, but the level of support services may be more aligned with the objectives of the City relative to the objectives of an airport authority.

THE ASHEVILLE REGIONAL AIRPORT TAKEOVER

While the CLT takeover attempt was highly visible and originally attracted attention as an unusual and pathway case for studying the role of costs in airport takeovers, it quickly became apparent that part of the reason for the switch to the authority was based on past precedent. The State's earlier airport takeover was of the Asheville

Regional Airport (AVL) in 2012.[6] This case was not nearly as big because AVL is a smaller regional airport, and the City of Asheville did not put up a fight to stop the North Carolina General Assembly from transferring the airport to an authority. Interestingly, indirect and support costs were an issue in the Asheville case as well.

Background on AVL and the Asheville Airport Authority

On June 28, 2012 the North Carolina General Assembly passed the law to take the control of Asheville Regional Airport (AVL) away from the City of Asheville (Barrett, 2012). This had been after two prior attempts to take the airport control from the City of Asheville had failed in 2011 and 2008 (Burgess 2011). Instead of local control primarily from the city and the county, a commission that consisted of a seven-member board was created, and this commission stripped control of the airport away from the city and the county.

AVL was created by a 1958 referendum vote by the City of Asheville, which led to the approval of a $1.2 million bond for the purchase of land and for the construction of AVL (Hillier, 2011). By the time of the AVL takeover by the state in 2012 it had accumulated $60 million worth of assets (Hillier, 2011). Over the 60 years that AVL was under the control of the city there were some key developments such as the creation of the Asheville Regional Airport authority board in 1979. This board moved AVL from being run as a city department to a joint governmental agency. This joint governmental agency consisted of the City of Asheville and Buncombe County. In the 1990s, Buncombe County passed $8 million-dollar bond referenda for the expansion of the airport (Hillier, 2011). At the time of the takeover the airport had an operating and capital budget of over $8 million and had more than 330,000 annual passengers (Burgess, 2012).

A key difference regarding the debate for airport control between Asheville and the General Assembly, and the fight that took place between Charlotte and the General Assembly is that there appeared to be a resignation by the City of Asheville leadership that the airport takeover was inevitable. Vice Mayor Esther Manheimer was quoted in the press as saying "Of course, we oppose the bill ... but since it's going to happen no matter what we say, why expend great political capital on this issue?" (Burgess, 2011).

Some of the facts that the legislation outlined were that AVL had been for decades reliant upon funds from the Federal Aviation Authority (FAA) to expand and improve the infrastructure. In the unlikely event that AVL would cease to exist the legislation had a provision that property ownership would revert back to the City of Asheville. AVL was a self-sustaining enterprise, and the legislation shared the belief that the proposed arrangement would help streamline grant acceptances and long-term leasing processes (Hillier, 2011).

As for the actual voting, the bill passed the House with 110-0 majority with, one Democrat, Susan Fisher, abstaining (Burgess, 2012). When it came time for the Senate vote the issue had become more contentious with the final vote tally being 31-16, a vote split along party lines in a Republican controlled State Senate (North Carolina General Assembly, 2012).

This sentiment of contention was lingering even after the passage of the legislation as was demonstrated by actions of the Asheville Mayor at the time, Terry Bellamy. The legislation amended the size of AVL's governing board to increase from five to seven. City of Asheville, Buncombe County and Henderson County would each appoint two board members and then the seventh would be voted on by the other six members. Mayor Bellamy decided to serve the remainder of Vice Mayor Manheimer's tenure on the airport board, and the subsequent term, after the Councilwoman resigned from the board two days before the legislation was passed. The legislation stated that no elected official could be appointed to the board, but after the incident with the Mayor and Vice Mayor, the General Assembly passed a technical corrections bill on July 3rd, 2012 to clarify that no elected official could serve on a board under any circumstance. At this point, the City of Asheville had lost control of the airport, and it was being run by a completely independent authority and board.

Costs in the AVL Takeover

The issue of cost was relevant to the takeover debate mainly because of the City of Asheville demanded reimbursement for their initial $1.2 million investment that required a bond referendum in 1958. The incorporation documents of the airport indicate that the property should be returned to the City of Asheville if the airport ceased to operate. Because that bond was paid by taxpayers, the City originally demanded that it be repaid. However, the City chose not to fight the airport authority bill and was not repaid. On the side of the authority, the proponents of the AVL authority noted that it would provide greater contractual and grant flexibility and "current inefficiencies would be removed" (Hiller, 2011).

The issue of indirect costs came to light shortly before the AVL takeover. The debate regarded whether to build a fire station at the airport. That would mean that City firefighters, who were unionized, would operate out of the same building as the airport's public safety staff. The airport was concerned that having their public safety staff work alongside the City's unionized firefighters could cause morale issues if public safety staff learned how much less they were being paid compared to the firefighters. Having public safety staff and firefighters in the same location would allow for a more streamlined service delivery that would save the City of Asheville an estimated $4-5 million and could potentially decrease property insurance premiums around the airport. The AVL Airport Director, Lew Bleiweis, also said that the City was not willing to pay its fair share of the construction costs. The City tried to get the architect to split out the cost of the fire station for the city and the airport, but Director Bleiweis noted that the FAA has strict rules on not allowing the airport to pay for non-airport related functions. City Councilman Cecil Bothwell said "Lew is willing to sacrifice the safety of his air travelers for his political bias" and "to have U.S. taxpayers and Asheville taxpayers pay extra in order to have two separate facilities" (Barrett, 2011).

While costs were not nearly as extreme in the Asheville case, costs were noted on both sides. The major difference between the Asheville case and the Charlotte case seems to be in the level of development of the cost arguments. Asheville was fairly resigned to losing the airport. However, it would have liked to recoup its initial investment in the airport. The difficult to untangle indirect costs were clearly an issue for the City of Asheville and AVL prior to the Airport becoming an independent authority. While the creation of an independent authority means that the airport can now easily reject the City's attempts to coordinate services, it is not unreasonable that cities and airports coordinate these services.

DISCUSSION AND CONCLUSION

This chapter started with some basic questions about costs in attempts to takeover airports and switch governance from municipally owned airports to regional authorities. In both the Charlotte case and the Asheville case, costs were a noted issue on both the sides. The costs that were found to matter include the initial investment in the property, bond and interest costs, and indirect and support costs among others. The development of the arguments around cost appear to be strongly related to the intentions of both sides and, particularly if the city is going to resist the takeover. Clearly, the cost arguments in the CLT case were much more developed; however, these cost arguments had time to develop as the City of Charlotte fought the takeover and possibly in anticipation as they watched the State takeover in Asheville.

In both cases, indirect and shared resources appear to be a management and accounting struggle. While the case of Devonte Tisdale shows how concerns about security can force municipal owners of airports to increase security, the case of Asheville shows that even in this relatively peaceful transfer of an airport the use of shared resources and the perception of their costs can be a challenge. This challenge seems to support the proponents of authorities because they point to higher costs. It is not readily apparent, though, that these costs are unreasonable and that they may be addressed through appropriate accounting means. However, the proponents of airport control are likely to see in the airport authority a way of escaping control by their municipal owners and thus minimizing costs. As the study by Craig et al (2012) notes, the preferences and objectives of municipalities may be different from that of their airports. Future work on airports may want to look specifically at whether indirect costs are statistically different for municipally owned airports in relation to airport authorities. Research on the best practices for splitting out shared costs at airports is especially encouraged.

This chapter also speaks to the struggle for reasonable cost accounting that will be supported by all of the parties involved (Mohr, 2017b). While the sustainability of airports may be supported by greater independence through authorities, it may be passing costs onto local governments. More research on cost accounting is needed in this area and in other areas of transportation.

One limitation of this chapter is that it only looks at two cases from North Carolina. Research in other states and particularly over time would be useful. Understanding if there is a trend in the use of airport authorities would be particularly helpful. Airport authorities may provide a more sustainable model of airport asset use and development, but financial management issues, like cost accounting and shared resource use, remains an under researched aspect of airport management and sustainability.

REFERENCES

Adler, N., & Liebert, V. (2014). Joint impact of competition, ownership form and economic regulation on airport performance and pricing. *Transportation Research Part A, Policy and Practice*, *64*, 92–109. doi:10.1016/j.tra.2014.03.008

Albalate, D., Bel, G., & Fageda, X. (2013). Joint versus single management of large transport infrastructures. *Ocean and Coastal Management*, *71*, 163–169. doi:10.1016/j.ocecoaman.2012.09.014

Assaf, A. G., & Gillen, D. (2012). Measuring the joint impact of governance form and economic regulation on airport efficiency. *European Journal of Operational Research*, *220*(1), 187–198. doi:10.1016/j.ejor.2012.01.038

Bacot, H. (2008). Civic culture as a policy premise: Appraising Charlotte's civic culture. *Journal of Urban Affairs*, *30*(4), 389–417. doi:10.1111/j.1467-9906.2008.00405.x

Bacot, H., & Christine, J. (2006). What's so "Special" about airport authorities? Assessing the administrative structure of US airports. *Public Administration Review*, *66*(2), 241–251. doi:10.1111/j.1540-6210.2006.00576.x

Barrett, M. (2011, August 6). Shouting match may derail city, airport deal-Fire station at airport *Asheville Citizen Times*.

Barrett, M. (2012, August 17). Airport board blocks mayor *Asheville Citizen Times*.

Burgess, J. (2011, June 16). No action likely on airport *Asheville Citizen Times*.

Burgess, J. (2012, June 21). City set to lose airport control *Asheville Citizen Times*.

Carney, M., & Mew, K. (2003). Airport governance reform: A strategic management perspective. *Journal of Air Transport Management*, *9*(4), 221–232. doi:10.1016/S0969-6997(03)00003-6

Craig, S. G., Airola, J., & Tipu, M. (2012). General purpose or special district governance? Technical efficiency versus rent dissipation in airport finances. *Public Finance Review*, *40*(6), 712–735. doi:10.1177/1091142112448415

Eger, R. J., & McDonald, B. D. (2017). Cost Accounting for Government Grants. In Z. T. Mohr (Ed.), *Cost Accounting in Government: Theory and Applications* (pp. 81–99). New York, NY: Routledge. doi:10.4324/9781315648897-5

FlightGlobal. (2017). *World Airline Rankings 2017*. Retrieved from https://www.flightglobal.com/asset/18223

Frazier, E. (2014, November 14). Harris: Costs more critical than control *Charlotte Observer*.

George, A. L., & Bennett, A. (2005). *Case studies and theory development in the social sciences*. The MIT Press.

Gerring, J. (2007). Is there a (viable) crucial-case method? *Comparative Political Studies*, *40*(3), 231–253. doi:10.1177/0010414006290784

Harrison, S. (2013, December 22). Orr out, airport rift remains *Charlotte Observer*.

Harrison, S., Rothacker, R., & Frazier, E. (2014, November 15). Charlotte asks FAA to act on airport. *Charlotte Observer*.

Hillier, D. (2011a, July 10). It's time for an independent airport authority *Asheville Citizen Times*.

Kutlu, L., & McCarthy, P. (2016). US airport ownership, efficiency, and heterogeneity. *Transportation Research Part E, Logistics and Transportation Review*, *89*, 117–132. doi:10.1016/j.tre.2016.03.003

Mohr, Z. (2017b). Cost Accounting at the Service Level: An Analysis of Transaction Cost Influences on Indirect Cost Measurement in the Cost Accounting Plans of Large US Cities. *Public Administration Quarterly*, *41*(1).

Mohr, Z. T. (2017a). *Cost Accounting in Government: Theory and Applications*. Taylor & Francis. doi:10.4324/9781315648897

North Carolina General Assembly. (2012, June 26). *North Carolina Senate*. Retrieved from North Carolina General Assembly: https://www.ncleg.net/gascripts/voteHistory/RollCallVoteTranscript.pl?sSession=2011&sChamber=S&RCS=1191

Orr, J. (2013, October 13). Let's move forward on CLT airport *Charlotte Observer*.

Oum, T. H., Yan, J., & Yu, C. (2008). Ownership forms matter for airport efficiency: A stochastic frontier investigation of worldwide airports. *Journal of Urban Economics*, *64*(2), 422–435. doi:10.1016/j.jue.2008.03.001

Perlmutt, D., & Portillo, E. (2013, April 21). City aims its anger at airport board. *Charlotte Observer*.

Pope, J. V., & Mohr, Z. T. (2017). Cost Accounting for Rates and User Fees. In Z. T. Mohr (Ed.), *Cost Accounting in Government: Theory and Applications* (pp. 66–80). New York, NY: Routledge. doi:10.4324/9781315648897-4

Portillo, E. (2013a, September 1). Airport dispute continues *Charlotte Observer*.

Portillo, E. (2013b, October 31). An eye on upgrades, not uproar *Charlotte Observer*.

Portillo, E. (2014a, May 11). Airport budget set to climb. *Charlotte Observer*.

Portillo, E. (2014b, July 13). Airport leader steps out of Orr's shadow *Charlotte Observer*.

Portillo, E. (2014c, August 28). Audit: Airport oversight lax *Charlotte Observer*.

Portillo, E. (2014d, March 6). Orr blamed for parking rates *Charlotte Observer*.

Portillo, E., & Harriosn, S. (2013, April 26). City-funded study recommends airport authority. *Charlotte Observer*.

Rivenbark, W. C. (2005). A historical overview of cost accounting in local government. *State & Local Government Review*, *37*(3), 217–227. doi:10.1177/0160323X0503700304

Roberts, D., & Morrill, J. (2013, March 06). Bond Questions Slow Airport Vote. *Charlotte Observer*.

Rothacker, R., & Henderson, B. (2013, July 27). Orr waits in wings as city sorts out new *Charlotte Observer*.

Whymann, O. (2013). *Charlotte Airport Governance Study Final Report*. Academic Press.

Zezima, K. (2010, Dec. 10). Body Fell From Plane, Authorities Say. *The New York Times*. Retrieved from https://www.nytimes.com/2010/12/11/us/11plane.html

KEY TERMS AND DEFINITIONS

Accounting: The process and system of recording, measuring, and communicating financial information. In this paper, accounting refers specifically to the public management processes related to the financial structure of airports.

Airport: Is defined in the law as any area of land or water used or intended for landing or takeoff of aircraft including appurtenant area used or intended for airport buildings, facilities, as well as rights of way together with the buildings and facilities. The use of Airport in this paper refers to a variety of airports which include municipally run airports (CLT), regionally governed airport authorities (AVL), and privately-operated airports.

Authority: The power to give orders, make decisions or enforce rules. In this paper Authority refers to the governing power exercised by an airport. This varies depending on how that particular airport is structured.

Cost: Anything that requires the resources of an organization. Cost in this context refer to both the direct operation costs and also the overhead costs that are associated with managing an airport.

Governance: The legal authority to manage an operation or policy. Governance in this paper refers to the legal authority to manage airports.

State: A sub-national political authority. In this paper State refers specifically to the North Carolina government.

Takeover: Is the usurping and supplanting of authority to manage the affairs of an organization and this is typically done by an entity with superior control such as a larger corporation or government authority. In this paper Takeover refers to the usurping of a city or region's authority over an airport by the State government.

ENDNOTES

1 Bacot and Christine (2006) note that there are four categories of management of commercial airports: city, county, state and independent authorities. They note that authorities are "distinct government units" (p. 243) and the others fall under the direction or control of another general purpose government. As these authors define it, we note that airport authorities are airports that are not under the specific control or can not be controlled as in the case of a component unit of some other general purpose government.

2 A small percentage of airports (2.5%) are operated by states (Bacot and Christine 2006). However, these entities would not pertain to our discussion of airport takeovers by the state and given to a local authority because the state would presumably have the power to divest its control of a state owned airport to an independent authority. The costs of the airport and supporting services would be known and likely not contested.

3 As far as regulations around safety go, the FAA specifically through the Aircraft Rescue Fire-Fighting (ARFF) Index Parts 139.315, has a number of safety requirements and guidelines. These guidelines include the type of vehicles and storage of firefighting chemicals on the airport premises. These regulations show that costs for operating airports also consist of safety measures required to remain in compliance with the FAA, but the airport owner may go above and beyond these requirements.

4 This chapter section is based upon a paper written by the authors presented at the 2015 North Carolina Political Science Association Conference and can be found at https://ssrn.com/abstract=2805233.

5 While the report seems to indicate that these costs could be avoided, it does not recognize that the City was providing significant support for the airport like payroll and much of the accounting. These support services would have had to be provided by the airport under the authority structure and it is not obvious that the cost of these services would be lower under an airport authority.

6 Which was also based on the precedent of the General Assembly taking Asheville's water department (Burgess 2012).

Chapter 10
Contracting Out in the Transit Industry:
Recent Perspectives of Transit Agency Managers

Suzanne Leland
University of North Carolina at Charlotte, USA

Olga Smirnova
East Carolina University, USA

ABSTRACT

Since the Government Accounting Office report "Transit Agencies' Use of Contracting to Provide Service," there is a growing interest in contracting out and any measures increasing efficiency and cost-savings. This chapter looks at the results of a unique national survey of transit agency managers conducted in 2017 for a modern snapshot of the transit industry in the United States. While there are specific factors that make transit contracting easier (e.g., competition in the provision of services), there are also factors that require contracting out but make monitoring of contracts more difficult (e.g., no capacity to provide services and monitoring in-house). The authors discuss these factors and provide illustrative examples of factors that may enhance efficiency.

DOI: 10.4018/978-1-5225-7396-8.ch010

INTRODUCTION

Historically, the post-World-War II period has resulted in large financial losses for the public transportation industry in the US (Nayan & Wang, 2017), especially with the increase in automobile use and the implementation of the interstate highway system in the 1950s. In the 1960s many private bus transportation companies were no longer profitable so many state and local government agencies stepped in. However, the environmental concerns over vehicle emissions, the cost of gasoline and increased congestion challenge the sustainability of reliance on cars as American's primary mode of transportation. Therefore, public transportation continues to be a subsidized industry in order to promote sustainable transportation (Wachs, 1989). The other chapters in this book discuss other ways to achieve financial sustainability such as toll roads or HOT lanes, earmarked sales taxes, or public private partnerships, this chapter examines a different aspect of the financial sustainability of public transportation in the United States - the current contracting practices of the public transportation industry. This chapter echoes the importance of political factors in the cost efficiency consideration that are discussed in other chapters as well. The cost savings are important as they can make transportation infrastructure and funding more sustainable.

This chapter highlights a survey of public transit industry managers' views on their current type of service provision arrangements. They were asked to provide a description of their governing structures, what modes of transportation they provide (bus, rail, ferry boat etc.), their contracting activities and current characteristics of their service areas. They were also asked questions concerning the performance indicators and monitoring and the size and length of current contracts to improve our understanding of the level of oversight and control over their contractual arrangements. This survey is needed because the Transportation Review Board (TRB) last studied this issue in 2001 in report #258 "Special Report: Contracting for Bus and Demand-Responsive Transit Services". Even the most recent Government Accountability Office report (GAO-13-782) "Transit Agencies" Use of Contracting to Provide Service," that showed a growing interest in contracting out a is over five years old. Also, both reports are not fully anonymous which may affect the way the administrators reply or phrase their replies. The GAO report was sanctioned by the Congress with the heightened interest in contracting out. "Political language is political reality" (Edelman 1985, p.10); hence, the beliefs or perceptions of politicians play an important role in contracting out decision-making. At the same time, the empirical literature (Leland and Smirnova, 2010; Zullo, 2008; Warner and Hebdon, 2001; Brown and Potoski 2005) finds little to no cost savings in contracting out.

This chapter looks at the factors that transit managers identify as the most important in the decision-making on contracting as well as the factors that influence their decisions.

BACKGROUND

Public Transportation, Efficiency, and Contracting Out

The cost of providing public transportation has risen rapidly across the United States, and there is growing concern for transit agencies' financial sustainability. Between 1997 and 2014, bus and rail service costs increased from $25 to $38 billion (adjusted for inflation). While service miles increased by approximately 16 percent, passenger miles have increased 35 percent. Some of this has been partly funded with increased user fees but the industry is still heavily subsidized by other types of taxation (Maroales Sarriera, Salvucci, & Zhao, 2017 p.3). Over half of cities, counties and special purpose governments have turned to contracting out as a way to reduce costs (Leland and Smirnova, 2009).

The evidence of whether this actually saves money has been mixed over the years. Early studies on contracting out public transit usually find some reduced cost in the US (Talley and Anderson 1986; Teal, 1988; Cox and Mundle, 1997). Such studies often have been based either on the evaluation of one company (Talley and Anderson 1986) or utilized cross-sectional data (Miranda 1994). As additional research emerges, and the cost of rail and bus transit has risen, the literature (Brown and Potoski 2005) finds fewer cases of cost-savings through contracting out services.

Earlier studies (Teal 1988) emphasized the potential of cost-savings from competition, assuming that contracting out would allow local governments to capitalize on those savings. The later literature has unmasked a more complex picture of transportation contracting markets (Smirnova and Leland 2015). For example, Perry and Babitsky (1986) and Leland and Smirnova (2009) 25 years later found no cost savings from contracting in the transit industry. However, if an agency contracts out a portion of its service (typically operations of some routes) and then keeps a portion of provided by government, some cost savings can be realized (Leland and Smirnova, 2010; Zullo, 2008; Warner and Hebdon 2001; Brown and Potoski 2005), especially in the area of operating expenses for public bus services. This is due to the agency's ability to monitor costs through benchmarking and minimize risk by keeping the capacity to deliver services.

Almost twenty years ago, the TRB (2001) surveyed approximately 500 agencies that received federal funding to determine the extent of contracting out for bus services and demand response. In this report, they found that larger agencies are more likely to contract out than smaller agencies and that agencies are more likely to contract out for demand response than for bus systems. According to this study, the main reasons transit systems contract for demand response and bus service are to reduce costs while increasing flexibility to introduce new services. Agencies typically keep services in house for increased control over services and operations. In the report's conclusion, the authors recommend further study be undertaken to determine the level of competition in the industry and understand the political and policy environments that affect the decision to contract.

While Warner and Habdon (2001) find that pragmatic reasons are the most important in contracting out for local public officials, Lu (2013) finds both political as well as programmatic factors accounting for variations in contracting out by state governments. While, the survey does not measure the attitudes of politicians, it does have questions for managers who may be influenced by the politicians. Also, the individual communications between private companies and agencies opens the pathways for private companies to influence policies; Kelleher & Yackee, (2009) call these "contract pathways".

Smirnova and Leland (2016) dig deeper into the question of efficiency, exploring its tentative link to agency contracting. They find that individual private companies (vendors) usually have greater market power (as measured by the number of contracts) as well as a higher market share. This may drive up costs of services. There is also a lack of competition in the actual provision of public transit services that also contributes to diminished savings (Smirnova and Leland, 2013). However, transit agencies may not fully realize why their contracting for services does not yield the expected cost savings. Therefore, this chapter gives us an important to look at underlying factors for the most recent trends and decisions about contracting for services in the industry.

During the Great Recession contracting out did not increase despite the political and economic pressure to reduce costs. Previously contracting out has been proposed as a potential cutback management mechanism during economic down turns for local governments (Cox and Mundle 1997; Gomez-Ibanez and Meyer 1993; Winston and Shirley 1998), as well as part of the popular movement of New Public Management which touted contracting out and privatization as a desired method for governments to reduce costs. Based on 2009 and 2013 industry wide contracting surveys, Smirnova and Leland (2015) discovered that most agencies could not realize such savings by switching to contracting out because of several constraints on transit agency managers such as state law, the level of competition in the bidding process, and length of contracts. The studies on other services (e.g., education Rho (2013)) also do not find freeing of resources. This chapter examines the differences in agency type and contracting out and how managers actually perceive contracting out vis-à-vis cutting costs and improving quality of services.

Therefore, this chapter focuses on the following questions 1) What type of agency is more likely to contract out all or some of their services? Large or small service areas? Special purpose or general purpose? Elected or appointed officials? 2) Do managers perceive contracting out as a way to cut costs or are their other underlying reasons? 3) How many contracts agencies and vendors have? 4) How are existing contracts monitored?

The findings indicate that while the form of government (special purpose vs. general purpose) does not seem to affect the rate of contracting out, the board composition (having elected officials) does seem to increase the likelihood of contracting out. The managers operating contracted out services tend to view contracting out as cost savings measure, while those who do not contract out do not view it that way. On average, a typical agency has about 2-3 bidders, but mostly 1 to 2 contracts, while their largest vendors usually have more than 5 contracts. Agencies tend to have a monitoring unit and use multiple methods to monitor contracting out.

SURVEY DESIGN AND METHODS

Transit managers were surveyed in the fall of 2017 to provide a recent snapshot of the industry, providing unique data not available from any other industry sources. The survey is distributed to all agencies that submit their information to the National Transit Database (NTD)[1] which is the largest source of industry-wide statistics covering over ninety percent of all transit services in the United States (American Public Transit Association, 2011). The Transit Research Board (2001) and the U.S. Government Accountability Office (2015) also use this method for sampling agency managers. Self-administered surveys are a good way to learn more about how managers perceive the industry and contracting out. The questions about contracting out may be viewed as intrusive involving the threat of disclosure to the third party (Tourangeau and Yan 2007), as such we go to length to ensure the confidentiality and anonymity of respondents. The self-administered surveys such as online surveys remove the interviewer, allowing the respondents to be candid in their replies. Both open- and closed-ended questions were used to provide a current portrait of the industry (Dillman, 2000). Since the respondents cannot clarify any items on the survey, all instructions and questions should be as clear as possible (Dillman, 2000).

The response rate was 31.4% (255 responses) with 813 surveys sent to agency managers. The responses by agency size are comparable to TRB (2001), with slightly smaller response from large agencies (over 250 VOMS) operating transit in-house. Incentives for the first 50 respondents were provided; about a one third of responders refused the survey incentives. The survey also offered a random drawing for a prize. The survey was administered electronically through SurveyShare to guarantee confidentiality to the respondents. The topics covered within the survey included: information on bus service that is contracted out, manager perceptions of transit operations the size of and number of contracts, and the fiscal characteristics of the contracts.

RESULTS

We organized our survey results by research questions noted earlier in this chapter.

What Type of Agency Contracts Out All or Some of Their Services?

Overall, larger agencies with elected officials on their boards (alone or in combination with appointed) report more contracting out. There is no difference in reporting of contracting out between special purpose and general-purpose governments.

The larger agencies in the sample, those over 250 vehicles[2] for the bus services and over 100 vehicles for the demand response services, are more likely to contract out on average than smaller agencies. Only about 43% of agencies with less than 50 busses contract out, while about 63% of agencies with over 500 vehicles contract out for bus services. If measured by either population served or service area size, smaller agencies report considerably less contracting activity (37% of those under 100 sq. mi. and 32% of agencies with under 100,000 population served contract out).

About 38% of agencies are special purpose governments, 16% are county, 31% are city agencies (for a combination of 47% of the agencies falling into the general-purpose government category). Fifteen percent said they were other, which includes state agencies, private companies (less than 1%), nonprofits (5%), universities (less than 2%), MPOs (about 1%), and COGs (about 1%) as well as multipurpose special authorities (less than 1%). Private agencies represent a little less than one percent likely due to the fact that the industry is not profitable unless only pieces of the service are contracted out.

While most of the city and county departments have elected boards (about 70%), most of the single purpose governments have appointed or combination boards (there are only about 29% of single purpose governments with elected boards). The remaining 30% of cities answered that either their agency has community representatives appointed by the city council on their board or a combination of elected/appointed officials. About 49% of single purpose governments have appointed boards.

Figure 1 represents the distribution of board composition (appointed, elected and mixed) for contracting out and in-house operations. The majority of agencies (52%) responding to the survey have elected officials on the board, followed by the appointed boards (29%), and mixed boards (16%). The other category includes volunteers, property owners' representatives, and they do not have a board of directors (4%). About 48% of those with elected officials on the board contract out, and 41% of those with appointed boards contract out.

About 80% of survey respondents provide bus services. And about 42% of bus service providers indicate that they contract out for bus services. The second most common services are paratransit and then demand response. Other services provided less frequently are light rail, heavy rail, trolley, ferryboat, incline, jitney, vanpool general public shuttles. Most agencies are small and about 69% operate less than 50 vehicles and only 5 agencies operate more than 500 vehicles.

Figure 1. Board composition
Source: Bi-annual Transit Survey, 2017

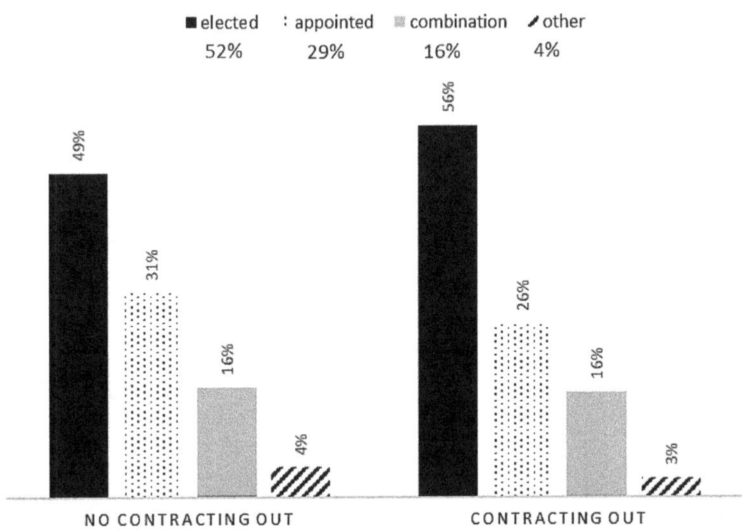

Do Managers Perceive Contracting Out as a Way to Cut Costs or are Their Other Underlying Reasons?

The survey contained a number of statements that the respondents could choose to strongly agree, agree, or disagree, and strongly disagree. Generally, agencies that contract out have more favorable perceptions of contracting out than the agencies that currently do not contract out. The exception is the statement that contracting out brings in additional revenues. Both groups tend to disagree with that statement, but those who currently contract out are even more negative towards this statement.

Nearly 69% disagree and strongly disagree with the statement that contracting out decreases efficiency. Among those respondents who agree with this statement (31%) the majority does not currently contract out for bus services (21%), only 9% currently contracting out and agree that contracting out decreases efficiency (see Figure 2).

At the same time about 38% somewhat agree that contracting brings greater efficiency. Among the respondents who agrees with this statement, only14% are represented by managers who do not contract out and 25% by those who contract out (see Figure 3). Overall, it may be that contracting out has the same efficiency, according to the respondents, as service delivered by employees in the agency.

Transit agency managers from agencies that contract out agree to a larger extent with the statement that contracting reduces risks (see Figure 4). About 18% those

Figure 2. Contracting out decreases efficiency
Source: Bi-annual Transit Survey, 2017

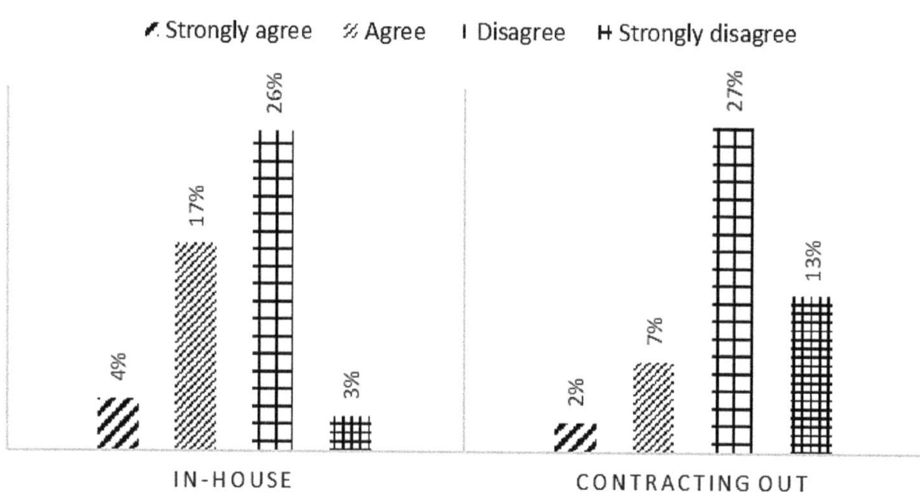

Figure 3. Contracting out increases efficiency
Source: Bi-annual Transit Survey, 2017

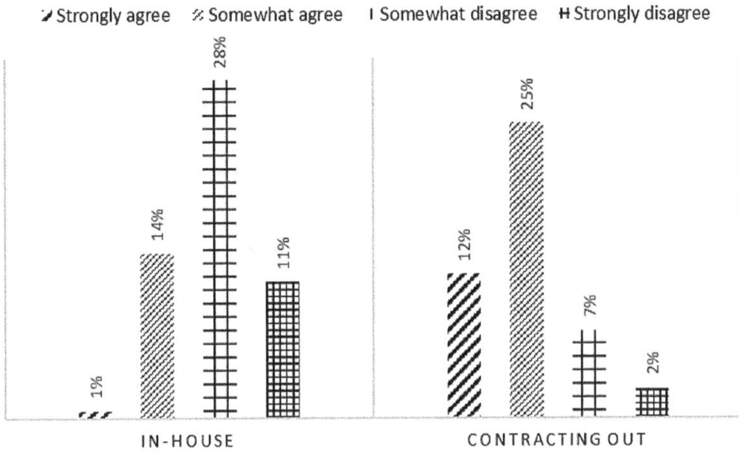

Figure 4. Contracting out reduces risks
Source: Bi-annual Transit Survey, 2017

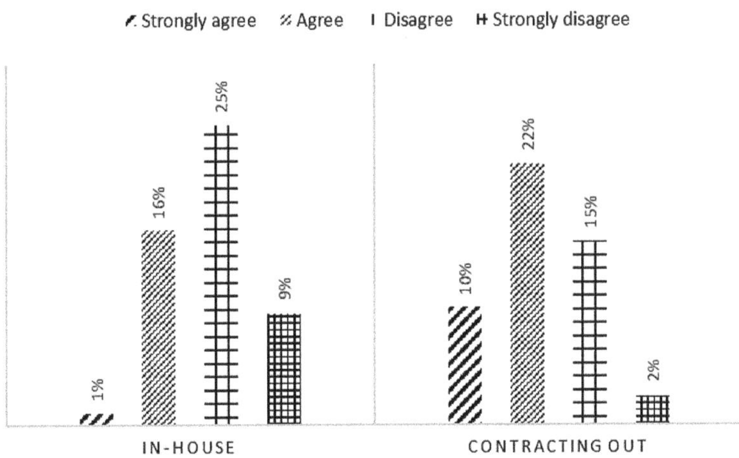

who contract out disagree and strongly disagree that contracting out reduces risks, while about 34% those with in-house operations disagree and strongly disagree that contracting out reduces risks. Only 17% managers of agencies providing services in-house agree and strongly agree (1%) that contracting out reduces risks, while about 31% managers of agencies contracting out agree and strongly agree (10%) with that statement. This could be a situation where agencies in the environment where contracting out may reduce risk take advantage of contracting out.

The agencies that currently do not contract out tend to disagree more with the statement that contracting out improves performance. About 12% managers in the agencies providing in-house operations agree that contracting out improves performance compared to about 24% managers of agencies that currently contract out who agree with this statement (see Figure 5).

When asked whether contracting out brings additional revenue, those who contract out tend to more often disagree with this statement. This is the only statement where agencies that contract out do not observe a positive side of contracting out. In fact, the managers who do not contract out transit operations seem to view contracting as bringing in more revenues (see Figure 6).

The split between those that contract out currently and those that do not continues for the improved quality of service. About 28% somewhat (20%) to strongly (8%) agree that contracting out improves quality of service are represented by those who contract out; only 12% (1% strongly) share similar opinion among those who do not contract out (see Figure 7).

There are certain statements where managers' perceptions from agencies with in-house operations and those that contract out are similar. Managers tend to disagree (64% of both groups) that contracting out reduces accountability. At the same time, 21% managers who oversee in-house operations agree that contracting out reduces accountability, while 15% managers who contract out agree with this statement.

Figure 5. Contracting out improves performance
Source: Bi-annual Transit Survey, 2017

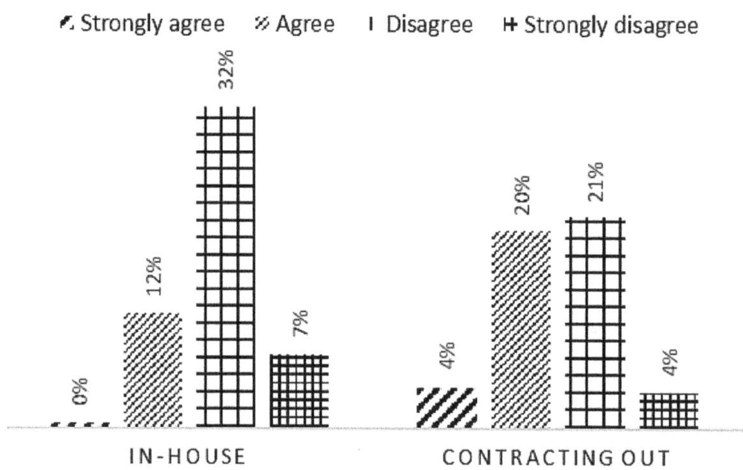

Figure 6. Contracting out brings in additional revenue
Source: Bi-annual Transit Survey, 2017

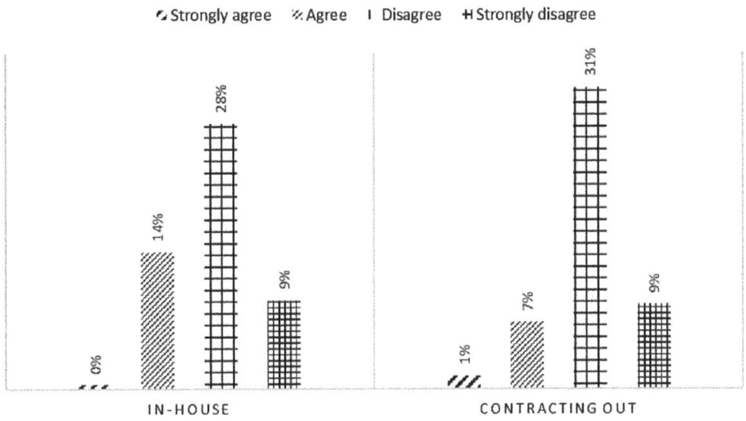

Figure 7. Contracting out improves quality
Source: Bi-annual survey Transit Survey, 2017

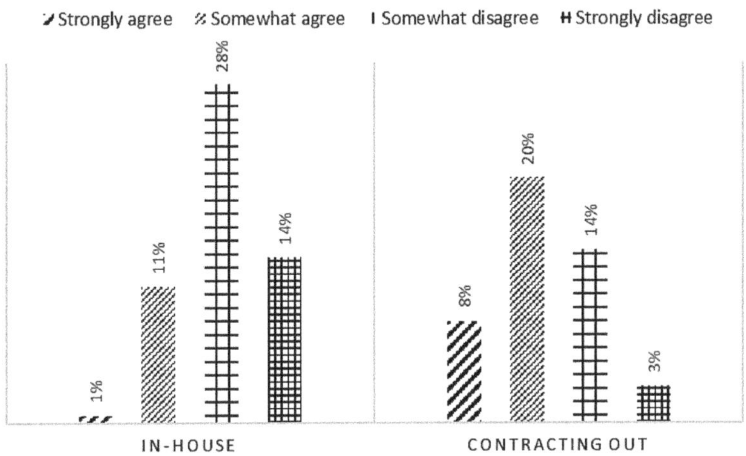

The bidding process again is viewed as fair and equitable by both groups with 86% of all respondents agreeing with the statement. None of managers who contract out disagree that the bidding process is fair, while 11% of those who do not contract out disagree that the process is fair.

There are also statements that both groups tend to disagree with. For example, 70% of respondents disagree and strongly disagree that contracting out decreases labor productivity. Only about 7% those who contract out agree with this statement, while about 23% those who provide in-house operations agree and strongly agree that contracting out reduces labor productivity (see Figure 8).

Also 81% disagree (or strongly disagree) that contracting out improves morale, but about 6% those who do not contract out somewhat agree with this statement and about 13% of those who contract out somewhat agree or strongly agree with this statement. About 67% disagree or strongly disagree that contracting out improves overall safety. About 8% those that do not contract out agree with this statement, while 25% those who contract out strongly agree and agree that contracting out improves overall safety. Nearly 75% of all respondents disagree or strongly disagree that contracting out increases ridership; 67% of all respondents disagree or strongly disagree that contracting out increases reliability.

We also asked transit agency manager to rank the reasons that they contract out (only of those agencies that contract out). The number one reason selected is cost efficiency, number two is that competition creates cost savings, and the third most important reason to contract out is to improve quality. This is consistent with the overall opinions of managers from agencies that contract out.

Figure 8. Labor productivity
Source: Bi-annual survey Transit Survey, 2017

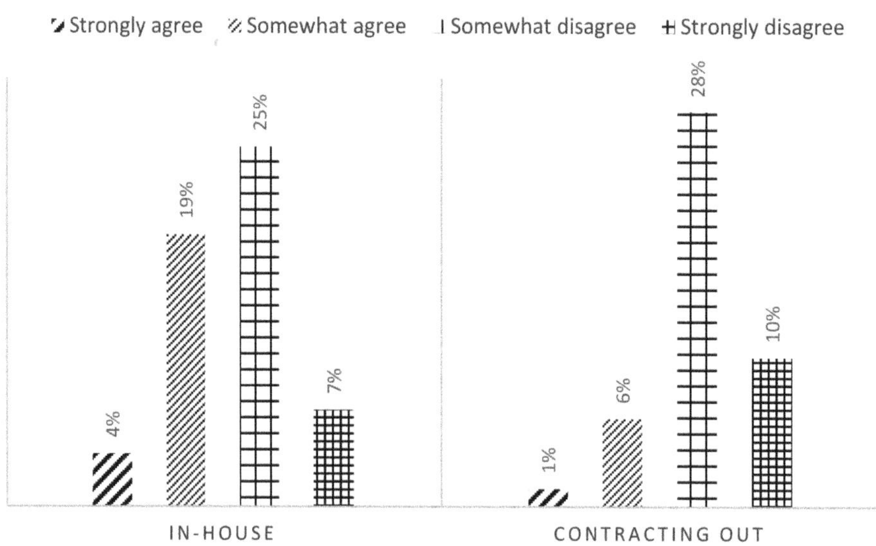

Figure 9. Number of contracts for the agency
Source: Bi-annual Transit Survey, 2017

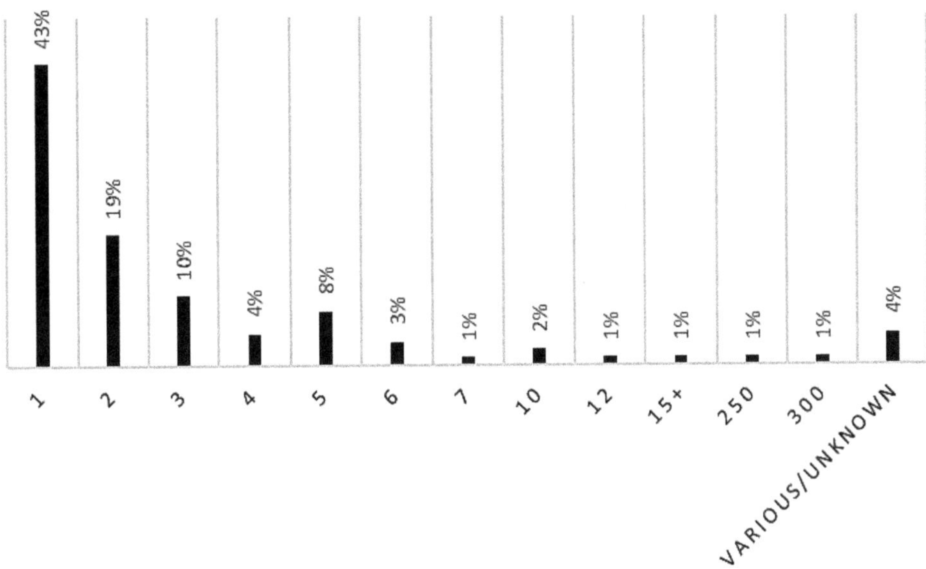

Figure 10. Largest vendors number of contracts (in addition to the contract with the agency)
Source: Bi-annual Transit Survey, 2017

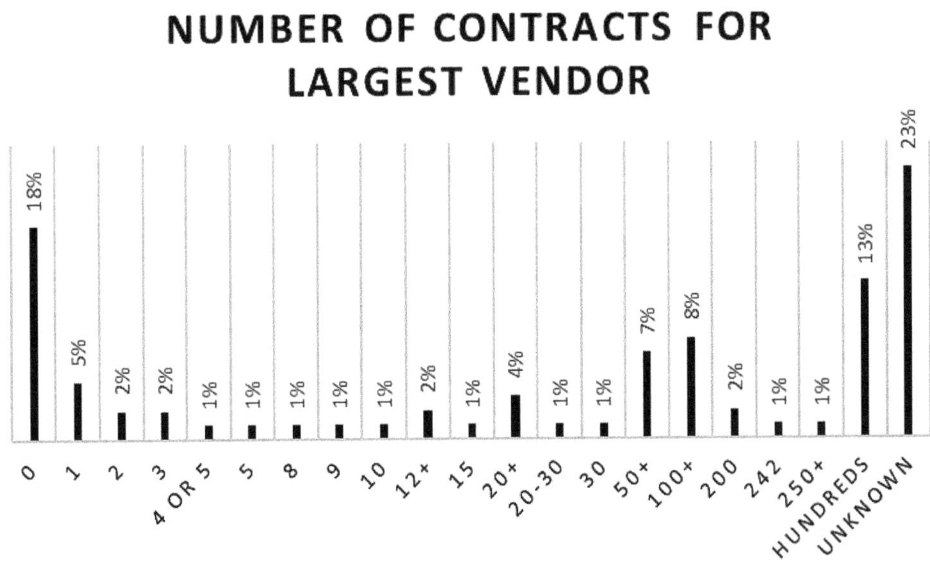

How Many Contracts and How Large is the Typical Contract in the Public Transit Industry?

Frequent rebidding and experience with contracting out is important for realizing cost savings from contracting out (Kettle 1993). In this regard, it is important to look at the number of bidders for contracts – competition for contracts. This does not guarantee the competition in the provision of services, but at least indicates some competition for the contracts and indirectly indicates the experience of agencies interacting with bidders. Our survey contained couple questions to study these issues: the number of bidders as well as number of contracts. Approximately 70% of agencies have three or less bidders per contract, with only 11% having just one bidder; Another 11% have more than five bidders. At the same time, 43% of agencies have only one vendor sometimes for couple different contracts. Figure 9 shows the number of contracts; the numbers go from 1 to 300; the unknown category may include smaller values than 300 contracts. These contract pathways (Kelleher & Yackee, 2009) may allow private and public entities to influence each other.

Besides asking questions about managerial experience, we also asked them for their views of their largest vendor. Figure 10 shows the number of contracts for the largest vendor. About 48% of an agency's largest vendors have over five contracts with somebody else. Only 18% have only one contracts (no additional contracts).

Hence, vendors, not agencies most likely have more experience with contracting out. Having multiple contracts may allow a vendor or an agency to easily switch from the least efficient contracts and even easier to monitor services since the benchmarks can be established on the multiple contracts over number of years. Also, Kelleher and Yackee (2009) stipulate that such "contract pathways" may allow private vendors control over public agencies' policies, including contracting out decision-making.

How Are Existing Contracts Monitored?

In the 2017 survey, 71% of agencies contracting out for bus services report having a monitoring unit. These findings differ from the previous waves of survey, potentially indicating a very positive trend towards more formal contracting oversight. Literature (Smirnova, Yusuf, and Leland 2016) shows that contracting oversight is very important for realizing savings.

Our survey asked respondents which methods of monitoring contracting out they use, and 77% of respondents use multiple methods of monitoring. The most frequent combinations include:

1. Customer satisfaction surveys, Monitoring level of service provision, Monitoring complaints, Secret shoppers (used by 25% of respondents); and
2. Monitoring level of service provision, Monitoring complaints (used by 15% of respondents).

Monitoring level of service provision is by far the most frequently used strategy (85% of respondents), usually in combination with other methods; 7% (5 respondents) use this strategy exclusively.

Ultimately, 83% of all respondents are satisfied with their current arrangements irrespective whether they contract out (34%) or provide services in house (49%). Only 8% plan to increase contracting out in the next five years (2% of those who do contract out currently and 6% of those who already do), and 10% plan to decrease contracting out in the next 5 years (6% of those who do not contract out for bus services now (that is, they are decreasing other services contracted out)) and 4% of those who do. At the same time, about 12% of agencies are somewhat and very dissatisfied with the current contracting arrangements, and about 32% are somewhat and very dissatisfied with the prior contracting arrangements.

CONCLUSION

In particular, this chapter reveals that the form of government does not impact the decision to contract out services which is contrary to some previous research on the topic (Perry and Babitsky 1986, Leland and Smirnova, 2009). Special purpose governments are not any more likely to contract out than general purpose governments. Instead this decision is driven by size of the agency and how the governing body of the transit agency is chosen. Larger agencies that have elected officials on their boards are more likely to contract out. Those with both elected and appointed also contract out more when compared to agencies with appointed boards. This suggests that political structure influences the choice for contracting out services. One explanation might be that elected officials may feel more pressure to cut costs and therefore turn to contracting without fully realizing it may not save money. Or they may be contracting out to obscure their accountability by having a third-party to blame.

The majority of transit agency managers are satisfied with their current contracting arrangements. Very few agencies plan to change their arrangements in the future which points to a level of stability and some might conclude even path-dependency.

However, such stability may be due to long contracts. There is increased performance monitoring and the number of oversight methods utilized in the industry compared to previous studies (Smirnova, Yusuf, and Leland 2016). This is encouraging from an accountability standpoint and may potentially lead to a more transparent process.

This study also reveals that a minority of managers reported some level of dissatisfaction with current arrangements and about a third of agency managers reported some dissatisfaction in the past. Among the managers that anticipate a change slightly more (53%) plan to bring operations in house in the future.

Previous studies establish that there is little evidence that contracting out transit services lowers costs and improves performance (Perry and Babitsky, 1986; Teal et. al. 1988; Leland and Smirnova, 2009). Yet, at the end of 2017, transit agencies still engaged in the practice of contracting out with managers and cited cost savings as the number one reason to contract out. But this chapter indicates that there is more to the story because those agencies with elected boards and a larger service area are more likely to contract out. Also contracting out may allow agencies to provide additional services to outlying areas. These issues need to be explored in the future by surveying board members across the industry.

REFERENCES

American Public Transit Association. (2011). *Public Transportation Factbook*. Available online at http://www.apta.com/resources/statistics/Documents/FactBook/APTA_2011_Fact_Book.pdf

Brown, T., & Potoski, M. (2005, March). Transaction Costs and Contracting: The Practitioner Perspective. *Public Performance & Management Review*, *28*(3), 326–351.

Bureau of Transportation Statistics. (1999). *National Transit Database (NTD) and Safety Management Information System (SAMIS)*. Available online at http://www.bts.gov/programs/statistical_policy_and_research/source_and_accuracy_compendium/FTA_national_transit.html

Burns, N. (1994). *The Formation of American Local Governments: Private Values in Public Institutions*. New York: Oxford University Press.

Cox, W., & Mundle, S. (1997). *Denver Urban Transport Competitive Tendering Saves $88 Million*. Retrieved from http:/www.publicpurpose.com.ut-dencc.htm

Dillman, D. A. (2000). *Mail and Internet surveys: The tailored design method*. New York: Wiley and Sons, Inc.

Gomez-Ibanez, J. A., & Meyer, J. (1993). *Going Private: The International Experience with Transport Privatization*. Washington, DC: Brookings Institution.

Hirsch, B. T., & Macpherson, D. A. (2003). Union Membership and Coverage Database from the Current Population Survey: Note. *Industrial & Labor Relations Review*, *56*(2), 349–354. doi:10.1177/001979390305600208

Kettl, D. F. (1993). *Sharing Power: Public Governance and Private Markets*. Washington, DC: Brookings Institution.

Leland & Smirnova. (2009). Reassessing Privatization Strategies 25 Years Later: Revisiting Perry and Babitsky's Comparative Performance Study of Urban Bus Transit Services. *Public Administration Review.*

Miranda, R. (1994). Privatization and the budget-maximizing bureaucrat. *Public Productivity & Management Review*, 355-369.

Nayan, A., & Wang, D. Z. (2017). Optimal bus transit route packaging in a privatized contracting regime. *Transportation Research Part A, Policy and Practice*, *97*, 146–157. doi:10.1016/j.tra.2017.01.016

North, D. (1990). A transaction cost theory of politics. *Journal of Theoretical Politics*, *2*(4), 355–367. doi:10.1177/0951692890002004001

Perry, J., & Babitsky, T. (1986). Comparative Performance in urban bus transit: Assessing privatization strategies. *Public Administration Review*, *46*(1), 57–66. doi:10.2307/975443

Reja, R. (1999). *The Economies and Politics of Contracting out with the Private Sector: Evidence from the US Transit Industry*. Working paper. The World Bank.

Morales Sarriera, J., Salvucci, F. P., & Zhao, J. (2017). *What Drives the Costs of Transit Operations? The Implications of Labor Productivity, Contracting Out, and Unionization (No. 17-02927)*. Academic Press.

Smirnova, O. (2008). *Does Government Structure Really Matter? A Comparison of Efficiency and Effectiveness of Special Purpose Versus General Purpose government transit operations* (Dissertation). University of North Carolina at Charlotte.

Smirnova, O. V., & Leland, S. M. (2014). Cutback management during the great recession: The case of transit agencies and contracting out. *State & Local Government Review*, *46*(4), 272–281. doi:10.1177/0160323X14564777

Smirnova, O. V., & Leland, S. (2015). The Contracting Decision. *Government Contracting: A Public Solutions Handbook, 6*, 17.

Smirnova, O. V., & Leland, S. M. (2016). The role of power and competition in contracting out: An analysis of public transportation markets. *Administration & Society, 48*(4), 421–443. doi:10.1177/0095399713498748

Smirnova, Yusuf, & Leland. (2016). Managing for Performance: Measurement and Monitoring of Contracts in the Transit Industry. *Journal of Public Procurement, 16*(2), 208–242.

Talley, W., & Anderson, E. (1986). An Urban Transit Firm Providing Transit, Paratransit and Contracted-out Services: A Cost Analysis. *Journal of Transport Economics and Policy, 20*(3), 353–368.

Teal, R., Giulano, G., & Morlok, E. (1986). *Public Transit service contracting*. Washington, DC: US Department of Transportation, Urban Mass Transportation Administration.

Thompon, L., & Elling, R. C. (2000). Mapping Patterns of Support for Privatization in the Mass Public: The Case of Michigan. *Public Administration Review, 60*(4), 338–348. doi:10.1111/0033-3352.00096

Transportation Research Board. (2001). *Special Report 258: Contracting for bus and demand response services: a survey of U.S. practices and services.* Author.

Tourangeau, R., & Yan, T. (2007). Sensitive Questions in Surveys. *Psychological Bulletin, 133*(5), 859–883. doi:10.1037/0033-2909.133.5.859 PMID:17723033

Warner, M., & Hebdon, R. (2001). Local Government Restructuring: Privatization and its Alternatives. *Journal of Policy Analysis and Management, 20*(2), 315–336. doi:10.1002/pam.2027

Warner, M., & Hefetz, A. (2002). Applying Market Solutions to Public Services: An Assessment of Efficiency, Equity, and Voice. *Urban Affairs Review, 38*(1), 70–89. doi:10.1177/107808702401097808

Williamson, O. (1985). *The Economic Institution of Capitalism: Firms, Markets, Relational*. London: Contracting. Collier McMillan.

Winston & Shirley. (1998). *Alternative route: towards efficient urban transportation.* Washington, DC: Brookings Institution.

Zullo, R. (2007). Transit Contracting Re-examined: Determinants of Cost Efficiency and Resource Allocation. *Journal of Public Administration Research and Theory*.

KEY TERMS AND DEFINITIONS

Contract Pathway: The individual communications between private companies and agencies opens the pathways for private companies to influence policies.

FTA Formula Funding: FTA formula funding includes a number of programs and grants that funds public transit, and that are distributed according to a pre-specified formula, e.g. on the basis of population such as Urbanized Area Formula Grants program (49 U.S.C. 5307). Examples of other formula programs: https://www.transit.dot.gov/grant-type/formula.

General Purpose Governments: Organized local governments that provide more than one service such as counties, parishes, boroughs, municipalities and townships and have sufficient administrative and fiscal autonomy.

Great Recession: A world-wide economic decline during the late 2000s-early 2010s. The most significant financial downturn since the Great Depression.

Special Purpose Governments: Typically authorized by state law to provide a single service and/or work in a designated functional area of public policy such as transportation districts, school districts and utility districts. The agency also has sufficient administrative and fiscal autonomy to operate as a separate local government.

Transit Contracting Survey: A bi-annual survey conducted independently. The survey collects information on the key transit characteristics, with the focus on contracting out. The perceptions of performance measurement and contracting out are included as well.

Vehicles Operated in Maximum Service (VOMS): VOMS stands for a fleet size or number of vehicles that an agency has to support for the peak operations.

ENDNOTES

1 Agencies are required to report to the NTD if they receive FTA formula funding. In addition, agencies that have received federal funding must report for the next six years or as long as they operate capital assets purchased with the FTA funding.

Compilation of References

Adler, N., & Liebert, V. (2014). Joint impact of competition, ownership form and economic regulation on airport performance and pricing. *Transportation Research Part A, Policy and Practice*, *64*, 92–109. doi:10.1016/j.tra.2014.03.008

Afonso, W. (2018). *Modal Equity of Transportation Expenditures: The Impact of Collaborative Governance and the Use of Earmarked Local Sales Taxes*. Paper presented at the Symposium on implementing collaborative governance: Models, experiences and challenges to foster policy coordination, and to enhance sustainable community outcomes and public value generation conference, University of Palermo, Italy.

Afonso, W. (2013). Diversification toward Stability? The Effect of Local Sales Taxes on Own Source Revenue. *Journal of Public Budgeting, Accounting & Financial Management*, *25*(4), 649–674. doi:10.1108/JPBAFM-25-04-2013-B004

Afonso, W. (2014a). Local Sales Taxes as a Means of Increasing Revenues and Reducing Property Tax Burdens: An Analysis Using Propensity Score Matching. *Public Budgeting & Finance*, *34*(2), 24–43. doi:10.1111/pbaf.12039

Afonso, W. (2014b). "Fiscal Illusion in State and Local Finances: A Hindrance to Transparency." 2014. *State & Local Government Review*, *46*(3), 219–228. doi:10.1177/0160323X14550103

Afonso, W. (2017). State LST Laws: A Comprehensive Analysis of the Laws Governing Local Sales Taxes. *Public Budgeting & Finance*, *37*(4), 25–46. doi:10.1111/pbaf.12171

Afonso, W. B. (2015a). Leviathan or flypaper: Examining the fungibility of earmarked local sales taxes for transportation. *Public Budgeting & Finance*, *35*(3), 1–23. doi:10.1111/pbaf.12072

Afonso, W. B. (2015b). LOST and Found Tax Dollars: Local Option Sales Taxes, Property Taxes, and Own Source Revenue. *Journal of Public Budgeting, Accounting and Financial Management*, *27*(3), 318–351. doi:10.1108/JPBAFM-27-03-2015-B002

Afonso, W. B. (2016a). Time to Adoption of Local Option Sales Taxes: An Examination of Texas Municipalities. *Public Finance Review*. doi:10.1177/1091142116673147

Afonso, W. B. (2016b). The Equity of Local Sales Tax Distributions in Urban, Suburban, Rural, and Tourism-Rich Counties in North Carolina. *Public Finance Review*, *44*(6), 691–721. doi:10.1177/1091142115588976

Ahadzi, M., & Bowles, G. (2004). Public–private partnerships and contract negotiations: An empirical study. *Construction Management and Economics*, *22*(9), 967–978. doi:10.1080/0144619042000241471

Albalate, D., Bel, G., & Fageda, X. (2013). Joint versus single management of large transport infrastructures. *Ocean and Coastal Management*, *71*, 163–169. doi:10.1016/j.ocecoaman.2012.09.014

Albrecht, M., Brown, A., Lederman, J., Taylor, B., & Wachs, M. (2017). *The Equity Challenges and Outcomes of California County Transportation Sales Taxes. Report*. Center of Economic Competitiveness, University of California.

Altshuler, A. A., & Luberoff, D. E. (2004). *Mega-projects: The changing politics of urban public investment*. Washington, DC: Brookings Institution Press.

Amended And Restated Comprehensive Agreement Relating To The Route 495 Hot Lanes in Virginia Project, Dated as of December 19, 2007 by and among Virginia Department of Transportation, an Agency of the Commonwealth of Virginia and Capital Beltway Expressway LLC, a Delaware limited liability company. (2007). Retrieved from http://www.p3virginia.org/wp-content/uploads/2017/02/1.-ARCA-Execution-Copy.pdf

American Association of Port Authorities (AAPT). (n.d.). *Exports, jobs, and economic growth*. Available online at http://www.aapa-ports.org/advocating/content.aspx?ItemNumber=21150

American Association of State Highway and Transportation Officials. (2016). *Transportation Finance Governance and Finance*. Retrieved from www.financingtransportation.org/pdf/50_state_review_nov16.pdf

American Export Lines. (2017, May 8). *Understanding the 3 New Ocean Carrier Shipping Alliances | Learn How the New Carrier Alliances Affect Your Ocean Freight Contracts and/or Shipments*. Retrieved March 12, 2018, from https://www.shipit.com/archives/2017/05/08/understanding-the-3-new-ocean-carrier-shipping-alliances/

American Public Transit Association (APTA). (2017). *2016 Public Transportation Fact Book*. Available online at: http://www.apta.com/resources/statistics/Documents/FactBook/2016-APTA-Fact-Book.pdf

American Public Transit Association. (2011). *Public Transportation Factbook*. Available online at http://www.apta.com/resources/statistics/Documents/FactBook/APTA_2011_Fact_Book.pdf

American Public Transportation Association. (2018). *Mobility management*. Retrieved from http://www.apta.com/resources/mobility/Pages/default.aspx

American Public Transportation Association. (n.d.). *Public Transportation Use Is Growing—Here Are the Facts.* Retrieved from http://www.apta.com/mediacenter/ptbenefits/Pages/Public-Transportation-Use-is-Growing-.aspx

American Society of Civil Engineers (ASCE). (2017). *2017 Infrastructure Report Card.* Available online at https://www.infrastructurereportcard.org/americas-grades/

American Society of Civil Engineers. (2016). *Failure to Act: Closing the Infrastructure Investment Gap for America's Economic Future.* Retrieved from https://www.infrastructurereportcard.org/wp-content/uploads/2016/05/2016-FTA-Report-Close-the-Gap.pdf

American Society of Civil Engineers. (2017). *2017 Infrastructure Report Card.* Retrieved from https://www.infrastructurereportcard.org/

American Society of Civil Engineers. (2017). *2017 Infrastructure Report Card: Roads.* Retrieved from https://www.infrastructurereportcard.org/wp-content/uploads/2017/01/Roads-Final.pdf

Andersen, M. (2014). How Denver Got an Oil Company to Help Crowdfund a Protected Bike Lane. *People for Bikes.* Retrieved from http://www.peopleforbikes.org/blog/entry/how-denver-got-an-oil-company-to-help-crowdfund-a-protected-bike-lane?utm_source=twitterfeed&utm_medium=twitter

Anderson, M. (2013). *What Caused Portland's Biking Boom?* Retrieved from http://bikeportland.org/2013/07/02/what-caused-portlands-biking-boom-89491

Anderson, J. M., Nidhi, K., Stanley, K. D., Sorensen, P., Samaras, C., & Oluwatola, O. A. (2014). *Autonomous Vehicle Technology: A Guide for Policymakers.* Santa Monica, CA: Rand Corporation. Retrieved from https://www.rand.org/pubs/research_reports/RR443-2.html

Arentze, T., Hofman, F., & Timmermans, H. (2004). Predicting multi-faceted activity-travel adjustment strategies in response to possible congestion pricing scenarios using an Internet-based stated adaptation experiment. *Transport Policy, 11*(1), 31–41. doi:10.1016/S0967-070X(03)00016-7

Arizona State Treasurer. (2018). *Revenue Distributions List.* Retrieved from https://aztreasury.gov/local-govt/revenue-distributions/revenue-distributions-list/

Assaf, A. G., & Gillen, D. (2012). Measuring the joint impact of governance form and economic regulation on airport efficiency. *European Journal of Operational Research, 220*(1), 187–198. doi:10.1016/j.ejor.2012.01.038

Bacot, H. (2008). Civic culture as a policy premise: Appraising Charlotte's civic culture. *Journal of Urban Affairs, 30*(4), 389–417. doi:10.1111/j.1467-9906.2008.00405.x

Bacot, H., & Christine, J. (2006). What's so "Special" about airport authorities? Assessing the administrative structure of US airports. *Public Administration Review, 66*(2), 241–251. doi:10.1111/j.1540-6210.2006.00576.x

Barrett, M. (2011, August 6). Shouting match may derail city, airport deal-Fire station at airport *Asheville Citizen Times.*

Barrett, M. (2012, August 17). Airport board blocks mayor *Asheville Citizen Times.*

Bartle, J. R., & Chen, C. (2014). Future Issues in State Transportation Finance. In Sustaining the States: The Fiscal Viability of American State Governments (pp. 211-233). New York: Taylor & Francis Group Press. doi:10.1201/b17267-12

Bartle, J. (2006). The sustainable development of U.S. air transportation: The promise and challenge of institutional reform. *Public Works Management & Policy, 10*(3), 214–224. doi:10.1177/1087724X06287494

Bartle, J. R., & Devan, J. (2006). Sustainable highways: Destination or mirage? *Public Works Management & Policy, 10*(3), 225–234. doi:10.1177/1087724X06287493

Beadle, A. D. (2004, June 27). *Is there life to chassis legislation?* Retrieved March 3, 2018, from https://www.joc.com/trucking-logistics/there-life-chassis-legislation_20040627.html

Bennett, J., & Iossa, E. (2006). Building and managing facilities for public services. *Journal of Public Economics, 90*(10–11), 2143–2160. doi:10.1016/j.jpubeco.2006.04.001

Black, W. R. (2010). *Sustainable transportation: problems and solutions.* Guilford Press.

Black, W. R. (2010). *Sustainable transportation: Problems and solutions.* Guilford Press.

Blackwell, C., Crotts, J. C., Litvin, S. W., & Styles, A. K. (2006). Local Government Compliance with Earmarked Tax Regulation. *Public Finance Review, 34*(2), 212–228. doi:10.1177/1091142105284213

Blair, R., & Janousek, C. (2013). Collaborative mechanisms in interlocal cooperation: A longitudinal examination. *State & Local Government Review, 45*(4), 268–282. doi:10.1177/0160323X13511647

Boarnet, M. G., & Dimento, J. F. (2004). The private sector's role in highway finance: Lessons from SR 91. *Access, 25*, 26–31.

Bonney, J. (2012a, January 16). *"Open Pools" for Chassis.* Retrieved March 14, 2018, from https://www.joc.com/maritime-news/open-pools-chassis_20120116.html

Bonney, J. (2012b, April 23). *Fixing Broken Chassis System Proves Elusive.* Retrieved May 22, 2018, from https://www.joc.com/maritime-news/international-freight-shipping/fixing-broken-chassis-system-proves-elusive_20120423.html

Bonney, J. (2014, September 23). *DOJ won't challenge LA-LB 'gray chassis' plan.* Retrieved November 21, 2017, from https://www.joc.com/port-news/us-ports/port-los-angeles/doj-won%E2%80%99t-challenge-la-lb-%E2%80%98gray-chassis%E2%80%99-plan_20140923.html

Bonney, J. (2015, December 7). *NY-NJ 'gray' chassis pool awaits deal with ILA*. Retrieved October 5, 2017, from https://www.joc.com/port-news/us-ports/port-new-york-and-new-jersey/ny-nj-%E2%80%98gray%E2%80%99-chassis-pool-awaits-deal-ila_20151207.html

Bonney, J. (2016a, November 23). *Lessors: THE Alliance members shouldn't jointly buy equipment*. Retrieved March 12, 2018, from https://www.joc.com/maritime-news/container-lines/alliance/lessors-alliance-members-shouldnt-jointly-buy-equipment_20161123.html

Bonney, J. (2016b, November 29). *FMC's Doyle warns of alliance's contracting power*. Retrieved March 12, 2018, from https://www.joc.com/regulation-policy/transportation-regulations/us-transportation-regulations/fmc-doyle-warns-alliance-contracting-power_20161129.html

Bonney, J. (2017, December 4). *Trucker-led pool adds chassis, plans more expansion*. Retrieved March 7, 2018, from https://www.joc.com/trucking-logistics/trucking-equipment/trucker-led-pool-adds-chassis-plans-more-expansion_20171204.html

Brennan, G., & Buchanan, J. (1980). *The Power to Tax: Analytical Foundations of a Fiscal Constitution*. Cambridge, UK: Cambridge University Press.

Broaddus, A., Browne, M., & Allen, J. (2015). Sustainable Freight: Impacts of the London Congestion Charge and Low Emissions Zone. In *Transportation Research Board Annual Conference*. Retrieved from http://andreabroaddus.com/wp-content/uploads/2015/01/Impacts-of-Sust-Transport-Policies-on-Freight-TRB-2015.pdf

Broaddus, A. (2014). Sustainable Transportation: Lessons from London. *Focus: Journal of the City and Regional Planning Department*, *11*(1), 10.

Brock, T. J., & Souleyrette, R. R. (2013). *An Overview of US Commuter Rail*. Lexington, KY: University of Kentucky, Kentucky Transportation Center.

Brock, T. J., & Souleyrette, R. R. (2014). Commuter Rail. In M. Garrett (Ed.), *The Encyclopedia of Transportation: Social Science and Policy* (pp. 392–394). Thousand Oaks, CA: Sage.

Brown, D., Cabbage, M., McCarthy, L., & Norton, K. (2016). *NASA, NOAA Analyses Reveal Record-Shattering Global Warm Temperatures in 2015*. Press Release. Available at http://www. nasa. gov/press-release/nasanoaa-analyses-reveal-record-shattering-global-warm-temperatures-in-2015

Brownstone, D., Ghosh, A., Golob, T. F., Kazimi, C., & Van Amelsfort, D. (2003). Drivers' willingness-to-pay to reduce travel time: Evidence from the San Diego I-15 congestion pricing project. *Transportation Research Part A, Policy and Practice*, *37*(4), 373–387. doi:10.1016/S0965-8564(02)00021-6

Brown, T., & Potoski, M. (2005, March). Transaction Costs and Contracting: The Practitioner Perspective. *Public Performance & Management Review*, *28*(3), 326–351.

Brundtland, G. H. (1987). World Commission on Environment and Development. *Our Common Future*, 8–9.

Brunori, D. (2007). *Local Tax Policy: A Federalist Perspective* (2nd ed.). Washington, DC: Urban Institute Press.

Bureau of Transportation Statistics (BTS). (2015). *Freight Facts and Figures 2015*. Available online at https://www.transtats.bts.gov/TRAFFIC/

Bureau of Transportation Statistics (BTS). (2017). *US Carrier Traffic Statistics through May 2017*. Available online at https://www.transtats.bts.gov/TRAFFIC/

Bureau of Transportation Statistics. (1999). *National Transit Database (NTD) and Safety Management Information System (SAMIS)*. Available online at http://www.bts.gov/programs/statistical_policy_and_research/source_and_accuracy_compendium/FTA_national_transit.html

Bureau of Transportation Statistics. (2017, October). *August 2017 North American Freight Numbers | Bureau of Transportation Statistics*. Retrieved October 30, 2017, from https://www.bts.gov/newsroom/august-2017-north-american-freight-numbers

Burge, G., & Rogers, C. (2011). Local option sales taxes and consumer spending patterns: Fiscal interdependence under multi-tiered local taxation. *Regional Science and Urban Economics*, *41*(1), 46–58. doi:10.1016/j.regsciurbeco.2010.08.001

Burgess, J. (2011, June 16). No action likely on airport *Asheville Citizen Times*.

Burgess, J. (2012, June 21). City set to lose airport control *Asheville Citizen Times*.

Burkhardt, J. E., Koffman, D., & Murray, G. (2003). *Economic benefits of coordinating human service transportation and transit services*. Transportation Research Board of the National Academies. Retrieved from http://onlinepubs.trb.org/onlinepubs/tcrp/tcrp_rpt_91.pdf

Burkhardt, J., & McLary, J. (n.d.). *The business case for mobility management*. American Public Transportation Association. Retrieved from https://www.apta.com/resources/mobility/Documents/Business-Case-for-Mobility-Management.pdf

Burns, N. (1994). *The Formation of American Local Governments: Private Values in Public Institutions*. New York: Oxford University Press.

Callahan, R., & Pisano, M. (2014). Aligning Fiscal and Environmental Sustainability. In D. Mazmanian & H. Blanco (Eds.), The Elgar Companion to Sustainable Cities: Strategies, Methods and Outlook (pp. 154–165). Academic Press.

Callahan, R., M. Pisano, & A. Linder. (2010). Leadership and Strategy: A Comparison of the Outcomes and Institutional Designs of the Alameda Corridor and Alameda Corridor East Projects. *Public Works Policy and Management*, (14), 263-287.

Callahan, R. (2007). Governance: The Collision of Politics and Cooperation. *Public Administration Review*, (67), 299–301.

Callahan, R., & Pisano, M. (n.d.). Fiscal Sustainability, Demographics, and the Social Determinants of Health Driving Intergenerational Equity. In D. L. S. María (Ed.), *Financial Sustainability and Intergenerational Equity in Local Governments. IGI Global*. doi:10.4018/978-1-5225-3713-7.ch006

Carney, M., & Mew, K. (2003). Airport governance reform: A strategic management perspective. *Journal of Air Transport Management*, *9*(4), 221–232. doi:10.1016/S0969-6997(03)00003-6

Carroll, D. A. (2009). Diversifying Municipal Government Revenue Structures: Fiscal Illusion or Instability? *Public Budgeting & Finance*, *29*(1), 27–48. doi:10.1111/j.1540-5850.2009.00922.x

Carroll, D. A., & Stater, K. (2009). Revenue Diversification in Nonprofit Organizations: Does It Lead to Financial Stability? *Journal of Public Administration: Research and Theory*, *19*(4), 947–966. doi:10.1093/jopart/mun025

Center for Transportation Excellence. (2018). *Transportation Ballot Measures*. Retrieved from http://www.cfte.org/elections

Cervero, R. (2007). Models for Change: Lessons for Creating Active Living Communities. *Planning Magazine*, 1.

Cervero, R. (2012). *Integrating Transit and Urban Development: Lessons and Challenges for Developing Countries*. Presented at the Urban Transport Pathways: Do the Experiences of the US and the EU Offer Lessons for the Developing World? Berkeley, CA.

Cervero, R., & Duncan, M. (2003). Walking, Bicycling, and Urban Landscapes: Evidence from the San Francisco Bay Area. *American Journal of Public Health*, *93*(9), 1478–1483. doi:10.2105/AJPH.93.9.1478 PMID:12948966

Cervero, R., & Kockelman, K. (1997). Travel Demand and the 3Ds: Density, Diversity, and Design. *Transportation Research Part D, Transport and Environment*, *2*(3), 199–219. doi:10.1016/S1361-9209(97)00009-6

Chapman, J. I. (2008). State and local fiscal sustainability: The challenges. *Public Administration Review*, *68*, S115–S131. doi:10.1111/j.1540-6210.2008.00983.x

Chatman, D. G. (2013). Does TOD need the T? On the importance of factors other than rail access. *Journal of the American Planning Association*, *79*(1), 17–31. doi:10.1080/01944363.2013.791008

Cheetah Chassis. (n.d.). *Standard international chassis image*. Retrieved from http://www.cheetahchassis.com/media/1189/blue-20-40-45-tandem-ps.jpg

Chen, C., & Afonso, W. (2017). *A First Look at the Time to Adoption of Local Option Fuel Taxes: Evidence from Florida Counties*. Paper presented at the Association of Budgeting and Financial Management conference, Washington, DC.

Chen, C. (2014). Measuring state highway sustainability: Taking the fiscal dimension into account. *Public Works Management & Policy*, *19*(3), 255–276. doi:10.1177/1087724X14528475

Chen, Z., Daito, N., & Gifford, J. L. (2014). Do State Fiscal Constraints Affect Implementation of Highway Public-Private Partnerships? A Panel Fixed Logit Assessment. *Journal of the Transportation Research Forum*, *53*(2), 111–128.

Chen, Z., & Haynes, K. (2015). Regional Impact of Public Transportation Infrastructure: A Spatial Panel Assessment of the U.S. Northeast Megaregion. *Economic Development Quarterly*, *29*(3), 275–291. doi:10.1177/0891242415584436

Chesley, R. (2018, February 28). Settling-out process reduces howls about NOVA tolls -- and maybe on I-64, too. *The Virginian-Pilot.* Retrieved from https://pilotonline.com/news/local/columnist/roger-chesley/article_398fa1b9-1e64-5c3e-917f-dc12ab400039.html

Christopher Newport University Judy Ford Wason Center for Public Policy. (2014a). *South Hampton Roads Midtown and Downtown Tunnels Tolls Survey Part I: Pre-Tolling Report and Findings.* Retrieved from http://www.hrtpo.org/uploads/docs/South%20Hampton%20Roads%20Midtown%20and%20Downtown%20Tunnels%20Tolls%20Survey-Part%20II.pdf

Christopher Newport University Judy Ford Wason Center for Public Policy. (2014b). *South Hampton Roads Midtown and Downtown Tunnels Tolls Survey Part II: Post-Tolling Report and Findings.* Retrieved from http://www.hrtpo.org/uploads/docs/South%20Hampton%20Roads%20Midtown%20and%20Downtown%20Tunnels%20Tolls%20Survey-Part%20II.pdf

Chung, D., Hensher, D. A., & Rose, J. M. (2010). Toward the betterment of risk allocation: Investigating risk perceptions of Australian stakeholder groups to public–private-partnership tollroad projects. *Research in Transportation Economics*, *30*(1), 43–58. doi:10.1016/j.retrec.2010.10.007

City and County of Denver. (2014). *Transit Oriented Denver: Transit Oriented development Strategic Plan.* Retrieved from www.denvergov.org/content/dam/denvergov/portals/193/documents/tod_plan/Tod_strategic_plan_final.pdf

City of Portland. (2018). *Shared Electric Scooter Pilot.* Retrieved from https://www.portlandoregon.gov/transportation/77294

City of San Francisco. (2013). *San Francisco Parklet Manual.* Retrieved from http://pavementtoparks.sfplanning.org/docs/SF_P2P_Parklet_Manual_1.0_FULL.pdf

Clewlow, R. R., & Mishra, G. S. (2017). Disruptive Transportation: The Adoption, Utilization, and Impacts of Ride-Hailing in the United States. In *Research Report UCD-ITS-RR-17-07.* Davis, CA: University of California, Davis, Institute of Transportation Studies. Retrieved from https://itspubs.ucdavis.edu/wp-content/themes/ucdavis/pubs/download_pdf.php?id=2752

Comprehensive Agreement Relating To The I-95 HOV/HOT Lanes Project, Dated as of July 31, 2012 by and between Virginia Department of Transportation, an Agency of The Commonwealth of Virginia and 95 Express Lanes LLC, a Delaware limited liability company. (2012). Retrieved from http://www.virginiadot.org/projects/resources/NorthernVirginia/Express_Lanes_Comprehensive_Agreement.pdf

Connors, P. (2017, November 13). *Shippers need "open choice" for chassis*. Retrieved March 1, 2018, from https://www.joc.com/port-news/port-equipment/evolution-chassis-business-model-must-continue_20171113.html

Cook, M. (2014). Toll Poll: Voters opposed to 241 extension. *Orange County Register*. Retrieved from https://www.ocregister.com/2014/09/20/toll-poll-voters-opposed-to-241-extension/

Copeland, L., & Overberg, P. (2012, December 4). *Controversial HOT lanes spread nationally*. Retrieved from https://www.usatoday.com/story/news/nation/2012/12/04/controversial-hot-lanes-spread-nationally/1747319/

Corburn, J. (2007). Reconnecting with Our Roots: American Urban Planning and Public Health in the Twenty-First Century. *Urban Affairs Review*, *42*(5), 688–713. doi:10.1177/1078087406296390

Cortright, J. (2014, October). *City Report: The Young and Restless and the Nation's Cities*. Retrieved from http://cityobservatory.org/wp-content/uploads/2014/10/YNR-Report-Final.pdf

Coté, M., & Cabanatuan, J. (2013). Task Force: Billions Needed to Fix Transit in S.F. *SFGate*. Retrieved from http://www.sfgate.com/bayarea/article/Task-force-billions-needed-to-fix-transit-in-S-F-5008878.php#src=fb

Cox, W., & Mundle, S. (1997). *Denver Urban Transport Competitive Tendering Saves $88 Million*. Retrieved from http:/www.publicpurpose.com.ut-dencc.htm

Crabbe, A., Hiatt, R., Poliwka, S., & Wachs, M. (2005). Local Transportation Sales Taxes: California's Experiment in Transportation Finance. *Public Budgeting & Finance*, *25*(3), 91–121. doi:10.1111/j.1540-5850.2005.00369.x

Cradock, A. L., Troped, P. J., Fields, B., Melly, S. J., Simms, S. V., Gimmler, F., & Fowler, M. (2009). Factors Associated with Federal Transportation Funding for Local Pedestrian and Bicycle Programming and Facilities. *Journal of Public Health Policy*, *30*(S1), S38–S72. doi:10.1057/jphp.2008.60 PMID:19190583

Craig, S. G., Airola, J., & Tipu, M. (2012). General purpose or special district governance? Technical efficiency versus rent dissipation in airport finances. *Public Finance Review*, *40*(6), 712–735. doi:10.1177/1091142112448415

Crowley, G., & Hoffer, A. (2012). Dedicating Tax Revenue: Constraining Government or Masking Its Growth? Working Paper. Mercatus Center at George Mason University, Arlington, VA.

Dadayan, L., & Ward, R. B. (2009). State and local finance: Increasing focus on fiscal sustainability. *Publius*, *39*(3), 455–475. doi:10.1093/publius/pjp014

Daito, N., Chen, Z., Gifford, J. L., Porter, T., & Gudgel, J. E. (2013). Implementing public private partnerships during challenging economic times: Case study of the 495 Express Lanes on the Virginia portion of the Washington Capital Beltway Project (USA). *Case Studies on Transport Policy*, *1*(1), 35–45. doi:10.1016/j.cstp.2013.07.001

Dalton, P. (2016). *Local Sales Taxes in Minnesota*. Information brief. Research Department, Minnesota House of Representatives.

DCLI. (2017). *DCLI to Purchase TRAC Intermodal's Domestic Chassis Fleet | Direct ChassisLink Inc*. Retrieved April 19, 2018, from https://dcli.com/about/news/dcli-to-purchase-trac-intermodals-domestic-chassis-fleet/

de Zeeuw, D., & Flusche, D. (2011). How a Bill Becomes a Bike Lane: Federal Legislation, Programs, and Requirements of Bicycling and Walking Projects. *Planning & Environmental Law*, *63*(8), 8–11. doi:10.1080/15480755.2011.604227

Dewatripont, M., & Legros, P. (2005). Public-private partnerships: Contract design and risk transfer. *EIB Papers*, *10*(1), 120-145.

Diaz, R. B. (1999). Impacts of Rail Transit on Property Values. In *American Public Transit Association Rapid Transit Conference Proceedings*. Retrieved from http://www.rtd-fastracks.com/media/uploads/nm/impacts_of_rail_transif_on_property_values.pdf

Dillman, D. A. (2000). *Mail and Internet surveys: The tailored design method*. New York: Wiley and Sons, Inc.

Dowall, D. E., & Landis, J. D. (1982). Land-Use Controls and Housing Costs: An Examination of San Francisco Bay Area Communities. *Real Estate Economics*, *10*(1), 67–93. doi:10.1111/1540-6229.00258

Dunham-Jones, E., & Williamson, J. (2011). *Retrofitting Suburbia, Updated Edition: Urban Design Solutions for Redesigning Suburbs*. John Wiley & Sons. Retrieved from http://books.google.com/books?hl=en&lr=&id=1xH4b4pQzOkC&oi=fnd&pg=PR6&dq=Dunham+2011+complete+streets&ots=VQxXk-4eTP&sig=-j4ZvAz2R8Rk_Bsvugby_qqQmQg

Dunham, M. (2011). Where the Shoe Leather Meets the Road: Learning From Experience in Crafting a Complete Streets Ordinance. *Planning & Environmental Law*, *63*(8), 3–8. doi:10.1080/15480755.2011.604226

Dunn, J. (2018). Sales Tax By State: Are Grocery Items Taxable? *TaxJar*. Retrieved from https://blog.taxjar.com/states-grocery-items-tax-exempt/

Dyble, L. N. (2011). Tolls and control: The Chicago Skyway and the Pennsylvania Turnpike. *Journal of Planning History*.

Dye, R., & McGuire, T. (1992). The Effect of Earmarked Revenues on the Level and Composition of Expenditures. *Public Finance Review*, *20*(4), 543–556. doi:10.1177/109114219202000410

Eger, R. J., & McDonald, B. D. (2017). Cost Accounting for Government Grants. In Z. T. Mohr (Ed.), *Cost Accounting in Government: Theory and Applications* (pp. 81–99). New York, NY: Routledge. doi:10.4324/9781315648897-5

Egilmez, G., & Tatari, O. (2012). A dynamic modeling approach to highway sustainability: Strategies to reduce overall impact. *Transportation Research Part A, Policy and Practice*, *46*(7), 1086–1096. doi:10.1016/j.tra.2012.04.011

Ehrenhalt, A. (2012). *The great inversion and the future of the American city*. New York: Knopf Doubleday Publishing Group.

El-Gohary, N. M., Osman, H., & El-Diraby, T. E. (2006). Stakeholder management for public private partnerships. *International Journal of Project Management*, *24*(7), 595–604. doi:10.1016/j.ijproman.2006.07.009

Elizabeth River Tunnels. (2018). *Toll Rates Information*. Retrieved from https://www.driveert.com/toll-info/toll-rates/

Elkind, E. N. (2014). *Railtown: The Fight for the Los Angeles Metro Rail and the Future of the City*. Los Angeles, CA: University of California Press.

Engel, E., Fischer, R., & Galetovic, A. (2013). The basic public finance of public–private partnerships. *Journal of the European Economic Association*, *11*(1), 83–111. doi:10.1111/j.1542-4774.2012.01105.x

Environment, Department of. (1993). *Reducing Transport Emissions Through Planning*. London: HMSO.

Estache, A., Juan, E., & Trujillo, L. (2007). *Public-private partnerships in transport*. Policy Research Working Paper 4436. World Bank. Retrieved from https://openknowledge.worldbank.org/bitstream/handle/10986/7602/wps4436.pdf

Ewing, R. (2005). Can the Physical Environment Determine Physical Activity Levels? *Exercise and Sport Sciences Reviews*, *33*(2), 69–75. doi:10.1097/00003677-200504000-00003 PMID:15821427

Ewing, R., Schmid, T., Killingsworth, R., Zlot, A., & Raudenbush, S. (2003). Relationship between Urban Sprawl and Physical Activity, Obesity, and Morbidity. *Urban Ecology*, 567–582. PMID:13677962

Fagnant, D. J., & Kockelman, K. M. (2014). The Travel and Environmental Implications of Shared Autonomous Vehicles, Using Agent-Based Model Scenarios. *Transportation Research Part C, Emerging Technologies*, *40*, 1–13. doi:10.1016/j.trc.2013.12.001

Farmer, L. (2018, May). How cities fell out of love with sports stadiums. *Governing*, 24–29.

Federal Highway Administration (FHWA). (2017). *Fixing America's Surface Transportation Act or "FAST Act"*. Available online at https://www.fhwa.dot.gov/fastact/

Federal Highway Administration. (2011). *Summary of travel trends: 2009 National Household Travel Survey*. Washington, DC: US Department of Transportation.

Federal Highway Administration. (2017, February 1). *FHWA Freight and Land Use Handbook: Section 3.0 - FHWA Freight Management and Operations*. Retrieved September 19, 2018, from https://ops.fhwa.dot.gov/publications/fhwahop12006/sec_3.htm

Federal Maritime Commission Bureau of Trade Analysis. (2015). *US; Container Port Congestion & Related International Supply Chain Issues: Causes, Consequences & Challenges*. Retrieved from http://www.fmc.gov/assets/1/Page/PortForumReport_FINALwebAll.pdf

Fiorino, D. (2010). Sustainability as a Conceptual Focus for Public Administration. *Public Administration Review*, *70*(68), S78–S88. doi:10.1111/j.1540-6210.2010.02249.x

FlightGlobal. (2017). *World Airline Rankings 2017*. Retrieved from https://www.flightglobal.com/asset/18223

Florida, R., & Adler, P. (2018). The patchwork metropolis: The morphology of the divided postindustrial city. *Journal of Urban Affairs*, *40*(5), 609–624. doi:10.1080/07352166.2017.1360743

Forkenbrock, D. J. (2006). Financing local roads: Current problems and new paradigms. *Transportation Research Record: Journal of the Transportation Research Board*, (1960): 8–14.

Forrer, J., Kee, J. E., Newcomer, K. E., & Boyer, E. (2010). Public-Private Partnerships and the Public Accountability Question. *Public Administration Review*, *70*(3), 475–484. doi:10.1111/j.1540-6210.2010.02161.x

Forster, D. (2015, July 5). Drivers flock to Uber-style operations. *The Virginian-Pilot*. Retrieved from https://pilotonline.com/news/local/transportation/article_e5e2d0be-af03-5abc-abcd-5e42b3dd97d6.html

Forsyth, A., Hearst, M., Oakes, J. M., & Schmitz, K. H. (2008). Design and Destinations: Factors Influencing Walking and Total Physical Activity. *Urban Studies (Edinburgh, Scotland)*, *45*(9), 1973–1996. doi:10.1177/0042098008093386

Francsics, J., & Ingrey, M. (2000). *The EUROTOLL Project: Road user responses to transport demand management*. Paper presented at the Tenth International Conference on Road Transport Information and Control. 10.1049/cp:20000117

Frank, L. D., Andresen, M. A., & Schmid, T. L. (2004). Obesity Relationships with Community Design, Physical Activity, and Time Spent in Cars. *American Journal of Preventive Medicine*, *27*(2), 87–96. doi:10.1016/j.amepre.2004.04.011 PMID:15261894

Frank, L. D., Schmid, T. L., Sallis, J. F., Chapman, J., & Saelens, B. E. (2005). Linking Objectively Measured Physical Activity with Objectively Measured Urban Form: Findings from SMARTRAQ. *American Journal of Preventive Medicine*, *28*(2), 117–125. doi:10.1016/j.amepre.2004.11.001 PMID:15694519

Frank, L., Engelke, P., & Schmid, T. (2003). *Health and community design: The impact of the built environment on physical activity*. Washington, DC: Island Press.

Frank, L., Kerr, J., Chapman, J., & Sallis, J. (2007). Urban Form Relationships with Walk Trip Frequency and Distance among Youth. *American Journal of Health Promotion, 21*(4Suppl), 1–8. PMID:17465175

Frazier, E. (2014, November 14). Harris: Costs more critical than control *Charlotte Observer.*

Frederick, C., Riggs, W., & Gilderbloom, J. (2016). Multi-Modality and Public Health: A Multivariate Analysis of 148 U.S. Cities. Academic Press.

Frey, B. B., Lohmeier, J. H., Lee, S. W., & Tollefson, N. (2006). Measuring collaboration among grant partners. *The American Journal of Evaluation, 27*(3), 383–392. doi:10.1177/1098214006290356

Ganning, J., Beaudoin, M., Brewer, S., Kim, K., & Park, K. (2016, February 28). The Effects of Commuter Rail on Population Deconcentration and Commuting: A Salt Lake City Case Study. Portland, OR: National Institute for Transportation and Communities Report.

Garrett, M., & Taylor, B. (1999). Reconsidering social equity in public transit. *Berkeley Planning Journal, 13*(1), 6–27.

Garvin, A. (1996). *The American city: What works, what doesn't.* New York: McGraw-Hill.

Gen, S., & Kingsley, G. (2007). Effects of Contracting Out Engineering Services Over Time in a State Department of Transportation. *Public Works Management & Policy, 12*(1), 331–343. doi:10.1177/1087724X07302585

George, A., & Bennett, A. (2005). *Case studies and theory development in the social sciences.* Cambridge, MA: MIT Press.

Gerring, J. (2007). Is there a (viable) crucial-case method? *Comparative Political Studies, 40*(3), 231–253. doi:10.1177/0010414006290784

Gilderbloom, Riggs, & Meares. (2015). Does Walkability Matter? An Examination of Walkability's Impact on Housing Values, Foreclosures and Crime. *Cities, 42*(Part A), 13–24. doi:10.1016/j.cities.2014.08.001

Gillham, O. (2002). *The limitless city: A primer on the urban sprawl debate.* Washington, DC: Island Press.

Glaeser, E. L., Gottlieb, J. D., & Tobio, K. (2012). *Housing Booms and City Centers.* Retrieved from http://works.bepress.com/cgi/viewcontent.cgi?article=1007&context=joshua_gottlieb

Glaeser, E. (2011). *Triumph of the City: How Our Greatest Invention Makes Us Richer, Smarter, Greener, Healthier, and Happier.* New York: Penguin Press.

Glazier, R. H., Creatore, M. I., Weyman, J. T., Fazli, G., Matheson, F. I., Gozdyra, P., ... Booth, G. L. (2014). Density, Destinations or Both? A Comparison of Measures of Walkability in Relation to Transportation Behaviors, Obesity and Diabetes in Toronto, Canada. *PLoS One, 9*(1), e85295. doi:10.1371/journal.pone.0085295 PMID:24454837

Goldman, T., & Wachs, M. (2003). *A quiet revolution in transportation finance: The rise of local option transportation taxes*. University of California Transportation Center. Retrieved from https://escholarship.org/uc/item/2gp4m4xq

Gomez-Ibanez, J. A., & Meyer, J. (1993). *Going Private: The International Experience with Transport Privatization*. Washington, DC: Brookings Institution.

Gomez, J., Papanikolaou, A., & Vassallo, J. M. (2016). Measuring regional differences in users' perceptions towards interurban toll roads. *Journal of Transport Geography*, *54*, 22–33. doi:10.1016/j.jtrangeo.2016.05.001

Gray, B. (1989). *Collaborating: Finding common ground for multiparty problems*. San Francisco: Jossey-Bass Publishers.

Green, A. D., Neiman, M., Bockman, S., & Sirotnik, B. (2013). Public Support for Transportation Sales Taxes in California: A Two County Assessment. *California Journal of Politics and Policy*, *5*(4), 645–670. doi:10.5070/P2X886

Grossi, G., & Thomasson, A. (2015). Bridging the accountability gap in hybrid organizations: The case of Copenhagen Malmö Port. *International Review of Administrative Sciences*, *81*(3), 604–620. doi:10.1177/0020852314548151

Gross, M., & Garvin, M. (2011). Structuring PPP toll-road contracts to achieve public pricing objectives. *Engineering Project Organization Journal*, *1*(2), 143–156. doi:10.1080/21573727.2011.572256

Grout, P. A. (1997). The economics of the private finance initiative. *Oxford Review of Economic Policy*, *13*(4), 53–66. doi:10.1093/oxrep/13.4.53

Guerra, E. (2015). Planning for Cars That Drive Themselves Metropolitan Planning Organizations, Regional Transportation Plans, and Autonomous Vehicles. *Journal of Planning Education and Research*. doi:10.1177/0739456X15613591

Gurgun, A. P., & Touran, A. (2013). Public-private partnership experience in the international arena: Case of Turkey. *Journal of Management Engineering*, *30*(6), 04014029. doi:10.1061/(ASCE)ME.1943-5479.0000213

Hafner, K. (2018, January 10). HOT lanes on I-64 open today. *The Virginian-Pilot*. Retrieved from https://pilotonline.com/news/local/transportation/traffic/article_c7570800-6a65-575e-ae0d-8c2d8111b258.html

Hafner, K., & Pascale, J. (2018, January 5). HOT lanes are coming to I-64 Wednesday. Here's what you should know. *The Virginian-Pilot*. Retrieved from https://pilotonline.com/news/local/transportation/article_00b922b9-94e0-51b3-b50e-48c3270122c7.html

Hamideh, A., Oh, J. E., Labi, S., & Mannering, F. (2008). Public acceptance of local government transportation sales taxes: A statistical assessment. *State & Local Government Review*, *40*(3), 150–159. doi:10.1177/0160323X0804000302

Handy, S., Cao, X., & Mokhtarian, P. (2005). Correlation or Causality between the Built Environment and Travel Behavior? Evidence from Northern California. *Transportation Research Part D, Transport and Environment*, *10*(6), 427–444. doi:10.1016/j.trd.2005.05.002

Handy, S., Cao, X., & Mokhtarian, P. L. (2006). Self-Selection in the Relationship between the Built Environment and Walking: Empirical Evidence from Northern California. *Journal of the American Planning Association*, *72*(1), 55–74. doi:10.1080/01944360608976724

Handy, S., & McCann, B. (2010). The Regional Response to Federal Funding for Bicycle and Pedestrian Projects: An Exploratory Study. *Journal of the American Planning Association*, *77*(1), 23–38. doi:10.1080/01944363.2011.526537

Hannay, R., & Wachs, M. (2007). Factors influencing support for local transportation sales tax measures. *Transportation*, *34*(1), 17–35. doi:10.100711116-006-0006-4

Harb, M., Xiao, Y., Circella, G., Mokhtarian, P., & Walker, J. (2018). Projecting Travelers into a World of Self-Driving Cars: Naturalistic Experiment for Travel Behavior Implications. *Proceedings of the 97th Transportation Research Board.*

Harrison, S. (2013, December 22). Orr out, airport rift remains *Charlotte Observer.*

Harrison, S., Rothacker, R., & Frazier, E. (2014, November 15). Charlotte asks FAA to act on airport. *Charlotte Observer.*

Hart, O. (2003). Incomplete Contracts and Public Ownership: Remarks, and an Application to Public-Private Partnerships. *Economic Journal (London)*, *113*(486), C69–C76. doi:10.1111/1468-0297.00119

Hawaii Department of Taxation. (2018). *County Surcharge FAQs*. Retrieved from http://tax.hawaii.gov/geninfo/a2_b2_7csurchg_faq/

Heinrich, C. J., Hill, C., & Lynn, L. E., Jr. (2004). Governance as an organizing theme for empirical research. In *The Art of Governance: Analyzing Management and Administration* (pp. 3-19). London Financial Times.

Hendrick, R., & Crawford, J. (2014). Municipal Fiscal Policy Space and Fiscal Structure: Tools for Managing Spending Volatility. *Public Budgeting & Finance*, *34*(3), 24–50. doi:10.1111/pbaf.12042

Hillier, D. (2011a, July 10). It's time for an independent airport authority *Asheville Citizen Times.*

Hill, M., & Hupe, P. (Eds.). (2002). *Implementing Public Policy*. London: Sage Publications.

Hirsch, B. T., & Macpherson, D. A. (2003). Union Membership and Coverage Database from the Current Population Survey: Note. *Industrial & Labor Relations Review*, *56*(2), 349–354. doi:10.1177/001979390305600208

Hodge, G. A. (2004). The risky business of public–private partnerships. *Australian Journal of Public Administration*, *63*(4), 37–49. doi:10.1111/j.1467-8500.2004.00400.x

Hodge, G. A., & Greve, C. (2017). On public–private partnership performance: A contemporary review. *Public Works Management & Policy*, *22*(1), 55–78. doi:10.1177/1087724X16657830

Hodge, G., Boulot, E., Duffield, C., & Greve, C. (2017). After the ribbon cutting: Governing PPPs in the medium to long term. *Australian Journal of Public Administration*, *76*(3), 330–351. doi:10.1111/1467-8500.12239

Holeywell, R. (2013). Public. Private. Practical? *Governing*, *27*(2), 34–41.

Holtz-Eakin, D., & Wachs, M. (2011). *Strengthening connections between transportation investments and economic growth.* Bipartisan Policy Center National Transportation Policy Project. Retrieved from http://bipartisanpolicy.org/wp-content/uploads/sites/default/files/NTPP%20 Strengthening%20Connections%20Paper.pdf

Hu, F. B., Li, T. Y., Colditz, G. A., Willett, W. C., & Manson, J. A. E. (2003). Television Watching and Other Sedentary Behaviors in Relation to Risk of Obesity and Type 2 Diabetes Mellitus in Women. *Journal of the American Medical Association*, *289*(14), 1785. doi:10.1001/jama.289.14.1785 PMID:12684356

Hult, K. M., & Walcott, C. (1990). *Governing Public Organizations: Politics Structures, and Institutional Design*. Pacific Grove, CA: Brooks/Cole.

Hutchins, R. (2015, October 31). *Can Miami be a major gateway for Asian imports heading out of Florida?* Retrieved October 13, 2017, from https://www.joc.com/port-news/us-ports/port-miami/can-miami-be-major-gateway-asian-imports-heading-out-florida_20151031.html

Hutchins, R. (2017a, March 23). *Container lines defend against collusion suspicion.* Retrieved March 12, 2018, from https://www.joc.com/maritime-news/container-lines/container-lines-defend-against-collusion-suspicion_20170323.html

Hutchins, R. (2017b, May 3). *Fearing alliance collusion, Congress eyes US supplier protection.* Retrieved March 12, 2018, from https://www.joc.com/regulation-policy/transportation-regulations/us-transportation-regulations/fearing-alliance-collusion-us-congress-considers-oversight-expansion_20170503.html

Hutchins, R. (2017c, June 9). *Legislation would give FMC more alliance muscle and money.* Retrieved March 12, 2018, from https://www.joc.com/regulation-policy/transportation-regulations/us-transportation-regulations/legislation-would-give-fmc-more-alliance-muscle-and-money_20170609.html

Hwang, B.-G., Zhao, X., & Gay, M. J. S. (2013). Public private partnership projects in Singapore: Factors, critical risks and preferred risk allocation from the perspective of contractors. *International Journal of Project Management*, *31*(3), 424–433. doi:10.1016/j.ijproman.2012.08.003

Illinois Revenue. (2018). *Mass Transit District Taxes.* Retrieved from http://www.revenue.state.il.us/localgovernment/Overview/HowDisbursed/masstransit.htm

IllinoisD. O. T. (2015). *Transit system*. Retrieved from http://www.idot.illinois.gov/transportation-system/Network-Overview/transit-system/index

Indiana D. O. T. (2018a). *Rural Transit Assistance Programs*. Retrieved from http://www.in.gov/indot/2816.htm

Indiana D. O. T. (2018b). *Enhanced mobility of seniors and individuals with disabilities*. Retrieved from http://www.in.gov/indot/2817.htm

Iowa D. O. T. (n.d.). *Iowa mobility management*. Retrieved from https://iowadot.gov/iowamobilitymanagement/home

Jagers, S., Matti, S., & Nilsson, A. (2017). How exposure to policy tolls transforms the mechanisms behind public acceptability and acceptance - The case of the Gothenburg congestion tax. *International Journal of Sustainable Transportation*, *11*(2), 108–119. doi:10.1080/15568318.2016.1197348

Jang, K., Song, M. K., Choi, K., & Kim, D.-K. (2014). A bi-level framework for pricing of High-Occupancy Toll lanes. *Transport*, *29*(3), 317–325. doi:10.3846/16484142.2014.952248

Jeon, C. M., Amekudzi, A. A., & Guensler, R. L. (2013). Sustainability assessment at the transportation planning level: Performance measures and indexes. *Transport Policy*, *25*, 10–21. doi:10.1016/j.tranpol.2012.10.004

John A. Volpe National Transportation Systems Center. (2018). *Primer for Improved Urban Freight Mobility and Delivery - Operations, Logistics, and Technology Strategies* (Performimg Organization Report No. FHWA-HOP-18-020). Cambridge, MA: Federal Highway Administration.

Johnson, N., & Lav, I. (1998). *Should States Tax Food? Examining the Policy Issues and Options*. Center on Budget and Policy Priorities. Retrieved from https://www.cbpp.org/archiveSite/stfdtax98.pdf

Jordan, M., & Wagner, G. (2008). Revenue Diversification in Arkansas Cities: The Budgetary and Tax Effort Impacts. *Public Budgeting & Finance*, *28*(2), 68–82. doi:10.1111/j.1540-5850.2008.00911.x

Journal of Commerce. (2016, December 6). *THE Alliance strips joint-contracting language from VSA*. Retrieved March 12, 2018, from https://www.joc.com/maritime-news/container-lines/alliance/alliance-strips-joint-contracting-language-vsa_20161206.html

Journal of Commerce. (2018, March 7). *US and Global Trade: TPM 2018 Tracker - The latest news, insights, and analysis from the JOC's TPM 2018 conference in Long Beach*. Retrieved March 16, 2018, from https://www.joc.com/international-logistics/tpm-2018-tracker-news-insights-analysis_20180307.html

Jr, S. (2014). *The City and the Coming Climate: Climate Change in the Places We Live*. Retrieved from http://www.jurareview.ro/2014_6_1/p_103_108.pdf

Jr, S. (2005). Urban Heat and Air Pollution: An Emerging Role for Planners in the Climate Change Debate. *Journal of the American Planning Association*, *71*(1), 13–25. doi:10.1080/01944360508976402

Jung, C. (2001). Does the Local-Option Sales Tax Provide Property Tax Relief? The Georgia Case. *Public Budgeting & Finance*, *21*(1), 73–86. doi:10.1111/0275-1100.00037

Jung, C. (2002). The Effect of Local Earmarking on Capital Spending in Georgia Counties. *State & Local Government Review*, *34*(1), 29–37. doi:10.1177/0160323X0203400103

Kearns, K. P. (1996). *Managing for Accountability: Preserving the Public Trust in Public and Nonprofit Organizations*. San Francisco: Jossey-Bass.

Kellaway, K. (2014, April). *State of the Union in the Drayage Industry and the impact on Green Fleet Initiatives*. PowerPoint presented at the Port Stakeholders Summit, Baltimore, MD. Retrieved from https://www.epa.gov/sites/production/files/2014-07/documents/summit-kellaway.pdf

Kettl, D. F. (1993). *Sharing Power: Public Governance and Private Markets*. Washington, DC: Brookings Institution.

Keuleers, B., Chow, V., Thorpe, N., Timmermans, H., & Wets, G. (2006). Behavioural change in activity-travel patterns in response to road user charging. *Journal of Transport Economics and Policy*, *40*(1), 119–134.

Kidder, B. (2006). *The challenges of rural transportation*. Logan, UT: Western Rural Development Center.

Kimberlin, J. (2012, March 4). Tolls, tolls, tolls. How did we get into this mess? *The Virginian-Pilot*. Retrieved from http://pilotonline.com/news/local/transportation/tolls-tolls-tolls-how-did-we-get-into-this-mess/article_5bccaafd-87d5-5d68-b435-6eb9ccb74d30.html

King, D., Manville, M., & Shoup, D. (2007). The political calculus of congestion pricing. *Transport Policy*, *14*(2), 111–123. doi:10.1016/j.tranpol.2006.11.002

Kissel, C., Schwartz, B., James, S., & Dyer, S. (2016). *Regional rural transportation planning: State models for local consultation, regional coordination, and regional transportation planning organizations*. NADO Research Foundation.

Kneebone, E. (2013). *Job sprawl stalls: The Great Recession and metropolitan employment location*. Washington, DC: Metropolitan Policy Program at Brookings.

Komanoff, C. (2013). Lessons from London after 10 Years of the Congestion Charge. *Streetsblog*. Retrieved from http://www.streetsblog.org/2013/02/15/lessons-from-london-after-10-years-of-the-congestion-charge/

Kotkin, J. (2010, August 9). *Mass Transit: The Great Train Robbery*. Retrieved from https://www.forbes.com/2010/08/09/cities-transportation-class-opinions-columnists-joel-kotkin.html#a6aaac87208f

Krieger, N. (2000). Epidemiology and Social Sciences: Towards a Critical Reengagement in the 21st Century. *Epidemiologic Reviews*, *22*(1), 155–163. doi:10.1093/oxfordjournals.epirev.a018014 PMID:10939022

Krol, R. (2016). *Tolling the Freeway: Congestion Pricing and the Economics of Managing Traffic*. Academic Press.

Kutlu, L., & McCarthy, P. (2016). US airport ownership, efficiency, and heterogeneity. *Transportation Research Part E, Logistics and Transportation Review*, *89*, 117–132. doi:10.1016/j.tre.2016.03.003

Laing, K. (2013). Bill would eliminate federal transportation funding. *TheHill*. Retrieved from http://thehill.com/blogs/transportation-report/infrastructure/190402-bill-would-eliminate-federal-transportation

Laing, K. (2014). Poll: 79 percent supports tolling to pay for highways. *The Hill*. Retrieved from http://thehill.com/policy/transportation/217476-poll-79-percent-support-tolling-to-pay-for-highways

Landis, J., Guhathakurta, S., Huang, W., Zhang, M., & Fukuji, B. (1995). Rail Transit Investments, Real Estate Values, and Land Use Change: A Comparative Analysis of Five California Rail Transit Systems. *EScholarship*. Retrieved from http://escholarship.org/uc/item/4hh7f652

Lane, J. C. (2015, October). *NY-Appellate-Division-rules-Graves-Amendment-extends-to-intermodal-chassis*. Retrieved October 12, 2017, from http://www.thelanelawfirm.com/ny-appellate-division-rules-graves-amendment-extends-to-intermodal-chassis.html?no_redirect=true

Larco, N. (2017). *When Are AVs Coming? (10 Car Companies Say Within the Next 5 Years...) | Urbanism Next*. Retrieved from https://urbanismnext.uoregon.edu/2017/08/28/when-are-avs-coming-10-car-companies-say-within-the-next-5-years/

Lee, C., & Miller, J. S. (2016). *Conditions contributing to the attitudes for toll facilities in the United States with a specific focus on Virginia*. Academic Press.

Leland & Smirnova. (2009). Reassessing Privatization Strategies 25 Years Later: Revisiting Perry and Babitsky's Comparative Performance Study of Urban Bus Transit Services. *Public Administration Review*.

Lemp, J., & Kockelman, K. (2009). Understanding and accommodating risk and uncertainty in toll road projects: A review of the literature. *Transportation Research Record: Journal of the Transportation Research Board*, *2132*(1), 106–112. doi:10.3141/2132-12

Leuenberger, D. Z., & Bartle, J. R. (2009). *Sustainable development for public administration*. Armonk, NY: M.E. Sharpe.

Leuenberger, D. Z., & Bartle, J. R. (2009). *Sustainable Development for Public Administration*. Armonk, NY: M.E. Sharpe.

Leuenberger, D. Z., Bartle, J. R., & Chen, C. (2014). Sustainability and Transportation. *Public Works Management & Policy*, *19*(4), 316–321. doi:10.1177/1087724X14545540

Light, T., Patil, S., Erhardt, G. D., Tsang, F., Burge, P., Sorensen, P., & Zmud, M. (2015). *The Impact of Adopting Time-of-Day Tolling: Case Study of 183A in Austin, Texas*. Santa Monica, CA: RAND Corporation.

Lillis, R. (2013). Council Votes to Launch Bike Corrals, Plan Parklets in Sacramento. *The Sacramento Bee*. Retrieved from http://www.sacbee.com/2013/10/22/5844018/council-votes-launch-bike-corrals.html

Lindholm, M., & Behrends, S. (2012). Challenges in urban freight transport planning – a review in the Baltic Sea Region. *Journal of Transport Geography*, *22*, 129–136. doi:10.1016/j.jtrangeo.2012.01.001

Lipson, H., & Kurman, M. (2016). *Driverless: Intelligent Cars and the Road Ahead*. MIT Press.

Litman, T. (2006). London Congestion Pricing. *Implications for Other Cities*. Retrieved from http://www.mumbaidp24seven.in/reference/london_congestion_pricing.pdf

Little, R. G. (2011). The Emerging Role of Public-Private Partnerships in Megaproject Delivery. *Public Works Management & Policy*, *16*(3), 240–249. doi:10.1177/1087724X11409244

Liu, Z., Wang, S., Qu, X., & Shiwakoti, N. (2014). Congestion Pricing with Distance Tolls: A Review and New Developments. In *CICTP 2014: Safe, Smart and Sustainable Multimodal Transportation Systems*. American Society of Civil Engineers. doi:10.1061/9780784413623.331

Liyanage, C., & Villalba-Romero, F. (2015). Measuring success of PPP transport projects: A cross-case analysis of toll roads. *Transport Reviews*, *35*(2), 140–161. doi:10.1080/01441647.2014.994583

Lockwood, S. (1998). *The Changing State DOT*. Washington, DC: American Association of State Highway and Transportation Officials.

Loukaitou-Sideris, Brozen, Callahan, Brookover, LaMontagne, & Snehansh. (2012). *Reclaiming the Right-of-Way: A Toolkit for Creating and Implementing Parklets*. Academic Press.

Lucy, W. H., & Phillips, D. L. (2006). Tomorrow's cities, tomorrow's suburbs. Chicago, IL: American Planning Association.

Luetke, S. R., Gibbs, T., Pronier, C., Vandecar-Burdin, T., & Richman, J. (2012). *How is Life in Hampton Roads? 2012 Report for the 3rd Annual Life in Hampton Roads Survey*. Retrieved from http://images.bimedia.net/documents/odu_lifestyle_study.pdf

Luna, L. (2004). Local sales tax competition and the effect on county governments' tax rates and tax bases. *The Journal of the American Taxation Association*, *26*(1), 43–61. doi:10.2308/jata.2004.26.1.43

Lutte, R., & Bartle, J. (2016). Examining the new role of public administrators in transportation sustainability: A case in air navigation modernization. *Public Works Management & Policy*. doi:10.1177/1087724X16679845

Maantay, J. (2001). Zoning, Equity, and Public Health. *American Journal of Public Health*, *91*(7), 1033–1041. doi:10.2105/AJPH.91.7.1033 PMID:11441726

Mack, A., & Ruse, K. (n.d.). *Mobility management: Empirical evidence of fiscal benefits from multiple states*. Omaha: Nebraska Department of Roads and University of Nebraska at Omaha College of Public Affairs & Community Service. Retrieved from https://www.nebraskatransit.com/assets/pdf/Mobility%20Management%20White%20Paper.pdf

Magnier, M. (1988, June 16). *Container leasing enters new era*. Retrieved March 14, 2018, from https://www.joc.com/container-leasing-enters-new-era_19880616.html

Majumdar, S. R., Sen, L., Highsmith, M. K., & Cherrington, L. (2013). The case of performance measurement in mobility management programs. *Public Performance & Management Review*, *37*(2), 280–301. doi:10.2753/PMR1530-9576370205

Massey, D. S. (2004). Segregation and Stratification: A Biosocial Perspective. *Du Bois Review*, *1*(01), 7–25. doi:10.1017/S1742058X04040032

May, A. D., Liu, R., Shepherd, S. P., & Sumalee, A. (2002). The Impact of Cordon Design on the Performance of Road Pricing Schemes. *Transport Policy*, *9*(3), 209–220. doi:10.1016/S0967-070X(02)00031-8

Mazmanian, D. (2009). Los Angeles Clean Air Saga—Spanning the Three Epochs. In D. Mazmanian & M. Kraft (Eds.), *Toward Sustainable Communities: Transitions and Transformations in Environmental Policy* (pp. 88–113). MIT Press. doi:10.7551/mitpress/9780262134927.003.0004

Mediterranean Shipping Company. (2018a). *Intermodalism | MSC*. Retrieved March 5, 2018, from https://www.msc.com/gin/help-centre/guide-to-international-shipping/new-to-international-trade-intermodalism?lang=ru-ru

Mediterranean Shipping Company. (2018b). *Shipping Glossary of Terms | MSC*. Retrieved May 7, 2018, from https://www.msc.com/usa/help-centre/interactive-shipping-glossary

Mercier, J. (2009). Equity, Social Justice, and Sustainable Urban Transportation in the Twenty-First Century. *Administrative Theory & Praxis*, *31*(2), 145–163. doi:10.2753/ATP1084-1806310201

Merriam-Webster. (2018). *Definition of antitrust*. Retrieved April 19, 2018, from https://www.merriam-webster.com/dictionary/antitrust

Meyer, M. (1997). *A toolbox for alleviating traffic congestion and enhancing mobility* (I. T. Engineers, Ed.). Washington, DC: Institute of Transportation Engineers.

Milward, H. B., & Provan, K. (2003). Managing the hollow state: Collaboration and contracting. *Public Management Review, 5*(1), 1–18. doi:10.1080/1461667022000028834

Milward, H. B., & Provan, K. G. (2000). Governing the Hollow State. *Journal of Public Administration: Research and Theory, 10*(2), 359–380. doi:10.1093/oxfordjournals.jpart.a024273

Miranda, R. (1994). Privatization and the budget-maximizing bureaucrat. *Public Productivity & Management Review*, 355-369.

Mohr, Z. (2017b). Cost Accounting at the Service Level: An Analysis of Transaction Cost Influences on Indirect Cost Measurement in the Cost Accounting Plans of Large US Cities. *Public Administration Quarterly, 41*(1).

Mohr, Z. T. (2017a). *Cost Accounting in Government: Theory and Applications*. Taylor & Francis. doi:10.4324/9781315648897

Mongelluzzo, B. (2015a, May 5). *Drayage community praises new LA-LB gray chassis pool*. Retrieved October 5, 2017, from https://www.joc.com/port-news/us-ports/port-los-angeles/drayage-community-praises-new-la-lb-gray-chassis-pool_20150305.html

Mongelluzzo, B. (2015b, October 21). *ILWU chassis trouble likely to be resolved in court*. Retrieved October 5, 2017, from https://www.joc.com/port-news/longshoreman-labor/international-longshore-and-warehouse-union/ilwu-chassis-trouble-likely-be-resolved-court_20151021.html

Mongelluzzo, B. (2017, June 9). *New chassis business model would cut costs, trucker says*. Retrieved October 5, 2017, from https://www.joc.com/port-news/us-ports/new-chassis-business-model-would-cut-costs-trucker-says_20170609.html

Morales Sarriera, J., Salvucci, F. P., & Zhao, J. (2017). *What Drives the Costs of Transit Operations? The Implications of Labor Productivity, Contracting Out, and Unionization (No. 17-02927)*. Academic Press.

Morello, C., Keating, D., & Hendrix, S. (2011, May 5). Census: Young adults are responsible for most of D.C.'s growth in past decade. *Washington Post*. Retrieved from https://www.washingtonpost.com/local/census-young-adults-are-responsible-for-most-of-dcs-growth-in-past-decade/2011/05/04/AFJz5LtF_story.html

Morley, H. (2017a, September 14). *NY-NJ joint chassis pool collapses*. Retrieved October 5, 2017, from https://www.joc.com/trucking-logistics/drayage/ny-nj-joint-chassis-pool-collapses_20170914.html

Morley, H. (2017b, October 25). *DCLI snags TRAC's domestic chassis fleet*. Retrieved November 7, 2017, from https://www.joc.com/trucking-logistics/trucking-equipment/dcli-snags-trac%E2%80%99s-domestic-chassis-fleet_20171025.html

National Academy of Public Administration. (2018). *Federal Systems Redesign*. NAPA.

National Center for Mobility Management. (n.d.). *Performance measures for mobility management*. Retrieved from https://nationalcenterformobilitymanagement.org/wp-content/uploads/2014/09/Performance_Measures_Final.pdf

Nayan, A., & Wang, D. Z. (2017). Optimal bus transit route packaging in a privatized contracting regime. *Transportation Research Part A, Policy and Practice*, *97*, 146–157. doi:10.1016/j.tra.2017.01.016

Nebraska D. O. T. (2017). *Nebraska mobility management: Mobility manager (Southeast Region)*. Retrieved from https://www.nebraskatransit.com/assets/pdf/Newsletter%201_MM_Oct2017_Final.pdf

Nebraska Public Transit. (n.d.). *Mobility management statewide project*. Retrieved from https://nebraskatransit.com/mobility-1.php

Nebraska Public Transportation Act, Nebraska Revised Statutes, Chapter 13 §1201-1214 (1993).

Nelson, A. (1999). Transit Stations and Commercial Property Values: A Case Study with Policy and Land-Use Implications. *Journal of Public Transportation*, *2*(3), 77–93. doi:10.5038/2375-0901.2.3.4

Nesbit, T., & Kreft, S. (2009). Federal Grants, Earmarked Revenues, and Budget Crowd-Out: State Highway Funding. *Public Budgeting & Finance*, *29*(2), 94–110. doi:10.1111/j.1540-5850.2009.00930.x

Newman, P., & Jennings, I. (2008). *Cities as Sustainable Ecosystems: Principles and Practices*. Island Press.

Newman, P., & Kenworthy, J. (1999). *Sustainability and Cities: Overcoming Automobile Dependence*. Island Press.

Nichols, K. M. (2017). *The State of Transportation in Hampton Roads 2017*. Chesapeake, VA: Hampton Roads Transportation Planning Organization.

Nichols, K. M., & Belfield, S. S. (2016). Impact of New Tolls on Existing Roadway Facilities: Hampton Roads, Virginia, Experience. *Transportation Research Record: Journal of the Transportation Research Board*, *2597*(1), 28–36. doi:10.3141/2597-04

Nielsen, O. A. (2004). Behavioral Responses to Road Pricing Schemes: Description of the Danish AKTA Experiment. *Journal of Intelligent Transportation Systems: Technology, Planning, and Operations*, *8*(4), 233–251. doi:10.1080/15472450490495579

North American Chassis Pool Cooperative (NACPC). (2018). *About*. Retrieved March 16, 2018, from http://www.nacpc.org/about/

North Carolina Center for County Research. (2015). *Basics of North Carolina Local Option Sales Taxes*. Report. Author.

North Carolina General Assembly. (2012, June 26). *North Carolina Senate*. Retrieved from North Carolina General Assembly: https://www.ncleg.net/gascripts/voteHistory/RollCallVoteTranscript.pl?sSession=2011&sChamber=S&RCS=1191

North, D. (1990). A transaction cost theory of politics. *Journal of Theoretical Politics*, *2*(4), 355–367. doi:10.1177/0951692890002004001

O'Connell, L. (2005). Program accountability as an emergent property: The role of stakeholders in a program's field. *Public Administration Review*, *65*(1), 85–93. doi:10.1111/j.1540-6210.2005.00433.x

O'Connell, L. (2006). Emergent accountability in state-local relations: Some lessons from solid waste policy in Kentucky. *Administration & Society*, *38*(4), 500–513. doi:10.1177/0095399706290635

O'Connell, L. (2009). The Impact of Local Supporters on Smart Growth Policy Adoption. *Journal of the American Planning Association*, *75*(3), 281–291. doi:10.1080/01944360902885495

O'Connell, L., & Yusuf, J.-E. W. (2013). Improving Revenue Adequacy by Indexing the Gas Tax to Indicators of Need: A Simulation Analysis. *Public Works Management & Policy*, *18*(3), 229–243. doi:10.1177/1087724X12451575

O'Connell, L., Yusuf, J.-E., & Hackbart, M. (2009). Transportation Commissions as Accountability Structures A Review of their Statutory Roles and Other Attributes. *American Review of Public Administration*, *39*(4), 409–424. doi:10.1177/0275074008321892

Ocean Carrier Management Equipment Association. (2017). *Ocean carrier equipment management association* (4th ed.). Federal Maritime Commission Agreement.

Odeck, J., & Bråthen, S. (1998). The Planning of Toll Roads—Do Public Attitudes Matter? Case of the Oslo Toll Ring. *Transportation Research Record: Journal of the Transportation Research Board*, *1649*(1), 72–80. doi:10.3141/1649-09

Ohio D. O. T. (2018). *Ohio mobility management program guide*. Retrieved from http://www.dot.state.oh.us/Divisions/Planning/Transit/Documents/Mobility%20Management%20Program%20Guide.pdf

Oliff, P. (2015). *Funding Challenges in Highway and Transit: A federal-state-local analysis*. Fiscal Federalism Initiative, The PEW Charitable Trusts. Retrieved from http://www.pewtrusts.org/en/research-and-analysis/analysis/2015/02/24/funding-challenges-in-highway-and-transit-a-federal-state-local-analysis

Olszewski, P., & Xie, L. (2005). Modelling the effects of road pricing on traffic in Singapore. *Transportation Research Part A, Policy and Practice*, *39*(7), 755–772. doi:10.1016/j.tra.2005.02.015

Orr, J. (2013, October 13). Let's move forward on CLT airport *Charlotte Observer*.

Osei-Kyei, R., & Chan, A. P. (2015). Review of studies on the Critical Success Factors for Public–Private Partnership (PPP) projects from 1990 to 2013. *International Journal of Project Management*, *33*(6), 1335–1346. doi:10.1016/j.ijproman.2015.02.008

Oum, T. H., Yan, J., & Yu, C. (2008). Ownership forms matter for airport efficiency: A stochastic frontier investigation of worldwide airports. *Journal of Urban Economics*, *64*(2), 422–435. doi:10.1016/j.jue.2008.03.001

Palm, M., & Handy, S. (2016). Sustainable transportation at the ballot box: A disaggregate analysis of the relative importance of user travel mode, attitudes and self-interest. *Transportation*, 1–21.

Parker, S., Bush, C., Richman, J., & Vandecar-Burdin, T. (2014). *Life in Hampton Roads Report The Fifth Annual Life in Hampton Roads Survey*. Academic Press.

Parker, S., Close, J., Gainey, R., & Vandecar-Burdin, T. (2015). *Life in Hampton Roads Report: The Sixth Annual Life in Hampton Roads Survey*. Retrieved from http://www.odu.edu/content/dam/odu/col-dept/ssrc/docs/lihr-report-final.pdf

Parker, S., Close, J., Gainey, R., & Vandecar-Burdin, T. (2016). *Life in Hampton Roads Report The Seventh Annual Life in Hampton Roads Survey*. Academic Press.

Pascale, J. (2017, April 12). FAQ: VDOT officials explain how High Occupancy Tolls lanes will work on I-64. *The Virginian-Pilot*. Retrieved from https://pilotonline.com/news/local/transportation/faq-vdot-officials-explain-how-high-occupancy-tolls-lanes-will/article_b80f01e5-a2eb-546b-a53b-e5e70136c058.html

Peirce, N. (2007). Selling Our Toll Roads: Good or Retrograde Idea? *Government Finance Review*, *23*(3), 81–82.

Perez, B. G., Batac, T., & Vovsha, P. (2012). *Assessing Highway Tolling and Pricing Options and Impacts*. Retrieved from https://www.nap.edu/download/22701

Perez, B. G., Giordano, R., & Stamm, H. (2011). *Evaluation and Performance Measurement of Congestion Pricing Projects*. Washington, DC: Transportation Research Board.

Perlmutt, D., & Portillo, E. (2013, April 21). City aims its anger at airport board. *Charlotte Observer*.

Perry, J., & Babitsky, T. (1986). Comparative Performance in urban bus transit: Assessing privatization strategies. *Public Administration Review*, *46*(1), 57–66. doi:10.2307/975443

Peters, S. (2012). *Impact Fees for Complete Streets*. Los Angeles, CA: University of California. Retrieved from http://164.67.121.27/files/Lewis_Center/CompleteStreetsInitiative/Peters_report.pdf

Pisano, M., & Callahan, R. (2013, October). Developing effective mechanisms that promote fiscal sustainability. *Government Finance Review*, 74.

Pisano, M. (2017). *The Puzzle of the American Economy*. Santa Barbara, CA: Praeger.

Podgorski, K. V., & Kockelman, K. M. (2006). Public Perceptions of Toll Roads: A Survey of the Texas Perspective. *Transportation Research Part A, Policy and Practice*, *40*(10), 888–902. doi:10.1016/j.tra.2006.03.002

Polèse, M. (2013). Five principles of urban economics. *City Journal*. Retrieved from https://www.city-journal.org/html/five-principles-urban-economics-13531.html

Ponomariov, B., & Kingsley, G. (2008). Applicability of the Normative Model of Outsourcing in the Public Sector: The Case of a State Transportation Agency. *Public Organization Review*, *8*(3), 253–272. doi:10.100711115-008-0059-2

Poole, R. W. Jr, & Orski, C. K. (2000). HOT lanes: A better way to attack urban highway congestion. *Regulation*, *23*(1), 15–20.

Pope, J. V., & Mohr, Z. T. (2017). Cost Accounting for Rates and User Fees. In Z. T. Mohr (Ed.), *Cost Accounting in Government: Theory and Applications* (pp. 66–80). New York, NY: Routledge. doi:10.4324/9781315648897-4

Portillo, E. (2013a, September 1). Airport dispute continues *Charlotte Observer*.

Portillo, E. (2013b, October 31). An eye on upgrades, not uproar *Charlotte Observer*.

Portillo, E. (2014a, May 11). Airport budget set to climb. *Charlotte Observer*.

Portillo, E. (2014b, July 13). Airport leader steps out of Orr's shadow *Charlotte Observer*.

Portillo, E. (2014c, August 28). Audit: Airport oversight lax *Charlotte Observer*.

Portillo, E. (2014d, March 6). Orr blamed for parking rates *Charlotte Observer*.

Portillo, E., & Harriosn, S. (2013, April 26). City-funded study recommends airport authority. *Charlotte Observer*.

Principals, Intergovernmental Forum on Transportation Finance. (2008). *Financing Transportation in the 21st Century: An Intergovernmental Perspective*. National Academy of Public Administration. Retrieved from https://www.napawash.org/uploads/Academy_Studies/08-16.pdf

Puentes, R. (2014). The Indiana toll road. How did a good deal go bad? *Forbes*. Retrieved from https://www.forbes.com/sites/realspin/2014/10/03/the-indiana-toll-road-how-did-a-good-deal-go-bad/#4a33c8b62087

Puentes, R., & Prince, R. (2003). Fueling Transportation Finance: A Primer on the Gas Tax. *The Brookings Institution*. Retrieved from https://www.brookings.edu/research/fueling-transportation-finance-a-primer-on-the-gas-tax/

Puentes, R., & Istrate, E. (2011). *Moving Forward on Pubilc Private Partnerships: U.S. and International Experience with PPP Unites*. Brookings-Rockefeller.

Putnam, R. D. (2001). *Bowling Alone: The Collapse and Revival of American Community*. Simon and Schuster.

Rall, J., Wheet, A., Farber, N., & Reed, J. (2011). *Transportation Governance and Finance*. National Conference of State Legislatures. Retrieved from http://www.ncsl.org/documents/transportation/FULL-REPORT.pdf

Rashidi, T., Mohammadian, A., & Zhang, Y. (2010). Effect of Variation in Household Sociodemographics, Lifestyles, and Built Environment on Travel Behavior. *Transportation Research Record: Journal of the Transportation Research Board*, *2156*(1), 64–72. doi:10.3141/2156-08

Rasmussen Reports. (2014, May 7). *65% Oppose Tolls on the Interstate Highways*. Retrieved from http://www.rasmussenreports.com/public_content/politics/general_politics/may_2014/65_oppose_tolls_on_interstate_highways

Rassman, C. L. (2014). Regulating Rideshare without Stifling Innovation: Examining the Drivers, the Insurance Gap, and Why Pennsylvania Should Get on Board. *Pitt. J. Tech. L. & Pol'y*, *15*, 81.

Rayle, L., Dai, D., Chan, N., Cervero, R., & Shaheen, S. (2016). Just a Better Taxi? A Survey-Based Comparison of Taxis, Transit, and Ridesourcing Services in San Francisco. *Transport Policy*, *45*(January), 168–178. doi:10.1016/j.tranpol.2015.10.004

Reja, R. (1999). *The Economies and Politics of Contracting out with the Private Sector: Evidence from the US Transit Industry*. Working paper. The World Bank.

Rep Don Young (R) Alaska. Safe, Accountable, Flexible, Efficient Transportation Equity Act: A Legacy For Users, Pub. L. No. 109–59, § 4118. ROADABILITY, 1729. (2005). Retrieved from https://www.gpo.gov/fdsys/pkg/PLAW-109publ59/pdf/PLAW-109publ59.pdf

Resnick, S., Gibbs, T., Bush, C., Parker, S., Vandecar-Burdin, T., & Richman, J. (2013). *Life in Hampton Roads Report: The Fourth Annual Life in Hampton Roads Survey*. Retrieved from http://www.odu.edu/content/dam/odu/col-dept/al/docs/lihr-report2013withcomparisons.pdf

Rides, M. O. (2018). *Mobility management*. Retrieved from http://morides.org/mobility-management/

Riggs & Gilderbloom. (2015). Two-Way Street Conversion Evidence of Increased Livability in Louisville. *Journal of Planning Education and Research*. . doi:10.1177/0739456X15593147

Riggs & Gilderbloom. (2017). How Multi-Lane, One-Way Street Design Shapes Neighbourhood Life: Collisions, Crime and Community. *Local Environment*. Retrieved from http://www.tandfonline.com/eprint/qZNFa3D9zyn9RD5chhvE/full

Riggs, W. (2011). *Walkability and Housing: A Comparative Study of Income, Neighborhood Change and Socio-Cultural Dynamics in the San Francisco Bay Area* (Doctoral Dissertation). University of California, Berkeley, CA.

Riggs, W. (2014). Dealing with Parking Issues on an Urban Campus: The Case of UC Berkeley. *Case Studies on Transport Policy*. Retrieved from http://www.sciencedirect.com/science/article/pii/S2213624X14000431

Riggs, W. W., & Boswell, M. R. (2016a). *No Business as Usual in an Autonomous Vehicle Future.* Retrieved from https://works.bepress.com/williamriggs/53/

Riggs, W. W., & Boswell, M. R. (2016b). *Why Autonomous Vehicles Probably Won't Induce Sprawl.* Retrieved from https://works.bepress.com/williamriggs/60/

Riggs, W. W., Boswell, M. R., & Zoepf, S. (2017). A New Policy Agenda for Autonomous Vehicles: It's Time to Lead Innovation. *Planetizen.* Retrieved from https://works.bepress.com/williamriggs/75/

Riggs, W., & Boswell, M. R. (2016). *Thinking Beyond the (Autonomous) Vehicle: The Promise of Saved Lives.* Retrieved from https://works.bepress.com/williamriggs/71/

Riggs, W. (2018). *Disruptive Transport: Driverless Cars, Transport Innovation and the Sustainable City of Tomorrow. Routledge Equity, Justice and the Sustainable City Series.* London: Routledge.

Riggs, W., & McDade, E. (2016). Moving from Planning to Action: Exploring Best Practice Policy in the Finance of Local Bicycling and Pedestrian Improvements. *Case Studies on Transport Policy, 4*(3), 248–257. doi:10.1016/j.cstp.2016.06.004

Rivenbark, W. C. (2005). A historical overview of cost accounting in local government. *State & Local Government Review, 37*(3), 217–227. doi:10.1177/0160323X0503700304

Roberts, D., & Morrill, J. (2013, March 06). Bond Questions Slow Airport Vote. *Charlotte Observer.*

Rodrigue, J. P. (2012). *Guidebook for Assessing Evolving International Container Chassis Supply Models.* Transportation Research Board. Retrieved from http://www.trb.org/Publications/Blurbs/168158.aspx

Rodrigue, J.-P., & Booth, M. (2013, February 26). *Grounded and chassis container terminal operations.* Retrieved March 7, 2018, from https://people.hofstra.edu/jean-paul_rodrigue/downloads/Booth%20&%20Rodrigue_PT57_V5.pdf

Rodrigue, J. (2013). The Financing of Transportation Infrastructure. In *The Geography of Transport Systems.* New York: Routledge. doi:10.4324/9780203371183

Rodriguez, B., Pedro, M., Navarro Galera, A., Alcaide Munoz, L., & Deseada Lopez Subires, M. (2014). Factors influencing local government financial sustainability: An empirical study. *Lex Localis: Journal of Local Self-Government, 12*(1), 31–54. doi:10.4335/12.1.31-54(2014)

Rodriguez, B., Pedro, M., Navarro Galera, A., Alcaide Munoz, L., & Deseada Lopez Subires, M. (2016). Analyzing forces to the financial contribution of local governments to sustainable development. *Sustainability, 8*(9), 925–943. doi:10.3390u8090925

Rogers, C. (2004). Local option sales tax (LOST) policy on the urban fringe. *The Journal of Regional Analysis & Policy, 34*(1), 25–50.

Rothacker, R., & Henderson, B. (2013, July 27). Orr waits in wings as city sorts out new *Charlotte Observer*.

Rouhani, O. M., Gao, H. O., & Geddes, R. R. (2015). Policy lessons for regulating public–private partnership tolling schemes in urban environments. *Transport Policy*, *41*, 68–79. doi:10.1016/j.tranpol.2015.03.006

Roumboutsos, A., & Pantelias, A. (2015). Allocating revenue risk in transport infrastructure public private partnership projects: How it matters. *Transport Reviews*, *35*(2), 183–203. doi:10.1080/01441647.2014.988306

Rusch, W. A. (1984). *Toll Highway Financing*. Washington, DC: Transportation Research Board.

Saelens, B., Sallis, J., & Frank, L. (2003). Environmental Correlates of Walking and Cycling: Findings from the Transportation, Urban Design, and Planning Literatures. *Annals of Behavioral Medicine*, *25*(2), 80–91. doi:10.1207/S15324796ABM2502_03 PMID:12704009

Safirova, E., Gillingham, K., Harrington, W., & Nelson, P. (2003). *Are HOT lanes a hot deal? the potential consequences of converting HOV to HOT lanes in northern virginia*. RFF Issue Brief.

Sallis, J. F., Frank, L. D., Saelens, B. E., & Kraft, M. K. (2004). Active Transportation and Physical Activity: Opportunities for Collaboration on Transportation and Public Health Research. *Transportation Research Part A, Policy and Practice*, *38*(4), 249–268. doi:10.1016/j.tra.2003.11.003

Sallis, J. F., & Glanz, K. (2006). The Role of Built Environments in Physical Activity, Eating, and Obesity in Childhood. *The Future of Children*, *16*(1), 89–108. doi:10.1353/foc.2006.0009 PMID:16532660

Santos, G., & Fraser, G. (2006). Road pricing: Lessons from London. *Economic Policy*, *21*(46), 263–310. doi:10.1111/j.1468-0327.2006.00159.x

Schade, J., & Schlag, B. (2003). Acceptability of Urban Transport Pricing Strategies. *Transportation Research Part F: Traffic Psychology and Behaviour*, *6*(1), 45–61. doi:10.1016/S1369-8478(02)00046-3

Schaller, B. (2010). New York City's congestion pricing experience and implications for road pricing acceptance in the United States. *Transport Policy*, *17*(4), 266–273. doi:10.1016/j.tranpol.2010.01.013

Schank, J. (2011). *Public-Private Partnerships: Understanding the Tradeoffs*. EnoBrief.

Sciara, G.-C. (2017). Metropolitan Transportation Planning: Lessons From the Past, Institutions for the Future. *Journal of the American Planning Association*, *83*(3), 262–276. doi:10.1080/01944363.2017.1322526

Seskin, S., & McCann, B. (2012). Complete Streets. *Policy Analysis*, *2011*. Retrieved from http://trid.trb.org/view.aspx?id=1212459

Shaheen, S., & Chan, N. (2016). Mobility and the Sharing Economy: Potential to Facilitate the First-and Last-Mile Public Transit Connections. *Built Environment*, *42*(4), 573–588. doi:10.2148/benv.42.4.573

Shan, L., Garvin, M. J., & Kumar, R. (2010). Collar options to manage revenue risks in real toll public-private partnership transportation projects. *Construction Management and Economics*, *28*(10), 1057–1069. doi:10.1080/01446193.2010.506645

Shi, S., Chong, H.-Y., Liu, L., & Ye, X. (2016). Examining the Interrelationship among Critical Success Factors of Public Private Partnership Infrastructure Projects. *Sustainability*, *8*(12), 1313. doi:10.3390u8121313

Sjoquist, D. L., Walker, M. B., & Wallace, S. (2005). Estimating Differential Responses to Local Fiscal Conditions: A Mixture Model Analysis. *Public Finance Review*, *33*(1), 36–61. doi:10.1177/1091142104270656

Skelcher, C. (2010). Governance of public-private partnerships. In G. Hodge, C. Greve, & A. Boardman (Eds.), *International Handbook on Public-private Partnerships* (pp. 292–304). Cheltenham, UK: Edward Elgar. doi:10.4337/9781849804691.00021

Slone, S. (2016). *Evolution of the public-private partnership pipeline*. Retrieved from http://knowledgecenter.csg.org/kc/content/evolution-public-private-partnership-pipeline

Smirnova, O. (2008). *Does Government Structure Really Matter? A Comparison of Efficiency and Effectiveness of Special Purpose Versus General Purpose government transit operations* (Dissertation). University of North Carolina at Charlotte.

Smirnova, O. V., & Leland, S. (2015). The Contracting Decision. *Government Contracting: A Public Solutions Handbook, 6*, 17.

Smirnova, Yusuf, & Leland. (2016). Managing for Performance: Measurement and Monitoring of Contracts in the Transit Industry. *Journal of Public Procurement*, *16*(2), 208–242.

Smirnova, O. V., & Leland, S. M. (2014). Cutback management during the great recession: The case of transit agencies and contracting out. *State & Local Government Review*, *46*(4), 272–281. doi:10.1177/0160323X14564777

Smirnova, O. V., & Leland, S. M. (2016). The role of power and competition in contracting out: An analysis of public transportation markets. *Administration & Society*, *48*(4), 421–443. doi:10.1177/0095399713498748

Sobel, R. S., & Holcombe, R. J. (1996). Measuring the Growth and Variability of Tax Bases Over the Business Cycle. *National Tax Journal*, *49*, 535–552.

South Norfolk Jordan Bridge. (n.d.). *Jordan Bridge History*. Retrieved from http://www.snjb.net/about-us/jordan-bridge-history

Southern California Association of Governments. (1994). *Regional Comprehensive Plan*. Los Angeles, CA: SCAG.

Southern California Association of Governments. (2004). *Compass: Charting the Course for a Sustainable Southland*. Los Angeles, CA: SCAG.

Southern California Association of Governments. (2008). *Comprehensive Regional Plan*. Los Angeles, CA: SCAG.

Spadafore, M. (2017). *Mobility management: Successes and future plans*. Retrieved from https://s3.amazonaws.com/v3-app_crowdc/assets/a/ab/ab5808057ca15bba/Coordination_StatewideMobilityManagementInitiatives_MikeSpadafore.original.1510071701.pdf?1510071705

Sperling, D. (2018). *Three Revolutions: Steering Automated, Shared, and Electric Vehicles to a Better Future*. Island Press. doi:10.5822/978-1-61091-906-7

Stich, B., & Miller, C. (2012). A New Public Philosophy for a New Railroad Era: Applying the Blacksburg Manifesto to Contemporary Transportation Issues. *Administrative Theory & Praxis*, *34*(4), 602–621. doi:10.2753/ATP1084-1806340405

Stone, B., Vargo, J., Liu, P., Habeeb, D., DeLucia, A., Trail, M., ... Russell, A. (2014). Avoided Heat-Related Mortality through Climate Adaptation Strategies in Three US Cities. *PLoS One*, *9*(6), e100852. doi:10.1371/journal.pone.0100852 PMID:24964213

Sturm, R., & Cohen, D. A. (2004). Suburban Sprawl and Physical and Mental Health. *Public Health*, *118*(7), 488–496. doi:10.1016/j.puhe.2004.02.007 PMID:15351221

Swift, E. (2011). *The Big Roads: The Untold Story of the Engineers, Visionaries, and Trailblazers who Created the American Superhighways*. Boston: Houghton Mifflin Harcourt.

Szoltysek, J. (Ed.). (2012). *Developing of Transportation Flows in 21st Century Supply Chains*. Wydawnictwo Uniwersytetu Ekonomicznego, Katowicach. Available online at: http://www.sbc.org.pl/Content/74125/SE_121.pdf

Talley, W., & Anderson, E. (1986). An Urban Transit Firm Providing Transit, Paratransit and Contracted-out Services: A Cost Analysis. *Journal of Transport Economics and Policy*, *20*(3), 353–368.

Tang, S. Y., Callahan, R. F., & Pisano, M. (2014). Using common-pool resource principles to design local government fiscal sustainability. *Public Administration Review*, *74*(6), 791–803. doi:10.1111/puar.12273

Taylor, B. (1995). Public Perceptions, Fiscal Realities, and Freeway Planning: The California Case. *Journal of the American Planning Association*, *61*(1), 43–56. doi:10.1080/01944369508975618

Teal, R., Giulano, G., & Morlok, E. (1986). *Public Transit service contracting*. Washington, DC: US Department of Transportation, Urban Mass Transportation Administration.

The Institute of International Container Lessors. (2017). *The Institute of International Container Lessors (IICL): About the Industry - Laws, Regulations, Conventions and Standards*. Retrieved November 21, 2017, from https://www.iicl.org/aboutIndustry/laws.cfm

Thompon, L., & Elling, R. C. (2000). Mapping Patterns of Support for Privatization in the Mass Public: The Case of Michigan. *Public Administration Review*, *60*(4), 338–348. doi:10.1111/0033-3352.00096

Thompson, D., Edelsberg, J., Colditz, G. A., Bird, A. P., & Oster, G. (1999). Lifetime Health and Economic Consequences of Obesity. *Archives of Internal Medicine*, *159*(18), 2177. doi:10.1001/archinte.159.18.2177 PMID:10527295

Tioga Group, Incorporated. (2011). *Truck Drayage Productivity Guide* (Vol. 11). Transportation Research Board. Retrieved from http://www.trb.org/Publications/Blurbs/165528.aspx

Tioga Group, Incorporated. (2012). SmartWay DrayFLEET, Truck Drayage Environment and Energy Model Version 2.0 User's Guide (Final Report No. EP 11H000338). Transportation and Climate Division Office of Transportation and Air Quality U.S. Environmental Protection Agency.

Tirschwell, P. (2017a, October 9). *Demise of chassis pool an ominous sign for NY-NJ*. Retrieved October 9, 2017, from https://www.joc.com/port-news/us-ports/demise-chassis-pool-ny-nj-has-many-shaking-heads_20171009.html

Tirschwell, P. (2017b, November 17). *It is time to give shippers a choice on chassis*. Retrieved March 1, 2018, from https://www.joc.com/maritime-news/container-lines/it%E2%80%99s-time-give-shippers-choice-chassis_20171117.html

Tourangeau, R., & Yan, T. (2007). Sensitive Questions in Surveys. *Psychological Bulletin*, *133*(5), 859–883. doi:10.1037/0033-2909.133.5.859 PMID:17723033

Transport for London. (2008). *Impacts Monitoring*. Retrieved from http://www.tfl.gov.uk/cdn/static/cms/documents/central-london-congestion-charging-impacts-monitoring-sixth-annual-report.pdf

Transportation Research Board (TRB). (2008). *Potential Impacts of Climate Change on U.S. Transportation*. Available online at http://onlinepubs.trb.org/onlinepubs/sr/sr290.pdf

Transportation Research Board. (2001). *Special Report 258: Contracting for bus and demand response services: a survey of U.S. practices and services*. Author.

U.S. Census Bureau. (2017). *2012-2016 American Community Survey 5-Year Estimates Table B2045 Tenure by Vehicles available by Age of Householder*. Retrieved from https://factfinder.census.gov

U.S. Census Bureau. (2018). *2012-2016 American Community Survey 5-Year Estimates Public Use Microdata Sample for Nebraska*. Retrieved from https:factfinder.census.gov

U.S. Customs and Border Protection. (2016, March 4). *CBP Releases Fiscal Year 2015 Trade and Travel Numbers- Trade Facilitation and Enforcement Supports Economic Prosperity*. Retrieved December 1, 2017, from https://www.cbp.gov/newsroom/national-media-release/cbp-releases-fiscal-year-2015-trade-and-travel-numbers

U.S. Department of Housing and Urban Development. (2016). Urban Revival? Not For Most Americans. *Gentrification, 18*(3). Retrieved from http://jedkolko.com/2016/03/30/urban-revival-not-for-most-americans/

U.S. Government Accountability Office (GAO). (2013). *State and Local Governments' Fiscal Outlook: April 2013 Update*. Retrieved from http://www.gao.gov/assets/660/654255.pdf

United States Congress. (2014). *U.S.C. Title 46 - SHIPPING*. Retrieved March 14, 2018, from https://www.gpo.gov/fdsys/pkg/USCODE-2014-title46/html/USCODE-2014-title46-subtitleIV-partA-chap403-sec40307.htm

United States Congress. The Shipping Act of 1984 as Modified by The Ocean Shipping Reform Act of 1998, Pub. L. No. S. 414, § 10. (1999). Retrieved from http://www.shippers.com/Shipping_Act.asp

Urban Land Institute. (2013). *When the Road Price Is Right: Land Use, Tolls, and Congestion Pricing*. Retrieved from http://www.uli.org/wp-content/uploads/2013/03/When-the-Road-Price-is-Right_web_F.pdf

Utah State Tax Commission. (2017). *Sales and Use Tax General Information*. Report. Publication 25. Author.

van Den Hurk, M., & Verhoest, K. (2017). On the fast track? Using standard contracts in public-private partnerships for sports facilities: A case study. *Sports Management Review, 20*, 226-238.

Vavrova, M., Chang, C., & Bina, L. (2017). *A Framework to Analyze the Feasibility of Vehicle Miles Traveled Fees to Finance A Sustainable Transportation System*. Academic Press.

Virginia Department of Motor Vehicles. (2016, December). *Transportation Network Companies, 2016 Report*. Retrieved from https://rga.lis.virginia.gov/Published/2016/RD524/PDF

Virginia Department of Transportation Office of Public-Private Partnership. (2014). *Elizabeth River Tunnels: Increasing travel options and improving safety*. Retrieved from https://www.p3virginia.org/projects/elizabeth-river-tunnels/

Virginia Department of Transportation. (2018, September 10). *Hampton Roads Tunnels and Bridges*. Retrieved from http://www.virginiadot.org/travel/hro-tunnel-default.asp

Voss, C., Winters, M., Frazer, A., & McKay, H. (2015). School-Travel by Public Transit: Rethinking Active Transportation. *Preventive Medicine Reports, 2*, 65–70. doi:10.1016/j.pmedr.2015.01.004 PMID:26793430

Wachs, M. (2003). A dozen reasons for raising gasoline taxes. *Public Works Management & Policy*, *7*(4), 235–242. doi:10.1177/1087724X03253152

Wachs, M. (2003a). Local Option Transportation Taxes: Devolution as Revolution. *Transportation Quarterly*, (22): 9–15.

Wachs, M. (2003b). *Improving efficiency and equity in transportation finance*. Brookings Institution, Center on Urban and Metropolitan Policy.

Wang, Y. (2015). Evolution of public–private partnership models in American toll road development: Learning based on public institutions' risk management. *International Journal of Project Management*, *33*(3), 684–696. doi:10.1016/j.ijproman.2014.10.006

Warner, M., & Hebdon, R. (2001). Local Government Restructuring: Privatization and its Alternatives. *Journal of Policy Analysis and Management*, *20*(2), 315–336. doi:10.1002/pam.2027

Warner, M., & Hefetz, A. (2002). Applying Market Solutions to Public Services: An Assessment of Efficiency, Equity, and Voice. *Urban Affairs Review*, *38*(1), 70–89. doi:10.1177/107808702401097808

Warne, T. R. (2003). *State DOT Outsourcing and Private-Sector Utilization: A Synthesis of Highway Practice*. Retrieved from http://onlinepubs.trb.org/onlinepubs/nchrp/nchrp_syn_313.pdf

Weiland, N. (2017, November 27). Synagogue's Mix of Arts and Religion Helps Shape Jewish Life in Washington. *New York Times*. Retrieved from https://www.nytimes.com/2017/11/27/us/sixth-i-historic-synagogue-jewish-culture.html

Weinreich, D. (2015). Thinking Big When Funding Is Local Assessing the Potential of Local Option Transportation Funding in a Multi-Jurisdictional Context. *Public Works Management & Policy*.

Whymann, O. (2013). *Charlotte Airport Governance Study Final Report*. Academic Press.

Willems, T., Verhoest, K., Voets, J., Coppens, T., Van Dooren, W., & Van den Hurk, M. (2017). Ten lessons from ten years PPP experience in Belgium. *Australian Journal of Public Administration*, *76*(3), 316–329. doi:10.1111/1467-8500.12207

Williams, R., & Hammond, P. (2015). *2015 Passenger Transportation Trends*. Retrieved from https://www.strategyand.pwc.com/trends/2015-transportation-trends

Williams, D. R., & Jackson, P. B. (2005). Social Sources of Racial Disparities in Health. *Health Affairs*, *24*(2), 325–334. doi:10.1377/hlthaff.24.2.325 PMID:15757915

Williamson, O. (1985). *The Economic Institution of Capitalism: Firms, Markets, Relational*. London: Contracting. Collier McMillan.

Wilson, K. (2017). Operator of Texas toll road with 85 mph speeed limit emerges from bankruptcy. *Texas Tribune*. Retrieved from https://www.texastribune.org/2017/06/28/sh-130-toll-operator-announces-new-post-bankruptcy-ownership-structure/

Winston & Shirley. (1998). *Alternative route: towards efficient urban transportation*. Washington, DC: Brookings Institution.

Wu, J., Liu, J., Jin, X., & Sing, M. C. (2016). Government accountability within infrastructure public–private partnerships. *International Journal of Project Management*, *34*(8), 1471–1478. doi:10.1016/j.ijproman.2016.08.003

Yin, R. K. (2009). *Case study research: Designs and methods* (4th ed.). Thousand Oaks, CA: Sage.

Yusuf, J.-E. W., & Leavitt, W. M. (2014). The Case for Performance Management in Public Works and Infrastructure. *Public Works Management & Policy*.

Yusuf, J.-E. W. (2014). Highway Trust Fund. In M. Garrett (Ed.), *The Encyclopedia of Transportation: Social Science and Policy* (Vol. 2, pp. 745–747). Thousand Oaks, CA: Sage Reference.

Yusuf, J.-E. W., & O'Connell, L. (2013). The Crisis in State Highway Finances: Its Roots, Current Effects, and Some Possible Remedies. *Journal of Public Budgeting, Accounting & Financial Management*, *25*(3), 502–521.

Yusuf, J.-E. W., & O'Connell, L. (2013). The Crisis in State Highway Finances: Its Roots, Current Effects, and Some Possible Remedies. *Journal of Public Budgeting, Accounting & Financial Management*, *25*(3), 502–521. doi:10.1108/JPBAFM-25-03-2013-B007

Yusuf, J.-E. W., & O'Connell, L. (2014). Outsourcing Expert Services by State Transportation Departments: A Look at Effects on Cost, Quality, and Changing Employment Levels. *American Review of Public Administration*, *44*(4), 477–492. doi:10.1177/0275074012469460

Yusuf, J.-E. W., O'Connell, L., & Abutabenjeh, S. (2011). Paying for Locally-owned Roads: A Crisis in Local Government Highway Finance. *Public Works Management & Policy*, *16*(3), 250–269. doi:10.1177/1087724X11402357

Yusuf, J.-E. W., O'Connell, L., & Anuar, K. A. (2014). For whom the tunnel be tolled: A four-factor model for explaining willingness-to-pay tolls. *Transportation Research Part A, Policy and Practice*, *59*, 13–21. doi:10.1016/j.tra.2013.10.021

Yusuf, J.-E. W., O'Connell, L., Hackbart, M., & Wallace, C. (2008). An Empirical Examination of the Statutory Characteristics and Effects of Highway and Transportation Commissions on DOT Capital Management Capacity. *Public Works Management & Policy*, *12*(3), 533–543. doi:10.1177/1087724X07311260

Zezima, K. (2010, Dec. 10). Body Fell From Plane, Authorities Say. *The New York Times*. Retrieved from https://www.nytimes.com/2010/12/11/us/11plane.html

Zhao, Z. J., & Hou, Y. (2008). Local option sales taxes and fiscal disparity: The case of Georgia counties. *Public Budgeting & Finance*, *28*(1), 39–57. doi:10.1111/j.1540-5850.2008.00896.x

Zhao, Z. J., & Jung, C. (2008). Does Earmarked Revenue Provide Property Tax Relief? Long-Term Budgetary Effects of Georgia's Local Option Sales Tax. *Public Budgeting & Finance*, *28*(4), 52–70. doi:10.1111/j.1540-5850.2008.00916.x

Zhao, Z., Guo, H., Coyle, D., Robinson, F., & Munnich, L. (2015). Revisiting the Fuel Tax–Based Transportation Funding System in the United States. *Public Works Management & Policy*, *20*(2), 105–126. doi:10.1177/1087724X14539139

Zmud, J. (2008). The Public Supports Pricing If... A Synthesis of Public Opinion Studies on Tolling and Road Pricing. *Connecting People, Places and Ideas*, 29-39.

Zmud, J., & Arce, C. (2008). *Compilation of Public Opinion Data on Tolls and Road Pricing*. Washington, DC: Transportation Research Board of the National Academies. doi:10.17226/14151

Zullo, R. (2007). Transit Contracting Re-examined: Determinants of Cost Efficiency and Resource Allocation. *Journal of Public Administration Research and Theory*.

Related References

To continue our tradition of advancing academic research, we have compiled a list of recommended IGI Global readings. These references will provide additional information and guidance to further enrich your knowledge and assist you with your own research and future publications.

Abed, S., Khir, T., & Ben Brahim, A. (2016). Thermodynamic and Energy Study of a Regenerator in Gas Turbine Cycle and Optimization of Performances. *International Journal of Energy Optimization and Engineering*, *5*(2), 25–44. doi:10.4018/IJEOE.2016040102

Abu Bakar, W. A., Abdullah, W. N., Ali, R., & Mokhtar, W. N. (2016). Polymolybdate Supported Nano Catalyst for Desulfurization of Diesel. In T. Saleh (Ed.), *Applying Nanotechnology to the Desulfurization Process in Petroleum Engineering* (pp. 263–280). Hershey, PA: IGI Global. doi:10.4018/978-1-4666-9545-0.ch009

Addo-Tenkorang, R., Helo, P., & Kantola, J. (2016). Engineer-To-Order Product Development: A Communication Network Analysis for Supply-Chain's Sustainable Competitive Advantage. In R. Addo-Tenkorang, J. Kantola, P. Helo, & A. Shamsuzzoha (Eds.), *Supply Chain Strategies and the Engineer-to-Order Approach* (pp. 43–59). Hershey, PA: IGI Global. doi:10.4018/978-1-5225-0021-6.ch003

Adebiyi, I. D., Popoola, P. A., & Pityana, S. (2016). Mitigation of Wear Damage by Laser Surface Alloying Technique. In E. Akinlabi, R. Mahamood, & S. Akinlabi (Eds.), *Advanced Manufacturing Techniques Using Laser Material Processing* (pp. 172–196). Hershey, PA: IGI Global. doi:10.4018/978-1-5225-0329-3.ch007

Ahmad, W. (2016). Sulfur in Petroleum: Petroleum Desulfurization Techniques. In T. Saleh (Ed.), *Applying Nanotechnology to the Desulfurization Process in Petroleum Engineering* (pp. 1–52). Hershey, PA: IGI Global. doi:10.4018/978-1-4666-9545-0.ch001

Ahmed, I., Ahmad, N., Mehmood, I., Haq, I. U., Hassan, M., & Khan, M. U. (2016). Applications of Nanotechnology in Transportation Engineering. In A. Khitab & W. Anwar (Eds.), *Advanced Research on Nanotechnology for Civil Engineering Applications* (pp. 180–207). Hershey, PA: IGI Global. doi:10.4018/978-1-5225-0344-6.ch006

Aikhuele, D. (2018). A Study of Product Development Engineering and Design Reliability Concerns. *International Journal of Applied Industrial Engineering*, *5*(1), 79–89. doi:10.4018/IJAIE.2018010105

Al-Najar, B. T., & Bououdina, M. (2016). Bioinspired Nanoparticles for Efficient Drug Delivery System. In M. Bououdina (Ed.), *Emerging Research on Bioinspired Materials Engineering* (pp. 69–103). Hershey, PA: IGI Global. doi:10.4018/978-1-4666-9811-6.ch003

Al-Shebeeb, O. A., Rangaswamy, S., Gopalakrishan, B., & Devaru, D. G. (2017). Evaluation and Indexing of Process Plans Based on Electrical Demand and Energy Consumption. *International Journal of Manufacturing, Materials, and Mechanical Engineering*, *7*(3), 1–19. doi:10.4018/IJMMME.2017070101

Alexakis, H., & Makris, N. (2016). Validation of the Discrete Element Method for the Limit Stability Analysis of Masonry Arches. In V. Sarhosis, K. Bagi, J. Lemos, & G. Milani (Eds.), *Computational Modeling of Masonry Structures Using the Discrete Element Method* (pp. 292–325). Hershey, PA: IGI Global. doi:10.4018/978-1-5225-0231-9.ch012

AlMegren, H. A., Gonzalez-Cortes, S., Huang, Y., Chen, H., Qian, Y., Alkinany, M., ... Xiao, T. (2016). Preparation of Deep Hydrodesulfurzation Catalysts for Diesel Fuel using Organic Matrix Decomposition Method. In H. Al-Megren & T. Xiao (Eds.), *Petrochemical Catalyst Materials, Processes, and Emerging Technologies* (pp. 216–253). Hershey, PA: IGI Global. doi:10.4018/978-1-4666-9975-5.ch009

Alshammari, A., Kalevaru, V. N., Bagabas, A., & Martin, A. (2016). Production of Ethylene and its Commercial Importance in the Global Market. In H. Al-Megren & T. Xiao (Eds.), *Petrochemical Catalyst Materials, Processes, and Emerging Technologies* (pp. 82–115). Hershey, PA: IGI Global. doi:10.4018/978-1-4666-9975-5.ch004

Amel, M. (2016). Synthesis, Characterizations, and Biological Effects Study of Some Quinoline Family. In M. Bououdina (Ed.), *Emerging Research on Bioinspired Materials Engineering* (pp. 160–196). Hershey, PA: IGI Global. doi:10.4018/978-1-4666-9811-6.ch006

Amna, T., Haasan, M. S., Khil, M., & Hwang, I. (2016). Impact of Electrospun Biomimetic Extracellular Environment on Proliferation and Intercellular Communication of Muscle Precursor Cells: An Overview – Intercellular Communication of Muscle Precursor Cells with Extracellular Environment. In M. Bououdina (Ed.), *Emerging Research on Bioinspired Materials Engineering* (pp. 247–265). Hershey, PA: IGI Global. doi:10.4018/978-1-4666-9811-6.ch009

Amuda, M. O., Lawal, T. F., & Akinlabi, E. T. (2017). Research Progress on Rheological Behavior of AA7075 Aluminum Alloy During Hot Deformation. *International Journal of Materials Forming and Machining Processes*, *4*(1), 53–96. doi:10.4018/IJMFMP.2017010104

An, M., & Qin, Y. (2016). Challenges of Railway Safety Risk Assessment and Maintenance Decision Making. In B. Rai (Ed.), *Handbook of Research on Emerging Innovations in Rail Transportation Engineering* (pp. 173–211). Hershey, PA: IGI Global. doi:10.4018/978-1-5225-0084-1.ch009

Anil, M., Ayyildiz-Tamis, D., Tasdemir, S., Sendemir-Urkmez, A., & Gulce-Iz, S. (2016). Bioinspired Materials and Biocompatibility. In M. Bououdina (Ed.), *Emerging Research on Bioinspired Materials Engineering* (pp. 294–322). Hershey, PA: IGI Global. doi:10.4018/978-1-4666-9811-6.ch011

Armutlu, H. (2018). Intelligent Biomedical Engineering Operations by Cloud Computing Technologies. In U. Kose, G. Guraksin, & O. Deperlioglu (Eds.), *Nature-Inspired Intelligent Techniques for Solving Biomedical Engineering Problems* (pp. 297–317). Hershey, PA: IGI Global. doi:10.4018/978-1-5225-4769-3.ch015

Arokiyaraj, S., Saravanan, M., Bharanidharan, R., Islam, V. I., Bououdina, M., & Vincent, S. (2016). Green Synthesis of Metallic Nanoparticles Using Plant Compounds and Their Applications: Metallic Nanoparticles Synthesis Using Plants. In M. Bououdina (Ed.), *Emerging Research on Bioinspired Materials Engineering* (pp. 1–34). Hershey, PA: IGI Global. doi:10.4018/978-1-4666-9811-6.ch001

Atik, M., Sadek, M., & Shahrour, I. (2017). Single-Run Adaptive Pushover Procedure for Shear Wall Structures. In V. Plevris, G. Kremmyda, & Y. Fahjan (Eds.), *Performance-Based Seismic Design of Concrete Structures and Infrastructures* (pp. 59–83). Hershey, PA: IGI Global. doi:10.4018/978-1-5225-2089-4.ch003

Aydin, A., Akyol, E., Gungor, M., Kaya, A., & Tasdelen, S. (2018). Geophysical Surveys in Engineering Geology Investigations With Field Examples. In N. Ceryan (Ed.), *Handbook of Research on Trends and Digital Advances in Engineering Geology* (pp. 257–280). Hershey, PA: IGI Global. doi:10.4018/978-1-5225-2709-1.ch007

Azevedo, N. M., Lemos, J. V., & Rocha de Almeida, J. (2016). Discrete Element Particle Modelling of Stone Masonry. In V. Sarhosis, K. Bagi, J. Lemos, & G. Milani (Eds.), *Computational Modeling of Masonry Structures Using the Discrete Element Method* (pp. 146–170). Hershey, PA: IGI Global. doi:10.4018/978-1-5225-0231-9.ch007

Bamufleh, H. S., Noureldin, M. M., & El-Halwagi, M. M. (2016). Sustainable Process Integration in the Petrochemical Industries. In H. Al-Megren & T. Xiao (Eds.), *Petrochemical Catalyst Materials, Processes, and Emerging Technologies* (pp. 150–163). Hershey, PA: IGI Global. doi:10.4018/978-1-4666-9975-5.ch006

Banerjee, S., Gautam, R. K., Gautam, P. K., Jaiswal, A., & Chattopadhyaya, M. C. (2016). Recent Trends and Advancement in Nanotechnology for Water and Wastewater Treatment: Nanotechnological Approach for Water Purification. In A. Khitab & W. Anwar (Eds.), *Advanced Research on Nanotechnology for Civil Engineering Applications* (pp. 208–252). Hershey, PA: IGI Global. doi:10.4018/978-1-5225-0344-6.ch007

Bas, T. G. (2017). Nutraceutical Industry with the Collaboration of Biotechnology and Nutrigenomics Engineering: The Significance of Intellectual Property in the Entrepreneurship and Scientific Research Ecosystems. In T. Bas & J. Zhao (Eds.), *Comparative Approaches to Biotechnology Development and Use in Developed and Emerging Nations* (pp. 1–17). Hershey, PA: IGI Global. doi:10.4018/978-1-5225-1040-6.ch001

Beale, R., & André, J. (2017). *Design Solutions and Innovations in Temporary Structures*. Hershey, PA: IGI Global. doi:10.4018/978-1-5225-2199-0

Behnam, B. (2017). Simulating Post-Earthquake Fire Loading in Conventional RC Structures. In P. Samui, S. Chakraborty, & D. Kim (Eds.), *Modeling and Simulation Techniques in Structural Engineering* (pp. 425–444). Hershey, PA: IGI Global. doi:10.4018/978-1-5225-0588-4.ch015

Ben Hamida, I., Salah, S. B., Msahli, F., & Mimouni, M. F. (2018). Distribution Network Reconfiguration Using SPEA2 for Power Loss Minimization and Reliability Improvement. *International Journal of Energy Optimization and Engineering*, *7*(1), 50–65. doi:10.4018/IJEOE.2018010103

Benjamin, S. R., de Lima, F., & Rathoure, A. K. (2016). Genetically Engineered Microorganisms for Bioremediation Processes: GEMs for Bioremediaton. In A. Rathoure & V. Dhatwalia (Eds.), *Toxicity and Waste Management Using Bioremediation* (pp. 113–140). Hershey, PA: IGI Global. doi:10.4018/978-1-4666-9734-8.ch006

Bhaskar, S. V., & Kudal, H. N. (2017). Effect of TiCN and AlCrN Coating on Tribological Behaviour of Plasma-nitrided AISI 4140 Steel. *International Journal of Surface Engineering and Interdisciplinary Materials Science, 5*(2), 1–17. doi:10.4018/IJSEIMS.2017070101

Bhowmik, S., Sahoo, P., Acharyya, S. K., Dhar, S., & Chattopadhyay, J. (2016). Effect of Microstructure Degradation on Fracture Toughness of 20MnMoNi55 Steel in DBT Region. *International Journal of Manufacturing, Materials, and Mechanical Engineering, 6*(3), 11–27. doi:10.4018/IJMMME.2016070102

Bhutto, A. W., Abro, R., Abbas, T., Yu, G., & Chen, X. (2016). Desulphurization of Fuel Oils Using Ionic Liquids. In H. Al-Megren & T. Xiao (Eds.), *Petrochemical Catalyst Materials, Processes, and Emerging Technologies* (pp. 254–284). Hershey, PA: IGI Global. doi:10.4018/978-1-4666-9975-5.ch010

Bhuyan, D. (2018). Designing of a Twin Tube Shock Absorber: A Study in Reverse Engineering. In K. Kumar & J. Davim (Eds.), *Design and Optimization of Mechanical Engineering Products* (pp. 83–104). Hershey, PA: IGI Global. doi:10.4018/978-1-5225-3401-3.ch005

Bouloudenine, M., & Bououdina, M. (2016). Toxic Effects of Engineered Nanoparticles on Living Cells. In M. Bououdina (Ed.), *Emerging Research on Bioinspired Materials Engineering* (pp. 35–68). Hershey, PA: IGI Global. doi:10.4018/978-1-4666-9811-6.ch002

Brunetti, A., Sellaro, M., Drioli, E., & Barbieri, G. (2016). Membrane Engineering and its Role in Oil Refining and Petrochemical Industry. In H. Al-Megren & T. Xiao (Eds.), *Petrochemical Catalyst Materials, Processes, and Emerging Technologies* (pp. 116–149). Hershey, PA: IGI Global. doi:10.4018/978-1-4666-9975-5.ch005

Bügler, M., & Borrmann, A. (2016). Simulation Based Construction Project Schedule Optimization: An Overview on the State-of-the-Art. In F. Miranda & C. Abreu (Eds.), *Handbook of Research on Computational Simulation and Modeling in Engineering* (pp. 482–507). Hershey, PA: IGI Global. doi:10.4018/978-1-4666-8823-0.ch016

Calderon, F. A., Giolo, E. G., Frau, C. D., Rengel, M. G., Rodriguez, H., Tornello, M., ... Gallucci, R. (2018). Seismic Microzonation and Site Effects Detection Through Microtremors Measures: A Review. In N. Ceryan (Ed.), *Handbook of Research on Trends and Digital Advances in Engineering Geology* (pp. 326–349). Hershey, PA: IGI Global. doi:10.4018/978-1-5225-2709-1.ch009

Carmona-Murillo, J., & Valenzuela-Valdés, J. F. (2016). Motivation on Problem Based Learning. In D. Fonseca & E. Redondo (Eds.), *Handbook of Research on Applied E-Learning in Engineering and Architecture Education* (pp. 179–203). Hershey, PA: IGI Global. doi:10.4018/978-1-4666-8803-2.ch009

Ceryan, N. (2016). A Review of Soft Computing Methods Application in Rock Mechanic Engineering. In P. Samui (Ed.), *Handbook of Research on Advanced Computational Techniques for Simulation-Based Engineering* (pp. 1–70). Hershey, PA: IGI Global. doi:10.4018/978-1-4666-9479-8.ch001

Ceryan, N., & Can, N. K. (2018). Prediction of The Uniaxial Compressive Strength of Rocks Materials. In N. Ceryan (Ed.), *Handbook of Research on Trends and Digital Advances in Engineering Geology* (pp. 31–96). Hershey, PA: IGI Global. doi:10.4018/978-1-5225-2709-1.ch002

Ceryan, S. (2018). Weathering Indices Used in Evaluation of the Weathering State of Rock Material. In N. Ceryan (Ed.), *Handbook of Research on Trends and Digital Advances in Engineering Geology* (pp. 132–186). Hershey, PA: IGI Global. doi:10.4018/978-1-5225-2709-1.ch004

Chandrasekaran, S., Silva, B., Patil, A., Oo, A. M., & Campbell, M. (2016). Evaluating Engineering Students' Perceptions: The Impact of Team-Based Learning Practices in Engineering Education. *International Journal of Quality Assurance in Engineering and Technology Education, 5*(4), 42–59. doi:10.4018/IJQAETE.2016100103

Chen, H., Padilla, R. V., & Besarati, S. (2017). Supercritical Fluids and Their Applications in Power Generation. In L. Chen & Y. Iwamoto (Eds.), *Advanced Applications of Supercritical Fluids in Energy Systems* (pp. 369–402). Hershey, PA: IGI Global. doi:10.4018/978-1-5225-2047-4.ch012

Chen, L. (2017). Principles, Experiments, and Numerical Studies of Supercritical Fluid Natural Circulation System. In L. Chen & Y. Iwamoto (Eds.), *Advanced Applications of Supercritical Fluids in Energy Systems* (pp. 136–187). Hershey, PA: IGI Global. doi:10.4018/978-1-5225-2047-4.ch005

Clementi, F., Di Sciascio, G., Di Sciascio, S., & Lenci, S. (2017). Influence of the Shear-Bending Interaction on the Global Capacity of Reinforced Concrete Frames: A Brief Overview of the New Perspectives. In V. Plevris, G. Kremmyda, & Y. Fahjan (Eds.), *Performance-Based Seismic Design of Concrete Structures and Infrastructures* (pp. 84–111). Hershey, PA: IGI Global. doi:10.4018/978-1-5225-2089-4.ch004

Cortés-Polo, D., Calle-Cancho, J., Carmona-Murillo, J., & González-Sánchez, J. (2017). Future Trends in Mobile-Fixed Integration for Next Generation Networks: Classification and Analysis. *International Journal of Vehicular Telematics and Infotainment Systems*, *1*(1), 33–53. doi:10.4018/IJVTIS.2017010103

Cui, X., Zeng, S., Li, Z., Zheng, Q., Yu, X., & Han, B. (2018). Advanced Composites for Civil Engineering Infrastructures. In K. Kumar & J. Davim (Eds.), *Composites and Advanced Materials for Industrial Applications* (pp. 212–248). Hershey, PA: IGI Global. doi:10.4018/978-1-5225-5216-1.ch010

Dalgıç, S., & Kuşku, İ. (2018). Geological and Geotechnical Investigations in Tunneling. In N. Ceryan (Ed.), *Handbook of Research on Trends and Digital Advances in Engineering Geology* (pp. 482–529). Hershey, PA: IGI Global. doi:10.4018/978-1-5225-2709-1.ch014

de la Varga, D., Soto, M., Arias, C. A., van Oirschot, D., Kilian, R., Pascual, A., & Álvarez, J. A. (2017). Constructed Wetlands for Industrial Wastewater Treatment and Removal of Nutrients. In Á. Val del Río, J. Campos Gómez, & A. Mosquera Corral (Eds.), *Technologies for the Treatment and Recovery of Nutrients from Industrial Wastewater* (pp. 202–230). Hershey, PA: IGI Global. doi:10.4018/978-1-5225-1037-6.ch008

del Valle-Zermeño, R., Chimenos, J. M., & Formosa, J. (2016). Flue Gas Desulfurization: Processes and Technologies. In T. Saleh (Ed.), *Applying Nanotechnology to the Desulfurization Process in Petroleum Engineering* (pp. 337–377). Hershey, PA: IGI Global. doi:10.4018/978-1-4666-9545-0.ch011

Delgado, J. M., Henriques, A. A., & Delgado, R. M. (2016). Structural Non-Linear Models and Simulation Techniques: An Efficient Combination for Safety Evaluation of RC Structures. In F. Miranda & C. Abreu (Eds.), *Handbook of Research on Computational Simulation and Modeling in Engineering* (pp. 540–584). Hershey, PA: IGI Global. doi:10.4018/978-1-4666-8823-0.ch018

Delgado, P. S., Arêde, A., Pouca, N. V., & Costa, A. (2016). Numerical Modeling of RC Bridges for Seismic Risk Analysis. In F. Miranda & C. Abreu (Eds.), *Handbook of Research on Computational Simulation and Modeling in Engineering* (pp. 457–481). Hershey, PA: IGI Global. doi:10.4018/978-1-4666-8823-0.ch015

Deng, Y., & Liu, S. (2016). Catalysis with Room Temperature Ionic Liquids Mediated Metal Nanoparticles. In H. Al-Megren & T. Xiao (Eds.), *Petrochemical Catalyst Materials, Processes, and Emerging Technologies* (pp. 285–329). Hershey, PA: IGI Global. doi:10.4018/978-1-4666-9975-5.ch011

Deperlioglu, O. (2018). Intelligent Techniques Inspired by Nature and Used in Biomedical Engineering. In U. Kose, G. Guraksin, & O. Deperlioglu (Eds.), *Nature-Inspired Intelligent Techniques for Solving Biomedical Engineering Problems* (pp. 51–77). Hershey, PA: IGI Global. doi:10.4018/978-1-5225-4769-3.ch003

Dias, G. L., Magalhães, R. R., Ferreira, D. D., & Vitoriano, F. A. (2016). The Use of a Robotic Arm for Displacement Measurements in a Cantilever beam. *International Journal of Manufacturing, Materials, and Mechanical Engineering*, *6*(3), 45–57. doi:10.4018/IJMMME.2016070104

Dimitratos, N., Villa, A., Chan-Thaw, C. E., Hammond, C., & Prati, L. (2016). Valorisation of Glycerol to Fine Chemicals and Fuels. In H. Al-Megren & T. Xiao (Eds.), *Petrochemical Catalyst Materials, Processes, and Emerging Technologies* (pp. 352–384). Hershey, PA: IGI Global. doi:10.4018/978-1-4666-9975-5.ch013

Dixit, A. (2018). Application of Silica-Gel-Reinforced Aluminium Composite on the Piston of Internal Combustion Engine: Comparative Study of Silica-Gel-Reinforced Aluminium Composite Piston With Aluminium Alloy Piston. In K. Kumar & J. Davim (Eds.), *Composites and Advanced Materials for Industrial Applications* (pp. 63–98). Hershey, PA: IGI Global. doi:10.4018/978-1-5225-5216-1.ch004

Drei, A., Milani, G., & Sincraian, G. (2016). Application of DEM to Historic Masonries, Two Case-Studies in Portugal and Italy: Aguas Livres Aqueduct and Arch-Tympana of a Church. In V. Sarhosis, K. Bagi, J. Lemos, & G. Milani (Eds.), *Computational Modeling of Masonry Structures Using the Discrete Element Method* (pp. 326–366). Hershey, PA: IGI Global. doi:10.4018/978-1-5225-0231-9.ch013

Dutta, S., Roy, P. K., & Nandi, D. (2016). Optimal Allocation of Static Synchronous Series Compensator Controllers using Chemical Reaction Optimization for Reactive Power Dispatch. *International Journal of Energy Optimization and Engineering*, *5*(3), 43–62. doi:10.4018/IJEOE.2016070103

Dutta, S., Roy, P. K., & Nandi, D. (2016). Quasi Oppositional Teaching-Learning based Optimization for Optimal Power Flow Incorporating FACTS. *International Journal of Energy Optimization and Engineering*, *5*(2), 64–84. doi:10.4018/IJEOE.2016040104

Eloy, S., Dias, M. S., Lopes, P. F., & Vilar, E. (2016). Digital Technologies in Architecture and Engineering: Exploring an Engaged Interaction within Curricula. In D. Fonseca & E. Redondo (Eds.), *Handbook of Research on Applied E-Learning in Engineering and Architecture Education* (pp. 368–402). Hershey, PA: IGI Global. doi:10.4018/978-1-4666-8803-2.ch017

Elsayed, A. M., Dakkama, H. J., Mahmoud, S., Al-Dadah, R., & Kaialy, W. (2017). Sustainable Cooling Research Using Activated Carbon Adsorbents and Their Environmental Impact. In T. Kobayashi (Ed.), *Applied Environmental Materials Science for Sustainability* (pp. 186–221). Hershey, PA: IGI Global. doi:10.4018/978-1-5225-1971-3.ch009

Ercanoglu, M., & Sonmez, H. (2018). General Trends and New Perspectives on Landslide Mapping and Assessment Methods. In N. Ceryan (Ed.), *Handbook of Research on Trends and Digital Advances in Engineering Geology* (pp. 350–379). Hershey, PA: IGI Global. doi:10.4018/978-1-5225-2709-1.ch010

Erinosho, M. F., Akinlabi, E. T., & Pityana, S. (2016). Enhancement of Surface Integrity of Titanium Alloy with Copper by Means of Laser Metal Deposition Process. In E. Akinlabi, R. Mahamood, & S. Akinlabi (Eds.), *Advanced Manufacturing Techniques Using Laser Material Processing* (pp. 60–91). Hershey, PA: IGI Global. doi:10.4018/978-1-5225-0329-3.ch004

Farag, H., & Kishida, M. (2016). Kinetic Models for Complex Parallel–Consecutive Reactions Assessment of Reaction Network and Product Selectivity. In H. Al-Megren & T. Xiao (Eds.), *Petrochemical Catalyst Materials, Processes, and Emerging Technologies* (pp. 330–351). Hershey, PA: IGI Global. doi:10.4018/978-1-4666-9975-5.ch012

Faroz, S. A., Pujari, N. N., Rastogi, R., & Ghosh, S. (2017). Risk Analysis of Structural Engineering Systems Using Bayesian Inference. In P. Samui, S. Chakraborty, & D. Kim (Eds.), *Modeling and Simulation Techniques in Structural Engineering* (pp. 390–424). Hershey, PA: IGI Global. doi:10.4018/978-1-5225-0588-4.ch014

Fernando, P. R., Hamigah, T., Disne, S., Wickramasingha, G. G., & Sutharshan, A. (2018). The Evaluation of Engineering Properties of Low Cost Concrete Blocks by Partial Doping of Sand with Sawdust: Low Cost Sawdust Concrete Block. *International Journal of Strategic Engineering*, *1*(2), 26–42. doi:10.4018/IJoSE.2018070103

Fragiadakis, M., Stefanou, I., & Psycharis, I. N. (2016). Vulnerability Assessment of Damaged Classical Multidrum Columns. In V. Sarhosis, K. Bagi, J. Lemos, & G. Milani (Eds.), *Computational Modeling of Masonry Structures Using the Discrete Element Method* (pp. 235–253). Hershey, PA: IGI Global. doi:10.4018/978-1-5225-0231-9.ch010

Gaines, T. W., Williams, K. R., & Wagener, K. B. (2016). ADMET: Functionalized Polyolefins. In H. Al-Megren & T. Xiao (Eds.), *Petrochemical Catalyst Materials, Processes, and Emerging Technologies* (pp. 1–21). Hershey, PA: IGI Global. doi:10.4018/978-1-4666-9975-5.ch001

Garg, H. (2016). Bi-Criteria Optimization for Finding the Optimal Replacement Interval for Maintaining the Performance of the Process Industries. In P. Vasant, G. Weber, & V. Dieu (Eds.), *Handbook of Research on Modern Optimization Algorithms and Applications in Engineering and Economics* (pp. 643–675). Hershey, PA: IGI Global. doi:10.4018/978-1-4666-9644-0.ch025

Gaspar, P. D., Dinho da Silva, P., Gonçalves, J. P., & Carneiro, R. (2016). Computational Modelling and Simulation to Assist the Improvement of Thermal Performance and Energy Efficiency in Industrial Engineering Systems: Application to Cold Stores. In F. Miranda & C. Abreu (Eds.), *Handbook of Research on Computational Simulation and Modeling in Engineering* (pp. 1–68). Hershey, PA: IGI Global. doi:10.4018/978-1-4666-8823-0.ch001

Ge, H., Tang, M., & Wen, X. (2016). Ni/ZnO Nano Sorbent for Reactive Adsorption Desulfurization of Refinery Oil Streams. In T. Saleh (Ed.), *Applying Nanotechnology to the Desulfurization Process in Petroleum Engineering* (pp. 216–239). Hershey, PA: IGI Global. doi:10.4018/978-1-4666-9545-0.ch007

Ghosh, S., Mitra, S., Ghosh, S., & Chakraborty, S. (2017). Seismic Reliability Analysis in the Framework of Metamodelling Based Monte Carlo Simulation. In P. Samui, S. Chakraborty, & D. Kim (Eds.), *Modeling and Simulation Techniques in Structural Engineering* (pp. 192–208). Hershey, PA: IGI Global. doi:10.4018/978-1-5225-0588-4.ch006

Gil, M., & Otero, B. (2017). Learning Engineering Skills through Creativity and Collaboration: A Game-Based Proposal. In R. Alexandre Peixoto de Queirós & M. Pinto (Eds.), *Gamification-Based E-Learning Strategies for Computer Programming Education* (pp. 14–29). Hershey, PA: IGI Global. doi:10.4018/978-1-5225-1034-5.ch002

Gill, J., Ayre, M., & Mills, J. (2017). Revisioning the Engineering Profession: How to Make It Happen! In M. Gray & K. Thomas (Eds.), *Strategies for Increasing Diversity in Engineering Majors and Careers* (pp. 156–175). Hershey, PA: IGI Global. doi:10.4018/978-1-5225-2212-6.ch008

Gopal, S., & Al-Hazmi, M. H. (2016). Advances in Catalytic Technologies for Selective Oxidation of Lower Alkanes. In H. Al-Megren & T. Xiao (Eds.), *Petrochemical Catalyst Materials, Processes, and Emerging Technologies* (pp. 22–52). Hershey, PA: IGI Global. doi:10.4018/978-1-4666-9975-5.ch002

Goyal, N., Ram, M., Bhardwaj, A., & Kumar, A. (2016). Thermal Power Plant Modelling with Fault Coverage Stochastically. *International Journal of Manufacturing, Materials, and Mechanical Engineering*, *6*(3), 28–44. doi:10.4018/IJMMME.2016070103

Goyal, N., Ram, M., & Kumar, P. (2017). Welding Process under Fault Coverage Approach for Reliability and MTTF. In M. Ram & J. Davim (Eds.), *Mathematical Concepts and Applications in Mechanical Engineering and Mechatronics* (pp. 222–245). Hershey, PA: IGI Global. doi:10.4018/978-1-5225-1639-2.ch011

Gray, M., & Lundy, C. (2017). Engineering Study Abroad: High Impact Strategy for Increasing Access. In M. Gray & K. Thomas (Eds.), *Strategies for Increasing Diversity in Engineering Majors and Careers* (pp. 42–59). Hershey, PA: IGI Global. doi:10.4018/978-1-5225-2212-6.ch003

Guha, D., Roy, P. K., & Banerjee, S. (2016). Application of Modified Biogeography Based Optimization in AGC of an Interconnected Multi-Unit Multi-Source AC-DC Linked Power System. *International Journal of Energy Optimization and Engineering*, *5*(3), 1–18. doi:10.4018/IJEOE.2016070101

Guha, D., Roy, P. K., & Banerjee, S. (2016). Grey Wolf Optimization to Solve Load Frequency Control of an Interconnected Power System: GWO Used to Solve LFC Problem. *International Journal of Energy Optimization and Engineering*, *5*(4), 62–83. doi:10.4018/IJEOE.2016100104

Gupta, A. K., Dey, A., & Mukhopadhyay, A. K. (2016). Micromechanical and Finite Element Modeling for Composites. In S. Datta & J. Davim (Eds.), *Computational Approaches to Materials Design: Theoretical and Practical Aspects* (pp. 101–162). Hershey, PA: IGI Global. doi:10.4018/978-1-5225-0290-6.ch005

Guraksin, G. E. (2018). Internet of Things and Nature-Inspired Intelligent Techniques for the Future of Biomedical Engineering. In U. Kose, G. Guraksin, & O. Deperlioglu (Eds.), *Nature-Inspired Intelligent Techniques for Solving Biomedical Engineering Problems* (pp. 263–282). Hershey, PA: IGI Global. doi:10.4018/978-1-5225-4769-3.ch013

Hansman, C. A. (2016). Developing Mentoring Programs in Engineering and Technology Education. *International Journal of Quality Assurance in Engineering and Technology Education*, *5*(2), 1–15. doi:10.4018/IJQAETE.2016040101

Hasan, U., Chegenizadeh, A., & Nikraz, H. (2016). Nanotechnology Future and Present in Construction Industry: Applications in Geotechnical Engineering. In A. Khitab & W. Anwar (Eds.), *Advanced Research on Nanotechnology for Civil Engineering Applications* (pp. 141–179). Hershey, PA: IGI Global. doi:10.4018/978-1-5225-0344-6.ch005

Hejazi, T., & Akbari, L. (2017). A Multiresponse Optimization Model for Statistical Design of Processes with Discrete Variables. In M. Ram & J. Davim (Eds.), *Mathematical Concepts and Applications in Mechanical Engineering and Mechatronics* (pp. 17–37). Hershey, PA: IGI Global. doi:10.4018/978-1-5225-1639-2.ch002

Hejazi, T., & Hejazi, A. (2017). Monte Carlo Simulation for Reliability-Based Design of Automotive Complex Subsystems. In M. Ram & J. Davim (Eds.), *Mathematical Concepts and Applications in Mechanical Engineering and Mechatronics* (pp. 177–200). Hershey, PA: IGI Global. doi:10.4018/978-1-5225-1639-2.ch009

Hejazi, T., & Poursabbagh, H. (2017). Reliability Analysis of Engineering Systems: An Accelerated Life Testing for Boiler Tubes. In M. Ram & J. Davim (Eds.), *Mathematical Concepts and Applications in Mechanical Engineering and Mechatronics* (pp. 154–176). Hershey, PA: IGI Global. doi:10.4018/978-1-5225-1639-2.ch008

Henao, J., & Sotelo, O. (2018). Surface Engineering at High Temperature: Thermal Cycling and Corrosion Resistance. In A. Paksersht (Ed.), *Production, Properties, and Applications of High Temperature Coatings* (pp. 131–159). Hershey, PA: IGI Global. doi:10.4018/978-1-5225-4194-3.ch006

Huirache-Acuña, R., Alonso-Nuñez, G., Rivera-Muñoz, E. M., Gutierrez, O., & Pawelec, B. (2016). Trimetallic Sulfide Catalysts for Hydrodesulfurization. In T. Saleh (Ed.), *Applying Nanotechnology to the Desulfurization Process in Petroleum Engineering* (pp. 240–262). Hershey, PA: IGI Global. doi:10.4018/978-1-4666-9545-0.ch008

Ilori, O. O., Adetan, D. A., & Umoru, L. E. (2017). Effect of Cutting Parameters on the Surface Residual Stress of Face-Milled Pearlitic Ductile Iron. *International Journal of Materials Forming and Machining Processes*, *4*(1), 38–52. doi:10.4018/IJMFMP.2017010103

Imam, M. H., Tasadduq, I. A., Ahmad, A., Aldosari, F., & Khan, H. (2017). Automated Generation of Course Improvement Plans Using Expert System. *International Journal of Quality Assurance in Engineering and Technology Education*, *6*(1), 1–12. doi:10.4018/IJQAETE.2017010101

Injeti, S. K., & Kumar, T. V. (2018). A WDO Framework for Optimal Deployment of DGs and DSCs in a Radial Distribution System Under Daily Load Pattern to Improve Techno-Economic Benefits. *International Journal of Energy Optimization and Engineering*, *7*(2), 1–38. doi:10.4018/IJEOE.2018040101

Ishii, N., Anami, K., & Knisely, C. W. (2018). *Dynamic Stability of Hydraulic Gates and Engineering for Flood Prevention*. Hershey, PA: IGI Global. doi:10.4018/978-1-5225-3079-4

J., J., Chowdhury, S., Goyal, P., Samui, P., & Dalkiliç, Y. (2016). Determination of Bearing Capacity of Shallow Foundation Using Soft Computing. In P. Saxena, D. Singh, & M. Pant (Eds.), *Problem Solving and Uncertainty Modeling through Optimization and Soft Computing Applications* (pp. 292-328). Hershey, PA: IGI Global. doi:10.4018/978-1-4666-9885-7.ch014

Jagan, J., Gundlapalli, P., & Samui, P. (2016). Utilization of Classification Techniques for the Determination of Liquefaction Susceptibility of Soils. In S. Bhattacharyya, P. Banerjee, D. Majumdar, & P. Dutta (Eds.), *Handbook of Research on Advanced Hybrid Intelligent Techniques and Applications* (pp. 124–160). Hershey, PA: IGI Global. doi:10.4018/978-1-4666-9474-3.ch005

Jayapalan, S. (2018). A Review of Chemical Treatments on Natural Fibers-Based Hybrid Composites for Engineering Applications. In K. Kumar & J. Davim (Eds.), *Composites and Advanced Materials for Industrial Applications* (pp. 16–37). Hershey, PA: IGI Global. doi:10.4018/978-1-5225-5216-1.ch002

Jeet, K., & Dhir, R. (2016). Software Module Clustering Using Bio-Inspired Algorithms. In P. Vasant, G. Weber, & V. Dieu (Eds.), *Handbook of Research on Modern Optimization Algorithms and Applications in Engineering and Economics* (pp. 445–470). Hershey, PA: IGI Global. doi:10.4018/978-1-4666-9644-0.ch017

Joshi, S. D., & Talange, D. B. (2016). Fault Tolerant Control for a Fractional Order AUV System. *International Journal of Energy Optimization and Engineering*, *5*(2), 1–24. doi:10.4018/IJEOE.2016040101

Julião, D., Ribeiro, S., de Castro, B., Cunha-Silva, L., & Balula, S. S. (2016). Polyoxometalates-Based Nanocatalysts for Production of Sulfur-Free Diesel. In T. Saleh (Ed.), *Applying Nanotechnology to the Desulfurization Process in Petroleum Engineering* (pp. 426–458). Hershey, PA: IGI Global. doi:10.4018/978-1-4666-9545-0.ch014

Kamthan, P. (2016). On the Nature of Collaborations in Agile Software Engineering Course Projects. *International Journal of Quality Assurance in Engineering and Technology Education*, *5*(2), 42–59. doi:10.4018/IJQAETE.2016040104

Karaman, O., Celik, C., & Urkmez, A. S. (2016). Self-Assembled Biomimetic Scaffolds for Bone Tissue Engineering. In M. Bououdina (Ed.), *Emerging Research on Bioinspired Materials Engineering* (pp. 104–132). Hershey, PA: IGI Global. doi:10.4018/978-1-4666-9811-6.ch004

Karkalos, N. E., Markopoulos, A. P., & Dossis, M. F. (2017). Optimal Model Parameters of Inverse Kinematics Solution of a 3R Robotic Manipulator Using ANN Models. *International Journal of Manufacturing, Materials, and Mechanical Engineering*, *7*(3), 20–40. doi:10.4018/IJMMME.2017070102

Kesimal, A., Karaman, K., Cihangir, F., & Ercikdi, B. (2018). Excavatability Assessment of Rock Masses for Geotechnical Studies. In N. Ceryan (Ed.), *Handbook of Research on Trends and Digital Advances in Engineering Geology* (pp. 231–256). Hershey, PA: IGI Global. doi:10.4018/978-1-5225-2709-1.ch006

Khanh, D. V., Vasant, P. M., Elamvazuthi, I., & Dieu, V. N. (2016). Multi-Objective Optimization of Two-Stage Thermo-Electric Cooler Using Differential Evolution: MO Optimization of TEC Using DE. In F. Miranda & C. Abreu (Eds.), *Handbook of Research on Computational Simulation and Modeling in Engineering* (pp. 139–170). Hershey, PA: IGI Global. doi:10.4018/978-1-4666-8823-0.ch004

Kim, D., Hassan, M. K., Chang, S., & Bigdeli, Y. (2016). Nonlinear Vibration Control of 3D Irregular Structures Subjected to Seismic Loads. In P. Samui (Ed.), *Handbook of Research on Advanced Computational Techniques for Simulation-Based Engineering* (pp. 103–119). Hershey, PA: IGI Global. doi:10.4018/978-1-4666-9479-8.ch003

Knoflacher, H. (2017). The Role of Engineers and Their Tools in the Transport Sector after Paradigm Change: From Assumptions and Extrapolations to Science. In H. Knoflacher & E. Ocalir-Akunal (Eds.), *Engineering Tools and Solutions for Sustainable Transportation Planning* (pp. 1–29). Hershey, PA: IGI Global. doi:10.4018/978-1-5225-2116-7.ch001

Kose, U. (2018). Towards an Intelligent Biomedical Engineering With Nature-Inspired Artificial Intelligence Techniques. In U. Kose, G. Guraksin, & O. Deperlioglu (Eds.), *Nature-Inspired Intelligent Techniques for Solving Biomedical Engineering Problems* (pp. 1–26). Hershey, PA: IGI Global. doi:10.4018/978-1-5225-4769-3.ch001

Kostić, S. (2018). A Review on Enhanced Stability Analyses of Soil Slopes Using Statistical Design. In N. Ceryan (Ed.), *Handbook of Research on Trends and Digital Advances in Engineering Geology* (pp. 446–481). Hershey, PA: IGI Global. doi:10.4018/978-1-5225-2709-1.ch013

Kumar, A., Patil, P. P., & Prajapati, Y. K. (2018). *Advanced Numerical Simulations in Mechanical Engineering*. Hershey, PA: IGI Global. doi:10.4018/978-1-5225-3722-9

Kumar, G. R., Rajyalakshmi, G., & Manupati, V. K. (2017). Surface Micro Patterning of Aluminium Reinforced Composite through Laser Peening. *International Journal of Manufacturing, Materials, and Mechanical Engineering*, *7*(4), 15–27. doi:10.4018/IJMMME.2017100102

Kumari, N., & Kumar, K. (2018). Fabrication of Orthotic Calipers With Epoxy-Based Green Composite. In K. Kumar & J. Davim (Eds.), *Composites and Advanced Materials for Industrial Applications* (pp. 157–176). Hershey, PA: IGI Global. doi:10.4018/978-1-5225-5216-1.ch008

Kuppusamy, R. R. (2018). Development of Aerospace Composite Structures Through Vacuum-Enhanced Resin Transfer Moulding Technology (VERTMTy): Vacuum-Enhanced Resin Transfer Moulding. In K. Kumar & J. Davim (Eds.), *Composites and Advanced Materials for Industrial Applications* (pp. 99–111). Hershey, PA: IGI Global. doi:10.4018/978-1-5225-5216-1.ch005

Lemos, J. V. (2016). The Basis for Masonry Analysis with UDEC and 3DEC. In V. Sarhosis, K. Bagi, J. Lemos, & G. Milani (Eds.), *Computational Modeling of Masonry Structures Using the Discrete Element Method* (pp. 61–89). Hershey, PA: IGI Global. doi:10.4018/978-1-5225-0231-9.ch003

Loy, J., Howell, S., & Cooper, R. (2017). Engineering Teams: Supporting Diversity in Engineering Education. In M. Gray & K. Thomas (Eds.), *Strategies for Increasing Diversity in Engineering Majors and Careers* (pp. 106–129). Hershey, PA: IGI Global. doi:10.4018/978-1-5225-2212-6.ch006

Macher, G., Armengaud, E., Kreiner, C., Brenner, E., Schmittner, C., Ma, Z., ... Krammer, M. (2018). Integration of Security in the Development Lifecycle of Dependable Automotive CPS. In N. Druml, A. Genser, A. Krieg, M. Menghin, & A. Hoeller (Eds.), *Solutions for Cyber-Physical Systems Ubiquity* (pp. 383–423). Hershey, PA: IGI Global. doi:10.4018/978-1-5225-2845-6.ch015

Maghsoodlou, S., & Poreskandar, S. (2016). Controlling Electrospinning Jet Using Microscopic Model for Ideal Tissue Engineering Scaffolds. *International Journal of Chemoinformatics and Chemical Engineering*, *5*(2), 1–16. doi:10.4018/IJCCE.2016070101

Mahendramani, G., & Lakshmana Swamy, N. (2018). Effect of Weld Groove Area on Distortion of Butt Welded Joints in Submerged Arc Welding. *International Journal of Manufacturing, Materials, and Mechanical Engineering*, *8*(2), 33–44. doi:10.4018/IJMMME.2018040103

Maiti, S. (2016). Engineered Gellan Polysaccharides in the Design of Controlled Drug Delivery Systems. In M. Bououdina (Ed.), *Emerging Research on Bioinspired Materials Engineering* (pp. 266–293). Hershey, PA: IGI Global. doi:10.4018/978-1-4666-9811-6.ch010

Majumdar, J. D., Weisheit, A., & Manna, I. (2016). Laser Surface Processing for Tailoring of Properties by Optimization of Microstructure. In E. Akinlabi, R. Mahamood, & S. Akinlabi (Eds.), *Advanced Manufacturing Techniques Using Laser Material Processing* (pp. 121–171). Hershey, PA: IGI Global. doi:10.4018/978-1-5225-0329-3.ch006

Maldonado-Macías, A. A., García-Alcaraz, J. L., Hernández-Arellano, J. L., & Cortes-Robles, G. (2016). An Ergonomic Compatibility Perspective on the Selection of Advanced Manufacturing Technology: A Case Study for CNC Vertical Machining Centers. In G. Alor-Hernández, C. Sánchez-Ramírez, & J. García-Alcaraz (Eds.), *Handbook of Research on Managerial Strategies for Achieving Optimal Performance in Industrial Processes* (pp. 137–165). Hershey, PA: IGI Global. doi:10.4018/978-1-5225-0130-5.ch008

Mamaghani, I. H. (2016). Application of Discrete Finite Element Method for Analysis of Unreinforced Masonry Structures. In V. Sarhosis, K. Bagi, J. Lemos, & G. Milani (Eds.), *Computational Modeling of Masonry Structures Using the Discrete Element Method* (pp. 440–458). Hershey, PA: IGI Global. doi:10.4018/978-1-5225-0231-9.ch017

Mansor, M. R., Sapuan, S. M., Salim, M. A., Akop, M. Z., Musthafah, M. T., & Shaharuzaman, M. A. (2016). Concurrent Design of Green Composites. In D. Verma, S. Jain, X. Zhang, & P. Gope (Eds.), *Green Approaches to Biocomposite Materials Science and Engineering* (pp. 48–75). Hershey, PA: IGI Global. doi:10.4018/978-1-5225-0424-5.ch003

Mansouri, I., & Esmaeili, E. (2016). Nanotechnology Applications in the Construction Industry. In A. Khitab & W. Anwar (Eds.), *Advanced Research on Nanotechnology for Civil Engineering Applications* (pp. 111–140). Hershey, PA: IGI Global. doi:10.4018/978-1-5225-0344-6.ch004

Manzoor, A. (2016). MOOCs for Enhancing Engineering Education. In D. Fonseca & E. Redondo (Eds.), *Handbook of Research on Applied E-Learning in Engineering and Architecture Education* (pp. 204–223). Hershey, PA: IGI Global. doi:10.4018/978-1-4666-8803-2.ch010

Martin, A., Kalevaru, V. N., & Radnik, J. (2016). Palladium in Heterogeneous Oxidation Catalysis. In H. Al-Megren & T. Xiao (Eds.), *Petrochemical Catalyst Materials, Processes, and Emerging Technologies* (pp. 53–81). Hershey, PA: IGI Global. doi:10.4018/978-1-4666-9975-5.ch003

Melnyczuk, J. M., & Palchoudhury, S. (2016). Introduction to Bio-Inspired Hydrogel and Their Application: Hydrogels. In M. Bououdina (Ed.), *Emerging Research on Bioinspired Materials Engineering* (pp. 133–159). Hershey, PA: IGI Global. doi:10.4018/978-1-4666-9811-6.ch005

Mitra-Kirtley, S., Mullins, O. C., & Pomerantz, A. E. (2016). Sulfur and Nitrogen Chemical Speciation in Crude Oils and Related Carbonaceous Materials. In T. Saleh (Ed.), *Applying Nanotechnology to the Desulfurization Process in Petroleum Engineering* (pp. 53–83). Hershey, PA: IGI Global. doi:10.4018/978-1-4666-9545-0.ch002

Moalosi, R., Uziak, J., & Oladiran, M. T. (2016). Using Blended Learning Approach to Deliver Courses in An Engineering Programme. *International Journal of Quality Assurance in Engineering and Technology Education*, *5*(1), 23–39. doi:10.4018/IJQAETE.2016010103

Mohammadzadeh, S., & Kim, Y. (2017). Nonlinear System Identification of Smart Buildings. In P. Samui, S. Chakraborty, & D. Kim (Eds.), *Modeling and Simulation Techniques in Structural Engineering* (pp. 328–347). Hershey, PA: IGI Global. doi:10.4018/978-1-5225-0588-4.ch011

Mohanty, I., & Bhattacherjee, D. (2016). Artificial Neural Network and Its Application in Steel Industry. In S. Datta & J. Davim (Eds.), *Computational Approaches to Materials Design: Theoretical and Practical Aspects* (pp. 267–300). Hershey, PA: IGI Global. doi:10.4018/978-1-5225-0290-6.ch010

Mohebkhah, A., & Sarhosis, V. (2016). Discrete Element Modeling of Masonry-Infilled Frames. In V. Sarhosis, K. Bagi, J. Lemos, & G. Milani (Eds.), *Computational Modeling of Masonry Structures Using the Discrete Element Method* (pp. 200–234). Hershey, PA: IGI Global. doi:10.4018/978-1-5225-0231-9.ch009

Molina, G. J., Aktaruzzaman, F., Soloiu, V., & Rahman, M. (2017). Design and Testing of a Jet-Impingement Instrument to Study Surface-Modification Effects by Nanofluids. *International Journal of Surface Engineering and Interdisciplinary Materials Science*, *5*(2), 43–61. doi:10.4018/IJSEIMS.2017070104

Montalvan-Sorrosa, D., de los Cobos-Vasconcelos, D., & Gonzalez-Sanchez, A. (2016). Nanotechnology Applied to the Biodesulfurization of Fossil Fuels and Spent Caustic Streams. In T. Saleh (Ed.), *Applying Nanotechnology to the Desulfurization Process in Petroleum Engineering* (pp. 378–389). Hershey, PA: IGI Global. doi:10.4018/978-1-4666-9545-0.ch012

Montillet, J., Yu, K., Bonenberg, L. K., & Roberts, G. W. (2016). Optimization Algorithms in Local and Global Positioning. In P. Vasant, G. Weber, & V. Dieu (Eds.), *Handbook of Research on Modern Optimization Algorithms and Applications in Engineering and Economics* (pp. 1–53). Hershey, PA: IGI Global. doi:10.4018/978-1-4666-9644-0.ch001

Moreira, F., & Ferreira, M. J. (2016). Teaching and Learning Requirements Engineering Based on Mobile Devices and Cloud: A Case Study. In D. Fonseca & E. Redondo (Eds.), *Handbook of Research on Applied E-Learning in Engineering and Architecture Education* (pp. 237–262). Hershey, PA: IGI Global. doi:10.4018/978-1-4666-8803-2.ch012

Mukherjee, A., Saeed, R. A., Dutta, S., & Naskar, M. K. (2017). Fault Tracking Framework for Software-Defined Networking (SDN). In C. Singhal & S. De (Eds.), *Resource Allocation in Next-Generation Broadband Wireless Access Networks* (pp. 247–272). Hershey, PA: IGI Global. doi:10.4018/978-1-5225-2023-8.ch011

Mukhopadhyay, A., Barman, T. K., & Sahoo, P. (2018). Electroless Nickel Coatings for High Temperature Applications. In K. Kumar & J. Davim (Eds.), *Composites and Advanced Materials for Industrial Applications* (pp. 297–331). Hershey, PA: IGI Global. doi:10.4018/978-1-5225-5216-1.ch013

Náprstek, J., & Fischer, C. (2017). Dynamic Stability and Post-Critical Processes of Slender Auto-Parametric Systems. In V. Plevris, G. Kremmyda, & Y. Fahjan (Eds.), *Performance-Based Seismic Design of Concrete Structures and Infrastructures* (pp. 128–171). Hershey, PA: IGI Global. doi:10.4018/978-1-5225-2089-4.ch006

Nautiyal, L., Shivach, P., & Ram, M. (2018). Optimal Designs by Means of Genetic Algorithms. In M. Ram & J. Davim (Eds.), *Soft Computing Techniques and Applications in Mechanical Engineering* (pp. 151–161). Hershey, PA: IGI Global. doi:10.4018/978-1-5225-3035-0.ch007

Nazir, R. (2017). Advanced Nanomaterials for Water Engineering and Treatment: Nano-Metal Oxides and Their Nanocomposites. In T. Saleh (Ed.), *Advanced Nanomaterials for Water Engineering, Treatment, and Hydraulics* (pp. 84–126). Hershey, PA: IGI Global. doi:10.4018/978-1-5225-2136-5.ch005

Nogueira, A. F., Ribeiro, J. C., Fernández de Vega, F., & Zenha-Rela, M. A. (2018). Evolutionary Approaches to Test Data Generation for Object-Oriented Software: Overview of Techniques and Tools. In M. Khosrow-Pour, D.B.A. (Ed.), Incorporating Nature-Inspired Paradigms in Computational Applications (pp. 162-194). Hershey, PA: IGI Global. doi:10.4018/978-1-5225-5020-4.ch006

Nunes, J. F., Moreira, P. M., & Tavares, J. M. (2016). Human Motion Analysis and Simulation Tools: A Survey. In F. Miranda & C. Abreu (Eds.), *Handbook of Research on Computational Simulation and Modeling in Engineering* (pp. 359–388). Hershey, PA: IGI Global. doi:10.4018/978-1-4666-8823-0.ch012

Ogunlaja, A. S., & Tshentu, Z. R. (2016). Molecularly Imprinted Polymer Nanofibers for Adsorptive Desulfurization. In T. Saleh (Ed.), *Applying Nanotechnology to the Desulfurization Process in Petroleum Engineering* (pp. 281–336). Hershey, PA: IGI Global. doi:10.4018/978-1-4666-9545-0.ch010

Ong, P., & Kohshelan, S. (2016). Performances of Adaptive Cuckoo Search Algorithm in Engineering Optimization. In P. Vasant, G. Weber, & V. Dieu (Eds.), *Handbook of Research on Modern Optimization Algorithms and Applications in Engineering and Economics* (pp. 676–699). Hershey, PA: IGI Global. doi:10.4018/978-1-4666-9644-0.ch026

Osho, M. B. (2018). Industrial Enzyme Technology: Potential Applications. In S. Bharati & P. Chaurasia (Eds.), *Research Advancements in Pharmaceutical, Nutritional, and Industrial Enzymology* (pp. 375–394). Hershey, PA: IGI Global. doi:10.4018/978-1-5225-5237-6.ch017

Padmaja, P., & Marutheswar, G. (2017). Certain Investigation on Secured Data Transmission in Wireless Sensor Networks. *International Journal of Mobile Computing and Multimedia Communications*, *8*(1), 48–61. doi:10.4018/IJMCMC.2017010104

Paixão, S. M., Silva, T. P., Arez, B. F., & Alves, L. (2016). Advances in the Reduction of the Costs Inherent to Fossil Fuels' Biodesulfurization towards Its Potential Industrial Application. In T. Saleh (Ed.), *Applying Nanotechnology to the Desulfurization Process in Petroleum Engineering* (pp. 390–425). Hershey, PA: IGI Global. doi:10.4018/978-1-4666-9545-0.ch013

Palmer, S., & Hall, W. (2017). An Evaluation of Group Work in First-Year Engineering Design Education. In R. Tucker (Ed.), *Collaboration and Student Engagement in Design Education* (pp. 145–168). Hershey, PA: IGI Global. doi:10.4018/978-1-5225-0726-0.ch007

Panneer, R. (2017). Effect of Composition of Fibers on Properties of Hybrid Composites. *International Journal of Manufacturing, Materials, and Mechanical Engineering*, *7*(4), 28–43. doi:10.4018/IJMMME.2017100103

Parker, J. (2016). Hubble's Expanding Universe: A Model for Quality in Technology Infused engineering and Technology Education. *International Journal of Quality Assurance in Engineering and Technology Education*, *5*(2), 16–29. doi:10.4018/IJQAETE.2016040102

Paul, S., & Roy, P. (2018). Optimal Design of Power System Stabilizer Using a Novel Evolutionary Algorithm. *International Journal of Energy Optimization and Engineering*, *7*(3), 24–46. doi:10.4018/IJEOE.2018070102

Pavaloiu, A. (2018). Artificial Intelligence Ethics in Biomedical-Engineering-Oriented Problems. In U. Kose, G. Guraksin, & O. Deperlioglu (Eds.), *Nature-Inspired Intelligent Techniques for Solving Biomedical Engineering Problems* (pp. 219–231). Hershey, PA: IGI Global. doi:10.4018/978-1-5225-4769-3.ch010

Peña, F. (2016). A Semi-Discrete Approach for the Numerical Simulation of Freestanding Blocks. In V. Sarhosis, K. Bagi, J. Lemos, & G. Milani (Eds.), *Computational Modeling of Masonry Structures Using the Discrete Element Method* (pp. 416–439). Hershey, PA: IGI Global. doi:10.4018/978-1-5225-0231-9.ch016

Penchovsky, R., & Traykovska, M. (2016). Synthetic Approaches to Biology: Engineering Gene Control Circuits, Synthesizing, and Editing Genomes. In M. Bououdina (Ed.), *Emerging Research on Bioinspired Materials Engineering* (pp. 323–351). Hershey, PA: IGI Global. doi:10.4018/978-1-4666-9811-6.ch012

Pieroni, A., & Iazeolla, G. (2016). Engineering QoS and Energy Saving in the Delivery of ICT Services. In P. Vasant & N. Voropai (Eds.), *Sustaining Power Resources through Energy Optimization and Engineering* (pp. 208–226). Hershey, PA: IGI Global. doi:10.4018/978-1-4666-9755-3.ch009

Pioro, I., Mahdi, M., & Popov, R. (2017). Application of Supercritical Pressures in Power Engineering. In L. Chen & Y. Iwamoto (Eds.), *Advanced Applications of Supercritical Fluids in Energy Systems* (pp. 404–457). Hershey, PA: IGI Global. doi:10.4018/978-1-5225-2047-4.ch013

Plaksina, T., & Gildin, E. (2017). Rigorous Integrated Evolutionary Workflow for Optimal Exploitation of Unconventional Gas Assets. *International Journal of Energy Optimization and Engineering*, *6*(1), 101–122. doi:10.4018/IJEOE.2017010106

Puppala, A. J., Bheemasetti, T. V., Zou, H., Yu, X., Pedarla, A., & Cai, G. (2016). Spatial Variability Analysis of Soil Properties using Geostatistics. In P. Samui (Ed.), *Handbook of Research on Advanced Computational Techniques for Simulation-Based Engineering* (pp. 195–226). Hershey, PA: IGI Global. doi:10.4018/978-1-4666-9479-8.ch008

Ramdani, N., & Azibi, M. (2018). Polymer Composite Materials for Microelectronics Packaging Applications: Composites for Microelectronics Packaging. In K. Kumar & J. Davim (Eds.), *Composites and Advanced Materials for Industrial Applications* (pp. 177–211). Hershey, PA: IGI Global. doi:10.4018/978-1-5225-5216-1.ch009

Ramesh, M., Garg, R., & Subrahmanyam, G. V. (2017). Investigation of Influence of Quenching and Annealing on the Plane Fracture Toughness and Brittle to Ductile Transition Temperature of the Zinc Coated Structural Steel Materials. *International Journal of Surface Engineering and Interdisciplinary Materials Science*, *5*(2), 33–42. doi:10.4018/IJSEIMS.2017070103

Razavi, A. M., & Ahmad, R. (2016). Agile Software Development Challenges in Implementation and Adoption: Focusing on Large and Distributed Settings – Past Experiences, Emergent Topics. In I. Ghani, D. Jawawi, S. Dorairaj, & A. Sidky (Eds.), *Emerging Innovations in Agile Software Development* (pp. 175–207). Hershey, PA: IGI Global. doi:10.4018/978-1-4666-9858-1.ch010

Reccia, E., Cecchi, A., & Milani, G. (2016). FEM/DEM Approach for the Analysis of Masonry Arch Bridges. In V. Sarhosis, K. Bagi, J. Lemos, & G. Milani (Eds.), *Computational Modeling of Masonry Structures Using the Discrete Element Method* (pp. 367–392). Hershey, PA: IGI Global. doi:10.4018/978-1-5225-0231-9.ch014

Ro, H. K., & McIntosh, K. (2016). Constructing Conducive Environment for Women of Color in Engineering Undergraduate Education. In U. Thomas & J. Drake (Eds.), *Critical Research on Sexism and Racism in STEM Fields* (pp. 23–48). Hershey, PA: IGI Global. doi:10.4018/978-1-5225-0174-9.ch002

Rodulfo-Baechler, S. M. (2016). Dual Role of Perovskite Hollow Fiber Membrane in the Methane Oxidation Reactions. In H. Al-Megren & T. Xiao (Eds.), *Petrochemical Catalyst Materials, Processes, and Emerging Technologies* (pp. 385–430). Hershey, PA: IGI Global. doi:10.4018/978-1-4666-9975-5.ch014

Rudolf, S., Biryuk, V. V., & Volov, V. (2018). Vortex Effect, Vortex Power: Technology of Vortex Power Engineering. In V. Kharchenko & P. Vasant (Eds.), *Handbook of Research on Renewable Energy and Electric Resources for Sustainable Rural Development* (pp. 500–533). Hershey, PA: IGI Global. doi:10.4018/978-1-5225-3867-7.ch021

Sah, A., Bhadula, S. J., Dumka, A., & Rawat, S. (2018). A Software Engineering Perspective for Development of Enterprise Applications. In A. Elçi (Ed.), *Handbook of Research on Contemporary Perspectives on Web-Based Systems* (pp. 1–23). Hershey, PA: IGI Global. doi:10.4018/978-1-5225-5384-7.ch001

Sahoo, P., & Roy, S. (2017). Tribological Behavior of Electroless Ni-P, Ni-P-W and Ni-P-Cu Coatings: A Comparison. *International Journal of Surface Engineering and Interdisciplinary Materials Science*, *5*(1), 1–15. doi:10.4018/IJSEIMS.2017010101

Sahoo, S. (2018). Laminated Composite Hypar Shells as Roofing Units: Static and Dynamic Behavior. In K. Kumar & J. Davim (Eds.), *Composites and Advanced Materials for Industrial Applications* (pp. 249–269). Hershey, PA: IGI Global. doi:10.4018/978-1-5225-5216-1.ch011

Sahu, H., & Hungyo, M. (2018). Introduction to SDN and NFV. In A. Dumka (Ed.), *Innovations in Software-Defined Networking and Network Functions Virtualization* (pp. 1–25). Hershey, PA: IGI Global. doi:10.4018/978-1-5225-3640-6.ch001

Saikia, P., Bharadwaj, S. K., & Miah, A. T. (2016). Peroxovanadates and Its Bio-Mimicking Relation with Vanadium Haloperoxidases. In M. Bououdina (Ed.), *Emerging Research on Bioinspired Materials Engineering* (pp. 197–219). Hershey, PA: IGI Global. doi:10.4018/978-1-4666-9811-6.ch007

Saladino, R., Botta, G., & Crucianelli, M. (2016). Advances in Nanotechnology Transition Metal Catalysts in Oxidative Desulfurization (ODS) Processes: Nanotechnology Applied to ODS Processing. In T. Saleh (Ed.), *Applying Nanotechnology to the Desulfurization Process in Petroleum Engineering* (pp. 180–215). Hershey, PA: IGI Global. doi:10.4018/978-1-4666-9545-0.ch006

Saleh, T. A., Danmaliki, G. I., & Shuaib, T. D. (2016). Nanocomposites and Hybrid Materials for Adsorptive Desulfurization. In T. Saleh (Ed.), *Applying Nanotechnology to the Desulfurization Process in Petroleum Engineering* (pp. 129–153). Hershey, PA: IGI Global. doi:10.4018/978-1-4666-9545-0.ch004

Saleh, T. A., Shuaib, T. D., Danmaliki, G. I., & Al-Daous, M. A. (2016). Carbon-Based Nanomaterials for Desulfurization: Classification, Preparation, and Evaluation. In T. Saleh (Ed.), *Applying Nanotechnology to the Desulfurization Process in Petroleum Engineering* (pp. 154–179). Hershey, PA: IGI Global. doi:10.4018/978-1-4666-9545-0.ch005

Salem, A. M., & Shmelova, T. (2018). Intelligent Expert Decision Support Systems: Methodologies, Applications, and Challenges. In T. Shmelova, Y. Sikirda, N. Rizun, A. Salem, & Y. Kovalyov (Eds.), *Socio-Technical Decision Support in Air Navigation Systems: Emerging Research and Opportunities* (pp. 215–242). Hershey, PA: IGI Global. doi:10.4018/978-1-5225-3108-1.ch007

Samal, M. (2017). FE Analysis and Experimental Investigation of Cracked and Un-Cracked Thin-Walled Tubular Components to Evaluate Mechanical and Fracture Properties. In P. Samui, S. Chakraborty, & D. Kim (Eds.), *Modeling and Simulation Techniques in Structural Engineering* (pp. 266–293). Hershey, PA: IGI Global. doi:10.4018/978-1-5225-0588-4.ch009

Samal, M., & Balakrishnan, K. (2017). Experiments on a Ring Tension Setup and FE Analysis to Evaluate Transverse Mechanical Properties of Tubular Components. In P. Samui, S. Chakraborty, & D. Kim (Eds.), *Modeling and Simulation Techniques in Structural Engineering* (pp. 91–115). Hershey, PA: IGI Global. doi:10.4018/978-1-5225-0588-4.ch004

Santhanakumar, M., Adalarasan, R., & Rajmohan, M. (2016). An Investigation in Abrasive Waterjet Cutting of Al6061/SiC/Al2O3 Composite Using Principal Component Based Response Surface Methodology. *International Journal of Manufacturing, Materials, and Mechanical Engineering*, *6*(4), 30–47. doi:10.4018/IJMMME.2016100103

Sareen, N., & Bhattacharya, S. (2016). Cleaner Energy Fuels: Hydrodesulfurization and Beyond. In T. Saleh (Ed.), *Applying Nanotechnology to the Desulfurization Process in Petroleum Engineering* (pp. 84–128). Hershey, PA: IGI Global. doi:10.4018/978-1-4666-9545-0.ch003

Sarhosis, V. (2016). Micro-Modeling Options for Masonry. In V. Sarhosis, K. Bagi, J. Lemos, & G. Milani (Eds.), *Computational Modeling of Masonry Structures Using the Discrete Element Method* (pp. 28–60). Hershey, PA: IGI Global. doi:10.4018/978-1-5225-0231-9.ch002

Sarhosis, V., Oliveira, D. V., & Lourenco, P. B. (2016). On the Mechanical Behavior of Masonry. In V. Sarhosis, K. Bagi, J. Lemos, & G. Milani (Eds.), *Computational Modeling of Masonry Structures Using the Discrete Element Method* (pp. 1–27). Hershey, PA: IGI Global. doi:10.4018/978-1-5225-0231-9.ch001

Satyam, N. (2016). Liquefaction Modelling of Granular Soils using Discrete Element Method. In P. Samui (Ed.), *Handbook of Research on Advanced Computational Techniques for Simulation-Based Engineering* (pp. 381–441). Hershey, PA: IGI Global. doi:10.4018/978-1-4666-9479-8.ch015

Sawant, S. (2018). Deep Learning and Biomedical Engineering. In U. Kose, G. Guraksin, & O. Deperlioglu (Eds.), *Nature-Inspired Intelligent Techniques for Solving Biomedical Engineering Problems* (pp. 283–296). Hershey, PA: IGI Global. doi:10.4018/978-1-5225-4769-3.ch014

Sezgin, H., & Berkalp, O. B. (2018). Textile-Reinforced Composites for the Automotive Industry. In K. Kumar & J. Davim (Eds.), *Composites and Advanced Materials for Industrial Applications* (pp. 129–156). Hershey, PA: IGI Global. doi:10.4018/978-1-5225-5216-1.ch007

Shah, M. Z., Gazder, U., Bhatti, M. S., & Hussain, M. (2018). Comparative Performance Evaluation of Effects of Modifier in Asphaltic Concrete Mix. *International Journal of Strategic Engineering*, *1*(2), 13–25. doi:10.4018/IJoSE.2018070102

Shah, V. S., Shah, H. R., & Samui, P. (2016). Application of Meta-Models (MPMR and ELM) for Determining OMC, MDD and Soaked CBR Value of Soil. In S. Bhattacharyya, P. Banerjee, D. Majumdar, & P. Dutta (Eds.), *Handbook of Research on Advanced Hybrid Intelligent Techniques and Applications* (pp. 454–482). Hershey, PA: IGI Global. doi:10.4018/978-1-4666-9474-3.ch015

Sharma, N., & Kumar, K. (2018). Fabrication of Porous NiTi Alloy Using Organic Binders. In K. Kumar & J. Davim (Eds.), *Composites and Advanced Materials for Industrial Applications* (pp. 38–62). Hershey, PA: IGI Global. doi:10.4018/978-1-5225-5216-1.ch003

Sharma, T. K. (2016). Application of Shuffled Frog Leaping Algorithm in Software Project Scheduling. In P. Saxena, D. Singh, & M. Pant (Eds.), *Problem Solving and Uncertainty Modeling through Optimization and Soft Computing Applications* (pp. 225–238). Hershey, PA: IGI Global. doi:10.4018/978-1-4666-9885-7.ch011

Shivach, P., Nautiyal, L., & Ram, M. (2018). Applying Multi-Objective Optimization Algorithms to Mechanical Engineering. In M. Ram & J. Davim (Eds.), *Soft Computing Techniques and Applications in Mechanical Engineering* (pp. 287–301). Hershey, PA: IGI Global. doi:10.4018/978-1-5225-3035-0.ch014

Shmelova, T. (2018). Stochastic Methods for Estimation and Problem Solving in Engineering: Stochastic Methods of Decision Making in Aviation. In S. Kadry (Ed.), *Stochastic Methods for Estimation and Problem Solving in Engineering* (pp. 139–160). Hershey, PA: IGI Global. doi:10.4018/978-1-5225-5045-7.ch006

Shukla, R., Anapagaddi, R., Singh, A. K., Allen, J. K., Panchal, J. H., & Mistree, F. (2016). Integrated Computational Materials Engineering for Determining the Set Points of Unit Operations for Production of a Steel Product Mix. In S. Datta & J. Davim (Eds.), *Computational Approaches to Materials Design: Theoretical and Practical Aspects* (pp. 163–191). Hershey, PA: IGI Global. doi:10.4018/978-1-5225-0290-6.ch006

Siero González, L. R., & Romo Vázquez, A. (2017). Didactic Sequences Teaching Mathematics for Engineers With Focus on Differential Equations. In M. Ramírez-Montoya (Ed.), *Handbook of Research on Driving STEM Learning With Educational Technologies* (pp. 129–151). Hershey, PA: IGI Global. doi:10.4018/978-1-5225-2026-9.ch007

Singh, R., & Dutta, S. (2018). Visible Light Active Nanocomposites for Photocatalytic Applications. In K. Kumar & J. Davim (Eds.), *Composites and Advanced Materials for Industrial Applications* (pp. 270–296). Hershey, PA: IGI Global. doi:10.4018/978-1-5225-5216-1.ch012

Singh, R., & Lou, H. H. (2016). Safety and Efficiency Enhancement in LNG Terminals. In H. Al-Megren & T. Xiao (Eds.), *Petrochemical Catalyst Materials, Processes, and Emerging Technologies* (pp. 164–176). Hershey, PA: IGI Global. doi:10.4018/978-1-4666-9975-5.ch007

Sözbilir, H., Özkaymak, Ç., Uzel, B., & Sümer, Ö. (2018). Criteria for Surface Rupture Microzonation of Active Faults for Earthquake Hazards in Urban Areas. In N. Ceryan (Ed.), *Handbook of Research on Trends and Digital Advances in Engineering Geology* (pp. 187–230). Hershey, PA: IGI Global. doi:10.4018/978-1-5225-2709-1.ch005

Stanciu, I. (2018). Stochastic Methods in Microsystems Engineering. In S. Kadry (Ed.), *Stochastic Methods for Estimation and Problem Solving in Engineering* (pp. 161–176). Hershey, PA: IGI Global. doi:10.4018/978-1-5225-5045-7.ch007

Strebkov, D., Nekrasov, A., Trubnikov, V., & Nekrasov, A. (2018). Single-Wire Resonant Electric Power Systems for Renewable-Based Electric Grid. In V. Kharchenko & P. Vasant (Eds.), *Handbook of Research on Renewable Energy and Electric Resources for Sustainable Rural Development* (pp. 449–474). Hershey, PA: IGI Global. doi:10.4018/978-1-5225-3867-7.ch019

Subburaman, D., Jagan, J., Dalkiliç, Y., & Samui, P. (2016). Reliability Analysis of Slope Using MPMR, GRNN and GPR. In F. Miranda & C. Abreu (Eds.), *Handbook of Research on Computational Simulation and Modeling in Engineering* (pp. 208–224). Hershey, PA: IGI Global. doi:10.4018/978-1-4666-8823-0.ch007

Sun, J., Wan, S., Lin, J., & Wang, Y. (2016). Advances in Catalytic Conversion of Syngas to Ethanol and Higher Alcohols. In H. Al-Megren & T. Xiao (Eds.), *Petrochemical Catalyst Materials, Processes, and Emerging Technologies* (pp. 177–215). Hershey, PA: IGI Global. doi:10.4018/978-1-4666-9975-5.ch008

Tüdeş, Ş., Kumlu, K. B., & Ceryan, S. (2018). Integration Between Urban Planning and Natural Hazards For Resilient City. In N. Ceryan (Ed.), *Handbook of Research on Trends and Digital Advances in Engineering Geology* (pp. 591–630). Hershey, PA: IGI Global. doi:10.4018/978-1-5225-2709-1.ch017

Tyukhov, I., Rezk, H., & Vasant, P. (2016). Modern Optimization Algorithms and Applications in Solar Photovoltaic Engineering. In P. Vasant & N. Voropai (Eds.), *Sustaining Power Resources through Energy Optimization and Engineering* (pp. 390–445). Hershey, PA: IGI Global. doi:10.4018/978-1-4666-9755-3.ch016

Ulamis, K. (2018). Soil Liquefaction Assessment by Anisotropic Cyclic Triaxial Test. In N. Ceryan (Ed.), *Handbook of Research on Trends and Digital Advances in Engineering Geology* (pp. 631–664). Hershey, PA: IGI Global. doi:10.4018/978-1-5225-2709-1.ch018

Umar, M. A., Tenuche, S. S., Yusuf, S. A., Abdulsalami, A. O., & Kufena, A. M. (2016). Usability Engineering in Agile Software Development Processes. In I. Ghani, D. Jawawi, S. Dorairaj, & A. Sidky (Eds.), *Emerging Innovations in Agile Software Development* (pp. 208–221). Hershey, PA: IGI Global. doi:10.4018/978-1-4666-9858-1.ch011

Üzüm, O., & Çakır, Ö. A. (2016). A Bio-Inspired Phenomena in Cementitious Materials: Self-Healing. In M. Bououdina (Ed.), *Emerging Research on Bioinspired Materials Engineering* (pp. 220–246). Hershey, PA: IGI Global. doi:10.4018/978-1-4666-9811-6.ch008

Valente, M., & Milani, G. (2017). Seismic Assessment and Retrofitting of an Under-Designed RC Frame Through a Displacement-Based Approach. In V. Plevris, G. Kremmyda, & Y. Fahjan (Eds.), *Performance-Based Seismic Design of Concrete Structures and Infrastructures* (pp. 36–58). Hershey, PA: IGI Global. doi:10.4018/978-1-5225-2089-4.ch002

Vasant, P. (2018). A General Medical Diagnosis System Formed by Artificial Neural Networks and Swarm Intelligence Techniques. In U. Kose, G. Guraksin, & O. Deperlioglu (Eds.), *Nature-Inspired Intelligent Techniques for Solving Biomedical Engineering Problems* (pp. 130–145). Hershey, PA: IGI Global. doi:10.4018/978-1-5225-4769-3.ch006

Vergara, D., Lorenzo, M., & Rubio, M. (2016). On the Use of Virtual Environments in Engineering Education. *International Journal of Quality Assurance in Engineering and Technology Education, 5*(2), 30–41. doi:10.4018/IJQAETE.2016040103

Verrollot, J., Tolonen, A., Harkonen, J., & Haapasalo, H. J. (2018). Challenges and Enablers for Rapid Product Development. *International Journal of Applied Industrial Engineering*, *5*(1), 25–49. doi:10.4018/IJAIE.2018010102

Wagner, C., & Ryan, C. (2016). Physical and Digital Integration Strategies of Electronic Device Supply Chains and Their Applicability to ETO Supply Chains. In R. Addo-Tenkorang, J. Kantola, P. Helo, & A. Shamsuzzoha (Eds.), *Supply Chain Strategies and the Engineer-to-Order Approach* (pp. 224–245). Hershey, PA: IGI Global. doi:10.4018/978-1-5225-0021-6.ch011

Wang, Z., Wu, P., Lan, L., & Ji, S. (2016). Preparation, Characterization and Desulfurization of the Supported Nickel Phosphide Catalysts. In H. Al-Megren & T. Xiao (Eds.), *Petrochemical Catalyst Materials, Processes, and Emerging Technologies* (pp. 431–458). Hershey, PA: IGI Global. doi:10.4018/978-1-4666-9975-5.ch015

Yardimci, A. G., & Karpuz, C. (2018). Fuzzy Rock Mass Rating: Soft-Computing-Aided Preliminary Stability Analysis of Weak Rock Slopes. In N. Ceryan (Ed.), *Handbook of Research on Trends and Digital Advances in Engineering Geology* (pp. 97–131). Hershey, PA: IGI Global. doi:10.4018/978-1-5225-2709-1.ch003

Zhang, L., Ding, S., Sun, S., Han, B., Yu, X., & Ou, J. (2016). Nano-Scale Behavior and Nano-Modification of Cement and Concrete Materials. In A. Khitab & W. Anwar (Eds.), *Advanced Research on Nanotechnology for Civil Engineering Applications* (pp. 28–79). Hershey, PA: IGI Global. doi:10.4018/978-1-5225-0344-6.ch002

Zindani, D., & Kumar, K. (2018). Industrial Applications of Polymer Composite Materials. In K. Kumar & J. Davim (Eds.), *Composites and Advanced Materials for Industrial Applications* (pp. 1–15). Hershey, PA: IGI Global. doi:10.4018/978-1-5225-5216-1.ch001

Zindani, D., Maity, S. R., & Bhowmik, S. (2018). A Decision-Making Approach for Material Selection of Polymeric Composite Bumper Beam. In K. Kumar & J. Davim (Eds.), *Composites and Advanced Materials for Industrial Applications* (pp. 112–128). Hershey, PA: IGI Global. doi:10.4018/978-1-5225-5216-1.ch006

About the Contributors

Olga Smirnova, Ph.D., is an associate professor at the MPA program, the Department of Political Science, East Carolina University. Her research interests include green transportation innovations, contracting out, program evaluation, economic development, and performance measurement. Her research appeared in Public Administration Review, Administration and Society, Journal of Public Transportation, State and local Government Review, Journal of Public Procurement, Industrial Relations Journal, American Behavioral Scientist, Public Works Management and Policy, and Municipal Finance Journal.

* * *

Whitney Afonso is an Associate Professor at the School of Government at the University of North Carolina at Chapel Hill. Her research examines the impact of revenue choices on state and local governments. Her research has appeared in journals such as Public Budgeting & Finance, Public Finance Review, Journal of Public Policy, and State and Local Government Review. Her research on local sales taxes earmarked for transportation was awarded the Burkhead Award for best article in Public Budgeting & Finance in 2015. She also serves on the executive committee for the Association for Budgeting and Financial Management.

James Amdal has extensive experience and expertise in transportation infrastructure development, urban planning, historic preservation and community design. He served as Associate Project Manager and private-sector NGO President for the planning, design, construction and operation of Phase 1 and 2 of the New Orleans Riverfront Streetcar. He also authored the Strategic Policy Plan for the New Orleans Riverfront, a project sponsored by the New Orleans City Planning Commission that encompassed the entire 26 mile riverfront corridor within Orleans Parish. Mr. Amdal served as a consultant on the development and implementation of the Statewide Intermodal Transportation Plan and the Louisiana Statewide Transportation Plan.

He has recently served on the UNO project team for various strategic plans and / or market assessments for Louisiana ports. Mr. Amdal currently serves as Chairman of the New Orleans Central Business District Historic District Landmarks Commission. Mr. Amdal also serves on the Board of Directors of the Louisiana Steam Train Association and the Old Algiers Main Street Corporation.

Khairul Anuar holds a PhD in Civil Engineering (Transportation) from Old Dominion University. His research and publication interests include traffic operations, transportation planning, modeling and simulation and data analytics.

Robert Blair is Professor of Public Administration and Urban Studies at the University of Nebraska at Omaha. He conducts research on local government and rural and community development.

Timothy Brock has eleven years of experience in a broad array of transportation and economic development planning and policy research. His expertise is in transportation and land use, specifically transit-oriented development, comparative transportation policy, policy transfer and economic development. He has conducted projects in multiple transportation modes, including transit, multimodal freight movements, highways and roadways and inland waterways. Currently a practicing land use planner and economic development professional, Dr. Brock previous experience includes senior planning consultant and assistant professor of urban planning and economic geography.

Richard Callahan's research, consulting, publications, and teaching focuses on leadership behaviors and strategy practices that are effective in complex, demanding and dynamic environments in the public in transportation, finance, and social determinants of health. Dr. Callahan is a Professor at the University of San Francisco, with a joint appointment in the USF School of Management and in the USF School of Nursing and Health Professions. Dr. Callahan is an elected Fellow the National Academy of Public Administration, a nonprofit academy chartered by Congress for the past 50 years to advise on complex public governance challenges. He is the current Editor in Chief of the International Journal of Public Leadership, a peer reviewed journal, published by Emerald Press, United Kingdom. He was a visiting researcher at Oxford University, 2016 and was the Visiting Scholar 2017-2018 for the Center for California Studies at the California State University, Sacramento. He received a Fulbright Specialist Program grant in 2011 for lectures on public institutions at Aydin University in Istanbul, Turkey. Callahan co-authored a national award-winning article, receiving the Brownlow award for best article from

Public Administration Review. He has been published in journals such as Public Management Review, Public Manager, Government Finance Officers Review, The National Civic Review. As well as book chapters published in the United States and Europe on public management, leadership, and sustainability.

David Chapman is the Graduate Program Director of the Master of Public Administration program and a Master Lecturer in the School of Public Service in the Strome College of Business at Old Dominion University. He received his Ph.D. in Public Administration & Urban Policy with a cognate in Instructional Design and Technology from Old Dominion University and his M.S. in the Management of Information Systems from the McIntire School of Commerce at the University of Virginia.

Donta Council is a Ph.D. Student at Old Dominion University and is currently studying Public Administration and Policy.

Jerome Deichert is Director of the Center for Public Affairs Research at the University of Nebraska at Omaha. His research interests include demographics and economic development.

Meagan Jordan, Ph.D., is an Associate Professor of Public Administration at Old Dominion University's School of Public Service. Her Ph.D. in Public Policy and Administration is from the University of Kentucky. Her research interests are in state and local government budgeting and budget theory, fiscal policy, revenue management, transparency, and citizen participation.

Suzanne Leland, Ph.D., is a Professor in the Department of Political Science and Public Administration at University of North Carolina at Charlotte. She received her Phd in Political Science from the University of Kansas and has published transportation policy articles in journals such as Public Administration Review, Journal of Urban Affairs, and Administration and Society.

Martin Mayer received his Ph.D. from the School of Public Service at Old Dominion University, where he currently teaches part-time. Prior to that time, Dr. Mayer earned a BA and MPA from the University of Akron. Dr. Mayer's research interests include healthcare policy, local government innovation, and environmental policy. He has published several articles and book chapters, and his work has been published in Social Science Quarterly, Politics and Policy, The Social Science Journal, and Politics and the Life Sciences, among other outlets.

Zachary Mohr, PhD, CGFM, is an Assistant Professor of Political Science and Public Administration at the University of North Carolina at Charlotte. His work on cost accounting in government has appeared in top public administration journals and he edited the book Government Cost Accounting: Theory and Applications, which was recently published by Routledge. While his work primarily revolves around technical issues of cost accounting, he has also looked at cost issues related to major defense acquisitions, elections, sustainability, and other issues regarding cost and resource control.

Tonderai Mushipe, MPA, is a PhD student in the Department of Geography and a Data Analyst for the College of Business at the University of North Carolina at Charlotte. He has worked in local government and has an interest in local economic development.

Mark Pisano is a Professor of Practice at the Sol Price School of Policy at University of Southern California. He is on the Board of the National Academy of Public Administration and Co-chairman of the Standing Panel on Intergovernmental Relations. Chairman of a national non-profit Infrastructure Funding Alliance. He is co-chairmen of the Infrastructure Working Group of California Forward. He recently published a book called the Puzzle of the American Economy: How Demography is changing America's Economy and Politics.

William (Billy) Riggs, Ph.D., AICP, LEED AP is a global expert and thought leader in the areas of future mobility and smart transportation, housing, economics and urban development. He is currently a professor at the University of San Francisco School of Management, Director of Strategy and Research at Sustinere Consulting, and a strategic advisor to multiple Silicon Valley technology firms and start-ups. This follows two decades of experience working as a transportation planner, economist, and engineer. He has been both a fellow with the National Science Foundation fellow and the University of California Transportation Center, is the founder of ReStreet.com (app.restreet.com)—an online tool for democratizing street design—and provides strategy and consulting to multiple companies on smart mobility and urban development. Dr. Riggs sits on the City of Palo Alto's Planning and Transportation Commission and is a member of the Transportation Research Board (TRB) Committee on Transportation Economics and Transportation Research Board's (TRB) Committee on Transportation Economics and the Standing Committee on Policy and Law.

Bethany Stich is the Department Chair of Planning and Urban Studies and the Director of the University of New Orleans Transportation Institute (UNOTI). Stich serves as the Co-Principal Investigator of UNOTI's two University Transportation Centers, the National Center for Strategic Transportation Policies, Investments and Decisions housed at the University of Maryland and the Maritime Transportation Research & Education Center housed at the University of Arkansas. Additionally, Stich serves as the Principal Investigator for the Department of Homeland Security Center of Excellence for Maritime Resiliency housed at Louisiana State University. Stich serves on the Transportation Research Board's Intermodal Freight Committee as the Subcommittee Chair and on the Committee on the Logistics of Disaster Relief and Business Continuity. She serves on the American Society for Public Administration's Section for Public Administration Research and is the Past President of the Section on Transportation Planning and Administration.

Tancy Vandecar-Burdin is the Associate Director of the Social Science Research Center at Old Dominion University (ODU) and has served as the Associate Director since the Center's inception in 1998. Dr. Vandecar-Burdin has over 20 years of experience with various forms of research methods and data collection including web and mail surveys, telephone surveys, face-to-face interviews, and focus groups. Dr. Vandecar-Burdin works with faculty, area businesses, and health and human service providers to determine their research and data collection needs, develop survey instruments, and develop evaluations of programs. Her research interests include nursing home policy, survey research methods, and issues surrounding the mentally ill and the criminal justice system. Dr. Vandecar-Burdin earned her Master of Arts in Applied Sociology with a certificate in Criminal Justice as well as a PhD in Public Administration and Urban Policy from Old Dominion University.

Morgan Vogel is a doctoral student at the University of Nebraska at Omaha in the School of Public Administration.

Peter Webb (b. 1967) is a native New Orleanian. He received his B.A. in Anthropology in May of 1994 from UNO. He received his M.A. in Anthropology in December of 1997 from the University of Memphis, with an applied urban focus, and a Master of Science in Urban Studies from UNO in May 2016. He earned his PhD in Urban Studies at UNO in 2018 with a concentration in Urban Anthropology, focusing on the social processes of last mile freight congestion at the Port of New Orleans. He has participated in research projects about electronic navigation

(E-nav), NAFTA freight trade corridors, megaregional development, the impact of Liquefied Natural Gas (LNG) on the Louisiana petrochemical industry, and the logistics of international shipping container chassis.

Juita-Elena (Wie) Yusuf, Ph.D., is an Associate Professor in the School of Public Service at Old Dominion University. Her research interests are in public budgeting and financial management, with a particular interest in transportation and infrastructure finance, and transparency, accountability and public participation.

Index

E

F

G

H

I

J

L

M

N

O

P

R

S

T

U

V

W

www.ingramcontent.com/pod-product-compliance
Lightning Source LLC
LaVergne TN
LVHW081336300926
849161LV00008B/1991